UNDER THE SURFACE

Tom Wilber

UNDER THE SURFACE

Fracking, Fortunes, and the Fate of the Marcellus Shale

CORNELL UNIVERSITY PRESS, ITHACA AND LONDON

First published 2012 by Cornell University Press
First printing, Cornell Paperbacks, updated with a new chapter, 2015

Printed in the United States of America

ISBN 978-0-8014-5654-1

The Library of Congress has catalogued the hardcover edition as follows:

Wilber, Tom, 1958–
Under the surface : fracking, fortunes and the fate of the Marcellus Shale / Tom Wilber.
    p. cm.
  Includes bibliographical references and index.
  ISBN 978-0-8014-5016-7 (cloth : alk. paper)
1. Shale gas industry—New York (State)   2. Shale gas industry—
Pennsylvania.   3. Hydraulic fracturing—New York (State)   4. Hydraulic fracturing—
Pennsylvania.   5. Marcellus Shale.   I. Title.
  HD9581.2.S53W55 2012
  333.8'23309747—dc23   2011047166

Paperback printing   10 9 8 7 6 5 4 3 2 1

To my wife, Julianne Boyd, and my mother,
Florence Leonard Wilber

CONTENTS

UNDER THE SURFACE

Cracks in the Rock

From a distance, the stratified band of black shale running along the east shore of Cayuga Lake is hard to pick out from an ensemble of natural features—rocks, trees, hills, and water—that compose the landscape of the Finger Lakes region of New York. Viewed from across the lake, the rock is a thin charcoal-colored border just above the shoreline, occasionally disappearing behind cottages and boathouses. A landscape artist might incorporate it into the broader picture as a shaded accent above the beach. To geologists, shale is a defining aspect of the landscape and the Marcellus layer, well below the visible Cayuga shoreline, is a resource at the heart of our national interests.

Terry Engelder, a professor of geosciences at Pennsylvania State University, is such a person. He pointed out the rock's features to me from windows of the Glenwood Pines Restaurant on the slope above the west shore of the lake. The promontory, a few miles north of Ithaca, is one of his favorite viewing spots to study these outcroppings. He has sought out this view many times, and he still finds this perspective worth the four-hour drive from his home in central Pennsylvania.

A compact, energetic man with a tanned and deeply lined face suggesting many days in the field, Engelder has built a distinguished academic career understanding the formation and physical characteristics of rocks. His career flourished in the early 1970s when the country faced a series of major energy crises. The price of oil tripled due to an embargo by major oil producing countries in the Middle East in 1973 and nearly doubled in 1979 during a period of political instability in Iran. At that time, Engelder was earning advanced degrees at Yale and Texas A&M Universities, after having worked as a hydrologist with the U.S. Geological Survey (USGS) and a geologist for Texaco. Policymakers and researchers began focusing—for a while, at least—on domestic energy sources to decrease dependence on unpredictable and uncontrollable supplies. That focus created opportunities for young researchers such as Engelder who had experience applying their work to both industry and government interests. The decades he has spent doing this since then have not diminished the romantic view he harbors about his vocation, which he defines as "a person who sleeps in the mud one night and the next night puts on a tux to tell people what he or she found."

The gray band of rock above the shoreline began as mud at the bottom of an inland sea between 300 million and 400 million years ago. Sediments, including decaying organic material, were compressed over time into layers of Devonian rock, named for the geological period in which it was formed. The various outcroppings in upstate New York represent only the visible tips of this compressed sea of silt, clay, and carbon that extends south from central New York through northern Tennessee, west to the Ohio Valley, and east to the Hudson Valley. Much of this area (with the exception of Tennessee) is spanned by one stratum, the Marcellus Shale, which outcrops in Marcellus, New York.

Black shales get their name from their dark hues, reflecting a rich mix of organic material. Engelder was among the first to assign a commercial value to this characteristic—an estimated 500 trillion cubic feet of natural gas in the Marcellus alone, enough to meet U.S. needs for decades. That calculation, which Engelder announced in 2008, caught the attention of industry executives, investors, and the media at a time when U.S. dependence on foreign oil was again a political issue.

There is another thing that makes the Marcellus special in addition to its content—its position. It sits over other gas-rich geological formations and under the infrastructure of a burgeoning natural gas distribution system to major metropolitan markets in the northeast. Although the upper Devonian layers visible along Cayuga Lake are too shallow to be effectively produced, the depth and thickness of the deeper Marcellus make it ideally suited for commercial production throughout southern New York, Pennsylvania, and parts of Ohio and West Virginia. In short, the rock holds one of the largest gas fields in the world in the middle of the largest energy markets, including the metropolitan areas of New York, Philadelphia, Pittsburgh, and Boston. At the time of Engelder's calculation of the Marcellus potential, natural gas, in addition to its traditional role in heating homes and businesses, was becoming an alternative to coal for some electricity-generation plants. It was also a proven fuel for internal combustion engines. With these facts, industry supporters began building the case that the exploitation of the Marcellus and other domestic shale gas reserves could lead the country to energy independence.

Throughout most of Engelder's career, industry lacked commercial incentives to develop technology to extract gas from shale. In the early 2000s, as the price of gas rose, prospectors began experimenting with unconventional methods in the Barnett Shale of northern Texas. Breakthroughs came when they began using two different techniques in tandem. The first was horizontal drilling, a process developed in the early 2000s using computer-generated models and steerable drilling mechanics to bore first vertical, then long horizontal wells through pancake-like strata. The second was hydraulic fracturing, developed in the late 1940s for vertical wells. Commonly known in the industry as "fracking," the process involves injecting the bore with a mix of water, sand, and chemicals under pressure great enough to split the rock and free gas embedded within. Applying the process to a horizontal well requires handling a much larger volume of solution under greater pressure compared to a vertical well and poses logistical and engineering challenges that weren't seriously pursued until the Barnett was developed.

By 2008, growing estimations of the Marcellus potential—lead by Engelder but also supported by early production results—triggered one of the biggest prospecting rushes in modern history. Speculators began bidding up the price of mineral leases; and the ensuing development of wells and infrastructure in Appalachia brought both a new source of capital and new concerns about the degradation of land, water, and air from which an environmental movement was born.

Engelder became an established source for reporters, including myself, covering Marcellus development. As the story progressed, he evolved from an expert within academic and industry circles to a controversial and outspoken public figure. His position as an authority and his eagerness to talk about his work made him a popular source for a growing corps of reporters who found the Marcellus story overlapping their daily beats on politics, environment, and local government. When I told him about my plans to write a book about shale gas issues, he proposed we meet for a day in the field.

Our day began at the Glenwood Pines. We roamed through the dining room, taking a moment to inform a waitress that we were not there to eat but to look at rock. "I bring my students here all the time," he reassured me. "They know me." He squeezed between tables filled with patrons to lead me to the exact spot at the windows where I could get a good view of the specimen across the lake. Tuning out the distractions of colored sails trimmed against a breeze, I located the band of rock above the far shore. Engelder pointed out a subtle sine wave pattern in the rock that was apparent only by examining a long cross section from a distance. Where I saw a backdrop for boats and cottages, Engelder saw features formed by geologic compression due to the movement of tectonic plates. He explained how this compression created stress fields that resulted in a pattern of fractures—largely unseen—in this shale layer and many layers below it. His insights transformed my view of a tranquil lake shore into a complex and dynamic geological formation.

Everything central to shale gas production—and the controversy surrounding it—involves understanding rock fractures. Gas cannot flow from shale, for the purposes of production, without cracks. The more cracks, the greater the flow. Consequently, wells are more productive by orders of magnitude if drilled to intersect the greatest number of natural fractures in the shortest distance. A uniform and predictable system of fractures allows gas to be optimally controlled and captured, whereas chaotic and unpredictable cracks or faults increase the risks of gas flowing to places where it's unwanted. The degree an operator is able to master the natural system of cracks, while systematically and effectively creating new ones through hydraulic fracturing, is the difference between boom and bust.

According to some arguments, exploiting this complicated network of stress points invites disaster as the industry pushes ahead with unproven technology and insufficient knowledge of natural systems. Publicized concerns include opening a path for highly pressurized gas, brine, heavy metals, and chemicals into water bearing zones—a risk the industry says is negligible. Engelder, too, has dismissed this risk; and this position has contributed to his reputation among some drilling opponents as a chief enabler of an exploitive and uncaring gas industry.

After leaving the Glenwood Pines, we climbed into my car and headed north. We were following a tour that Engelder, over the course of years, has led more than sixty times for students and colleagues. Noteworthy in this group is Tony Ingraffea, an engineering professor at Cornell University who shares academic interests with Engelder. The careers of both men are distinguished by their knowledge of rock fractures. Rather than focusing on how fractures were formed in nature, however, Ingraffea's field of study involves how those fractures can be induced through technology such as hydraulic fracturing. Engelder's geological expertise complements Ingraffea's engineering, and the two have coauthored several research papers. Ingraffea has also been the target of criticism due to his views on Marcellus development. In Ingraffea's case, however, the heat has come from industry advocates incensed by his public opposition—often stated at speaking engagements arranged by antidrilling groups—to shale gas development. Whereas Engelder is confident that extracting natural gas by means of horizontal drilling and hydrofracturing will lead to prosperity, Ingraffea believes pairing the application of these technologies for extracting shale gas poses unacceptable risks.

On our ride, Engelder was mostly interested in talking about how the tour relates to the work of another academic—Pearl Gertrude Sheldon, a structural geology student who earned her Ph.D. from Cornell in 1911. To explain Sheldon's involvement in the hands-on way he prefers to teach, Engelder directed me north a few miles on Route 89, flanking the west shore of the lake, to Taughannock State Park. We parked in the lot on the west side of the highway, near the opening of a forested gorge carved by Taughannock Creek. Terry shouldered his daypack, and we hiked away from the lake and along a stream cascading off tables of rock. Deeper in the gorge, as cliff walls eclipsed the horizon, I stopped and gazed up at roots of trees clinging to bedrock planes hundreds of feet above us. Observing these features against moving clouds gave the illusion of the ground moving, and a brief rush of vertigo.

"No matter how many times you return, you see something new," Engelder remarked, pointing out the geometric patterns in the cliff walls, many of which ran perpendicular to the stream bed and nearly at right angles to each other. He explained how his academic frame of reference was largely shaped by Sheldon's observations. After two summers charting the geology manifested in outcroppings along the lake and the gorge, Sheldon was the first person of record to see that upper Devonian shales were uniformly bisected by a grid of vertical planes corresponding to the dips and rises in the sine wave—the tectonic-induced stress field—that Engelder had shown me along the lake shore. She called them symmetrical joint sets, and Engelder now pointed them out in the blocky rock faces all around the gorge, exposed where complementing halves of the joints had fallen away. The pattern that Sheldon documented was the genesis of Engelder's belief that the joints are not products of the chaotic process of uplift and erosion that might have affected only the upper shale layers exposed during the glacier age. Rather, they are as old as the rock itself, a result of gas generated from organically rich sediment aging under high pressure and heat. The joints formed along the stress fields created throughout the entire Devonian layer by the action of tectonic collisions; and they reflect both the high organic content of the rock and an efficient means for it to escape—aspects that make it a more desirable target for commercial production.

Sheldon, according to Engelder, provided "the seminal observation that has turned the Marcellus into a supergiant gas field." In 1912, the *Journal of Geology* published Sheldon's twenty-seven-page paper, "Some Observations and Experiments on Joint Planes," detailing her observations about joints in the shale along the lake and gorge. In Engelder's estimation, this paper charted the direction of the petroleum industry into the twenty-first century and the era of shale gas development.

If this is true, there is irresistible irony that a field so dominated by men was influenced by the groundbreaking work of a woman prior to suffrage. Engelder speculates that Sheldon must have spent days on end wading up creek beds and hiking and camping along undeveloped shorelines as she took 3,000 readings over the course of a fall and two summers. Public transportation outside of cities was not well developed, and Engelder imagines she would have made the 15-mile trip from Ithaca to the gorge mostly or entirely on foot or perhaps riding a bicycle—a vision that fits his view of geologists as pioneers. At the time, the primary market for petroleum was producing kerosene for lamps, and the demand was falling in proportion to the rising popularity of electric light. The explosion of petroleum demand related to mass-produced motor vehicles and paved roads had yet to develop. Drawing from scant records of Sheldon's life, Engelder concludes that she was driven by a calling to advance knowledge about natural history rather than aiming for any particular commercial application. To promote Sheldon's as-yet-unrecognized influence on the multibillion-dollar hydraulic fracturing industry, Engelder proposed an award in her name from the American Association of Petroleum Geologists to honor women who have contributed to petroleum and natural gas development. Engelder, who is not reluctant to promote his role as an advisor to industry and investors, said he would welcome a personal role in giving the Sheldon award, the status of which was unresolved at the time of our field trip.

As we hiked along the gorge, retracing Sheldon's steps, Engelder found the spot he had been looking for. Here he left the path and stepped out onto a shelf of dry streambed—smooth dark rock dimpled by the wear of water over eons. He took chalk from the pocket of his daypack and knelt down. With his face close enough to feel the sun radiating from its surface,

Shale gas plays across the continental United States   *Source: United States Department of Energy*

he began drawing diagrams on the streambed to illustrate the broad concept he was pointing out in the landscape all around us: the orientation of stress fields created by the collision of the African plate with the North American plate nearly 380 million years ago and the resulting pattern of cracks from escaping gas. He then dug in his pack for a geology hammer and chipped off a piece of shale overhanging the creek. He brought it to his nose and sniffed. He chipped off another piece and held it to my nose, an exercise he had repeated with countless students over the years. I noted the musty, bituminous odor of the rock, and Engelder smiled approvingly.

This rock wasn't the Marcellus but one of the numerous shale layers above it that, based on Sheldon's field work and Engelder's deductions, serve as analogs to what lies beneath. "A good black shale like the Marcellus will knock your socks off," he said.

The gorge is a place where what's under the surface comes together with the land above, and the past merges with the future. Rock layers visible here extend out of view for thousands of miles. Above them, the waterways that created the gorge and exposed the rock flow past vineyards, orchards, bed and breakfasts, farms, businesses, and homes. A few miles to the south of Taughannock Gorge is Ithaca, a city of about 30,000 people that is home to Ithaca College and Cornell University. Further south are the industrial cities of Corning and Binghamton in the Susquehanna River watershed, and beyond that there are vast rural tracts of northern Pennsylvania. To the east is the Delaware River basin, encompassing a network of streams, reservoirs, and rivers that are the primary water source for New York City and Philadelphia. This is but a fraction of the Marcellus footprint, which encompasses 95,000 square miles of metropolitan, urban, suburban, rural, and wilderness land.

The destinies of communities over similar shale gas reserves—in Alabama, Louisiana, Wyoming, Arkansas, Texas, Colorado, and other places—are linked to the Marcellus region by local geology and global energy concerns. In all these shale gas regions, the relationships people have with the land, and with their neighbors, are as complicated and multidimensional as the topographical and geological terrain. Here, too, there are cracks. They are created by forces that sometimes pull in opposite directions, at other times collide with great force, and often are buried from view. The social parallels to the natural features of the Marcellus are the focus of this story, and they begin with a series of events unfolding in rural communities along the border of New York and Pennsylvania between 1999 and 2011.

1. AN AGENT OF DREAMS

Ron and Jeannie Carter were in their late sixties and on a fixed income when the landman parked his luxury sedan in front of their trailer home. He wore a cowboy hat, boots, and an oversized buckle on his belt. He was not a big man but, striding confidently up to the drive, he cut a striking figure; and he brought uncommon news: Cabot Oil and Gas—a company from Houston, Texas—was interested in leasing the Carters' land for natural gas exploration.

The year was 2006. In the Endless Mountains of northern Pennsylvania, where the Carters lived, visits from landmen were rare but not unheard of. Some property owners from previous generations had signed over mineral rights to land speculators for between $1 and $5 per acre, with little ever coming of it but a modest check and a little paperwork. The proposal this landman brought might also amount to nothing—or the 75-acre ancestral homestead could be sitting right on top of a source of income for Ron and future generations. Cabot was not a speculator but a fully capitalized drilling company, the landman explained. His name was James Underwood, and he quickly impressed Ron as experienced in these matters. The $25 per acre that Mr. Underwood was offering was far better than offers Ron's father had once entertained. It would be almost enough to cover the Carters' tax bill, even if nothing more came of it. If Cabot struck gas, the Carters would be guaranteed a piece of the action: 12.5 percent. That was in black and white on the lease Mr. Underwood brought. The way gas prices just kept on rising, that could amount to something.

Jeannie Carter, Ron's wife of fifty years, was decidedly against the idea. If the landman didn't immediately know this, he would soon learn that she required some convincing. Forty-five minutes into his introductory visit, when it was clear the Carters were not going to sign anything that afternoon, the landman announced he had taken enough of their time. He knew they would want to think about it, and he surely understood that. But they had better not wait too long, he added—and here Ron sensed genuine indifference in his voice. The neighbors were signing up, the landman told them; and with due respect, Cabot didn't need the Carters to move forward with its plans.

Drilling companies based in the south, like Cabot, were among the first to sense opportunity in Marcellus Shale formation underlying Appalachia in the first decade of the twenty-first century. These companies sent agents to acquire rights to the land, and they started with promising territories that could be leased cheaply. That brought them to Susquehanna County in the heart of the Endless Mountains.

When viewed heading north on Interstate 81 toward the New York State border, the Endless Mountains live up to their name. With each crest in the road, the northern Appalachian terrain appears and reappears, flowing in waves and channels to the horizon. The New Milford exit—one of the last before New York—drops into a place that could also be called the Timeless Mountains. Away from the traffic and commerce of the interstate, the northeastern Pennsylvania landscape appears essentially as it did a generation ago—farms and fields, agricultural supply and general stores, lunch counters and gas marts, and the occasional horse-drawn buggy venturing from one of the Amish settlements tucked away in the Susquehanna River Valley. Heading west, County Route 706 wanders over hill and dale before settling down in Montrose, a village of 1,660 residents and the seat of Susquehanna County. Here, roads merge into the village center, with a public green and a Greek revival courthouse facing a boulevard with broad sidewalks, period streetlamps, and tidy storefronts.

From this hub, Route 29 runs out of the village to the south, passing through Dimock Township and connecting with town roads meandering through forested hollows. Homesteads and trailers of generations of farmhands, quarry workers, and lumberjacks are mixed with nineteenth-century farmhouses and contemporary homes of artisans and organic farmers—newcomers pursuing ambitions of gentrified country living. Dry-laid stonewalls from settlement times mark boundaries. Some have buckled with time and exposure to the elements, swallowed by wild rose and elderberry bushes. Others, perhaps built by better craftsmen or restored by capable hands, remain plumb and square, at least for another generation. A section of this kind of sturdy wall ends in front of Ron and Jeannie Carter's home.

It's a neat white trailer with brown shutters, sitting on the edge of a wood-lot. Facing away from the woods, a canopied front porch, furnished with deck chairs, wind chimes, and hanging flower baskets overlooks an expanse of fallow field. This is where Ron reads Tom Clancy and Clive Cussler on summer days as hummingbirds and finches flit about feeders. He wears a cap tipped back on his head, exposing a face with earnest features, a small mustache, and sturdy metal glasses. Jeannie often sits with him doing cross-word puzzles. It had always been a quiet place, but that changed after 2007. Following the landman's visits, the unpaved and seldom-traveled road visible from the porch became a primary route for caravans of heavy equipment. Subsequent events drew people from far and wide: reporters from network television, national magazines, and newspapers; filmmakers; environmental advocates; politicians; and celebrity elites such as Robert F. Kennedy Jr.

I met the Carters while covering natural gas development in New York and Pennsylvania for Gannett's Central New York Newspaper Group. Their homestead was at the center of a rush to develop the Marcellus Shale—soon to be touted as the mother lode of domestic energy reserves and emblematic of a conflict over shale gas development that was taking shape across the county. On the surface, this was a story about ecology versus the economy, land use conflicts and policy, and the heady prospects of unheard of sums of money changing hands. These themes were compelling enough and easy for the mainstream media to grasp. But people soon sensed that the significance of events ran deeper. The energy future of the nation, global warming, energy in-dependence, and ecological sustainability—these were the larger issues of the day, and they were integral to the Marcellus story unfolding in out-of-the-way places such as Dimock.

The Marcellus, a mile underground and stretching from southern New York through Pennsylvania and parts of West Virginia and Ohio, is a black shale formation long known to hold an abundance of natural gas. That gas was essentially worthless, trapped in rock with no effective way to extract it using conventional drilling methods. The formation was a vault with no key;

and prior to the time I met the Carters, it was noteworthy only because of the technical challenges it posed to those drilling through it to reach proven gas-bearing formations deeper in the earth. The best known of these were the Oriskany Sandstone and the Trenton Black River Limestone. For years, drilling companies able to pinpoint porous, geographically sporadic pockets in these formations—mostly in parts of western New York, Pennsylvania and West Virginia—were rewarded with prolific and free-flowing production.

When landmen came calling to secure rights to property in Dimock in 2006, few people outside the industry had heard of the Marcellus. Residents had no way of knowing its worth to energy companies, which, with a series of breakthroughs pioneered in Texas, had found a key not only to the Marcellus vault but to black shale reserves worldwide. The discovery—involving a process that uses hydraulically applied force to shatter rock and free gas deep in the earth—held the potential to dramatically alter the lives of Dimock residents and change the global dynamics of energy production and consumption. By spring 2008, people living over the Marcellus from the Finger Lakes to the Ohio River Valley began catching on as Underwood and his colleagues made their rounds.

Newspaper reporting offered a front row seat from which to track daily developments in both states. I drove from upstate New York south through the Endless Mountains, where I met the Carters and other residents of Dimock, including Victoria Switzer, Norma Fiorentino, and Don Lockhart. The story of the Marcellus begins here, in the rural reaches of northeastern Pennsylvania, long before most people knew anything of shale gas or of hydraulic fracturing.

A HISTORY OF EXTRACTION

Carter Road is named after the Carters because they were the oldest family living there when the 911 emergency system was installed late in the last century. It's where beauty and utility have coexisted, not without tension, for generations. Primeval forests of hemlock, pine, maple, and beech were logged to exhaustion throughout northern Pennsylvania well before the Carter family arrived. By the early 1900s, a rush on resources had advanced with little technical or scientific experience, government oversight, or planning. Timber was a versatile and seemingly limitless resource for heating, building, burning, and tanning. Hillsides were stripped bare as industry took hold during the nineteenth century.

The trees provided timber for a vast network of mineshafts for a booming anthracite coal mining industry concentrated in five counties of northeastern Pennsylvania, including Susquehanna County. Just prior to World War I, the industry—extending south from the northern border of Lackawanna County through Dauphin County—employed 181,000 mine workers and led the growth, capitalization, and vitality of the region. The coal extracted in abundance from this region provided cheap, efficient fuel for factories, forging works, and homes. Railroads were built, canals dug, and an entire economic infrastructure was built on the back of the northeastern Pennsylvanian coal industry.

Economic success was accompanied by historic measures of calamity. Between 1869 and 1999, more than 31,100 miners died on the job in the anthracite region. Workplace safety improved over time, yet injury and fatality in the coal industry remain hazards to both workers and residents. Waste from mine shafts still fouls rivers and tributaries. The earth beneath entire communities spews fumes from inextinguishable fires. One of the most famous examples is found in the town of Centralia, once a vibrant borough with more than 2,000 residents in the southern part of the anthracite region. It was abandoned and condemned after an underground coal fire began spreading in 1962. The fire still burns today, and only a handful of people stubbornly hold on to their homes in what has become a ghost town.

Another disaster, less than 40 miles south of Dimock, accelerated the demise of the anthracite coal industry and highlighted the costs of resource extraction. The Knox Mining Company disregarded orders by the state Department of Mines to shut down an anthracite mine carving out a vein under the Susquehanna River in Port Griffith. Work continued even after an inspector found that the company was violating rules that prohibited mining within 35 feet of the bottom of a riverbed. On January 22, 1959, the river fell through the roof of the mine. A massive vortex formed over the hole as 10 billion gallons of water drained from the river and filled the interconnecting network of shafts below the Wyoming Valley. Twelve miners died, and mine operations were halted in the valley.

A few hundred miles to the west, in the Allegany region of Pennsylvania, oil and gas drilling witnessed a similar pattern of prosperity followed by neglect and decay. It began in 1859 with the pioneering venture of Edward L. Drake, a retired railway man hired by investors of the Pennsylvania Rock Oil Company to explore ways to extract oil, which had limited but growing value at the time. Drake was the first to employ drilling machinery to produce petroleum from wells rather than experimenting with other methods then in vogue (and proven ineffective) to collect it from surface seeps common in the area. His success in developing the first U.S. wells in the Titusville region sparked a rush of capital and workers to the region that marked the birth of the modern petroleum industry. For the next three decades, tens of thousands of wells—some clustered a dozen feet from one another—would be the hallmark of these boom towns. Coupled with advances in production and refinement as well as growth in applications for light, heat, manufacturing, and transportation, these Allegany wells laid the foundation for a fossil-fuel-based economy in a country then largely dependent on whale oil.

The petroleum boom in the Allegany region brought untold wealth to some investors and speculators, and gainful employment to teamsters, roughnecks, and laborers. When it was over, it left abandoned towns and memories of environmental degradation and human loss. A series of spectacular oil fires devastated entire communities in the Allegany region before the boom ended in the last part of the nineteenth century. The worst was on June 5, 1892, when petroleum tanks overturned in a flood, causing an inferno along Oil Creek. According to an account in the *New York Times,* "an explosion was heard up the stream, which was rapidly followed by two others, and quick as a flash of lightning the creek for a distance of two miles was filled with an awful mass of roaring flames and billows of smoke that rolled high above the creek and river hills." The fire raged for days from Titusville to Oil City and killed scores of men, women, and children.

Petroleum speculators began leaving western Pennsylvania to seek oil first in Ohio, and then after the turn of the century in Texas and Oklahoma, where oil fields were proving more productive than those of Pennsylvania. Electricity had begun displacing oil and kerosene as a preferred source of light, and automobile production had yet to begin driving the demand for oil exploration. It was about that time that Ron Carter's family legacy began in Dimock, just north of the still-bustling anthracite coal mines. By the 1920s, the virgin landscape once occupied by the Delaware and, later, Iroquois Indians was long gone, but the land was beginning to heal from decades of logging and was developing a rustic beauty that is at the base of its character today. From their porch, Jeannie can see three massive hickory trees that, like the stonewalls flanking the property, offer a tangible link to an earlier time. She vividly remembers being fifteen years old and picnicking with Ron under those trees the year they met during an outing organized by Ron's parents and other members of the Dimock Baptist Church.

At that time, Ron's family owned close to 200 acres. Ron was born and raised in a gabled, four-bedroom farmhouse perched beyond the hickory trees and across the field from Ron and Jeannie's current home. In the 1940s, Ron was a young boy, and his parents and grandparents were able to coax enough of a living from the land to comfortably support an extended family. His father, Ray Carter, tended the animals—cows, pigs, and chickens—and his grandfather, George Brown, logged and sold timber that provided infrastructure for coal-mine shafts being dug to the south of Dimock. In the house where Ron grew up, trips to the grocery store were infrequent. The family occasionally slaughtered a pig and a cow in the barn, and slabs of smoked and salted pork hung on the porch through the winter. The cellar was full of potatoes, canned fruit, produce, and bulk provisions for baking and cooking. The family made enough profit to buy the things they needed, and by today's standards, they didn't need much.

One thing they could not do without, however, was water. Crops and livestock need plenty, and in Dimock wells and streams are fed by one of the greatest freshwater circulatory systems in the country. Burdick Creek, where Ron fished as a boy, runs through the hollow to the west of Carter Road before connecting with Meshoppen Creek, which feeds the Susquehanna River. The Susquehanna River watershed and the adjoining Delaware River watershed to the east are the heart of a network of streams, ponds, lakes, wetlands, and rivers encompassing 41,000 square miles in parts of New York, Pennsylvania, New Jersey, and Maryland. In addition to sustaining a thriving ecosystem, the watersheds provide drinking water to millions of people (including residents in New York City) and support industries and commerce throughout the Northeast and mid-Atlantic.

The life that Ron knew as a young boy had begun to change by the time he was a teenager. The post–World War II economy grew with unprecedented levels of production and consumption. America's ever-growing demand for inexpensive commodities and processed food could be met more efficiently and profitably by production from other parts of the country and the world. New industry rose based on economies of scale, capital interests, mechanization, and cheap labor. Across the United States, agricultural practices changed accordingly, and those changes did not favor Appalachian family farms.

Lacking a single iconic attraction such as the Adirondack High Peaks, the clear and deep waters of the Finger Lakes, or the resorts of the Poconos, the Endless Mountains of the northern Appalachian countryside was largely forsaken by the tourism that came to the aid of other communities after World War II. Dreams here, from settlement times, were made from raw beauty difficult to measure yet easy to squander—timber, agricultural product, minerals, and stone valued in board feet, BTUs, and raw tonnage. This and hard labor fueled the industrial revolution and wartime economic booms. By the early 2000s, such dreams had largely played out, leaving a poverty rate greater than 12 percent and an economic void waiting to be filled.

Long before the arrival of the twenty-first century, it was common for those of Ron's generation to leave the farm after high school to pursue new dreams elsewhere. In 1957, Ron moved to Johnson City in the Southern Tier of New York, about 30 miles northwest of the family farm. Although technically a village then, it was growing into a bustling manufacturing hub, with the Endicott Johnson shoe empire still thriving and a promising information technology company called IBM expanding quickly. That same year, he married Jeannie, who had left her family farm in neighboring Brooklyn Township to work as a secretary in Endicott. Like the family farm, the Endicott Johnson operation soon fell amid shifting economic pressures involving free-market demand for more with less. In this case, shoes could be produced far more cheaply overseas. Ron and Jeannie and their three sons then moved back to the Carter homestead, and Ron found sporadic work at Bendix in South Montrose and, later, a steady job at Procter and Gamble near Scranton. He worked at Procter and Gamble for twenty years making disposable diapers until a heart attack forced him from the job before he became eligible for retirement benefits.

Ron tells me his family story matter-of-factly in a deep, slow voice, pausing to gather his thoughts now and then, and conveying his feelings through quick, forceful hand gestures. Jeannie mostly listens, but chimes in with details Ron may have forgotten. The heart attack, she offers, "made other things seem not so important."

Her husband calls her Jeannie or, sometimes, when not addressing her directly, "The Wife." She's a slight woman, with a sweet voice belying a formidable strength and a farmer's firmness with animals. She snags Brandee, the family's excitable terrier that likes to jump on visitors. The dog struggles unsuccessfully in her arms as she settles back into her chair.

Jeannie worked at various jobs over the years ranging from factory work at Allied Signal (formerly Bendix) to a greeter at Walmart in Tunkhannock after Allied Signal closed. She and Ron supplemented their income, eventually paying off the mortgage of their upgraded trailer home, by selling off pieces of the homestead (including the farmhouse where Ron grew up). The real estate market in northeastern Pennsylvania was generally too remote to attract developers of subdivisions, which tend to drive up prices of rural land situated near urban areas. In Dimock, the relatively low demand, low prices, and low cost of living attracted other kinds of buyers: those seeking acreage for wood lots, a country home, or perhaps a cheap alternative or supplement to long-term market investments.

In addition to using the real estate money to pay off debts, the Carters invested in machinery for a small quarry on the remaining land. Two of the Carter boys lived separately on the homestead, and both of them were in the stone business. Even as making a livelihood from coal, petroleum, and farming became a thing of the past, stone remained a significant industry in northeastern Pennsylvania. An abundant endowment of bluestone, coveted for upscale building projects, lies just under the surface of many fields and wood lots in this area, and it commands a good price when the building market is on. The old fieldstone walls also fetch a price on suburban markets; and it's not uncommon for farmers to sell their stone walls to out-of-town buyers, who have them trucked off and reassembled in affluent neighborhoods in New York and New Jersey.

But the bluestone market was flat, and Ron and Jeannie were down to 75 acres when the landman arrived.

THE LANDMEN

While gas reserves were found and produced to the west in New York and Pennsylvania throughout the twentieth century, prospecting to the east remained sporadic, low-key, and highly speculative. Breakthroughs in shale gas development changed that—by 2007, thousands of landmen were competing for rights to land over the Marcellus on both sides of the Pennsylvania–New York border. Many of them worked for national and multinational gas companies based in Texas and Oklahoma, including Cabot, Chief Oil and Gas, Range Resources, and Chesapeake Energy.

One of them was James Underwood, the landman who had approached the Carters. While I was researching developments in Dimock, I phoned him to arrange an interview. I wanted to ask him about his travels throughout the country, and his encounters with people in northeastern Pennsylvania. I was especially interested in what he told residents about the newly exploitable resource to which gas companies were seeking title. He had been living out of a hotel room at the Holiday Inn Express near Interstate 81, he told me. Ron and Jeannie Carter remembered that, in addition to his sedan, Underwood had a large pickup and a camper from which he sometimes worked. As I tried to arrange a place and time to meet, he became chronically unavailable.

The other name that often came up in my interviews with landowners was Frank Fletcher. Like Underwood, Fletcher dressed and talked like a cowboy, and he left a distinct impression on those he met. "He looked like a big Texan. He dressed real nice, real flashy, and wore gold," said Julie Sautner, one of the Carters' neighbors. She and her husband, Craig, signed their mineral rights over immediately after hearing Fletcher's pitch. In my phone interview with him, Fletcher fit their description. He was charming and polite, with an engaging southern drawl and a cowboyish manner. He was born in Massachusetts and had lived in Los Angeles, although, as he told me, he may have picked up his accent from having worked for decades in Texas and Oklahoma. We arranged to meet at the Flying J truck stop in New Milford. I was to look for the man in the cowboy hat, headgear that was increasingly in vogue throughout Susquehanna County. On the day we were to meet, I called Frank to tell him I was finishing an interview in Dimock and I was on my way. He informed me that he regretfully had to pass, "at least for now," after Cabot officials told him not to do the interview if he wanted to continue to work for them. "Sorry for having to back off, but I guess you understand," he said. "I hope you're at least sympathetic to our side of the story."

I was certainly curious. The business of Fletcher and his colleagues was integral to events unfolding in Susquehanna County.

Fletcher and Underwood are part of a brotherhood of peddlers better known in oil-rich communities in the Midwest and South than in the Northeast. But their approach and style fit well into the farm culture of any rural U.S. community. They're throwbacks to the day when door-to-door was a common way for men to do business and deals were made at the kitchen table rather than in law offices. They are versed in land-use and farming matters, and the types of things that landowners worry about: taxes, weather, crop yields, animal stock, and the art of extracting wealth from the land. They are charismatic enough to show up on people's doorsteps and gain access to their homes. Often they are invited in. If they aren't, they are persistent enough to prevail, sooner or later.

Understanding the risks and rewards inherent in a deal is essential to any negotiation. Here the gas companies had the clear advantage. They were privy to data collected from beneath residents' land: seismic studies and

geological surveys supporting the growing body of industry knowledge suggesting the Marcellus was the richest domestic energy source in the country. Northeastern Pennsylvania and southern New York straddled one of the
most promising sections. Few residents of Dimock had any idea of that, and
none of them had access to detailed scientific data.

Gas prospecting brought another interested party to the region in spring
2006—Jackie Root, a farmer and self-taught lease negotiator from Tioga
County, Pennsylvania. After having some success organizing residents to
deal collectively with companies in Jackson, about 70 miles west of Dimock,
she was recruited by the Penn State Cooperative Extension to help advise
people dealing with offers from landmen. She likened dealing with landmen
to "a poker game where they get to see your cards but you don't get to see
theirs." Mostly, people trusted landmen because they had high hopes and
little experience in such matters, she told me. It's natural to believe somebody "with a charming southern accent that looks like your grandfather," she
said, especially when that person conveys enticing information about your
land and "looks like he knows what he is talking about." It's hard to know
what to believe, she added, because along with good, honest landmen there
are some known to tell "untruths."

There was a way around the risks of this game, and it involved establishing solidarity with neighbors. Large tracts of contiguous land owned by
multiple individuals were made even more attractive if their mineral leases
could be dealt all together, eliminating patchwork acquisitions. Both owners and lessees could benefit from this, Root thought. This was especially
effective when landowners were knowledgeable about the resources they
possessed and united in their commitment to hold out to command the best
price. During a gas rush spurred by a proven resource, it was an especially
good way to deal with big companies with deep pockets. On the other hand,
when speculation was high and interest was low—as it had been for years in
Susquehanna County—there was little incentive for prospectors to play a
game in which the rules were established by a collection of holdouts.

In 2006, before things had developed very far in Dimock, Root was one of
the featured speakers at a meeting coordinated by the Penn State Cooperative
Extension in Montrose. Not wishing to alienate the industry, she invited
Frank Fletcher to participate in the spirit of public education. "He told
them 'we'll pay $25 an acre and never a nickel more,'" Root later recalled
about that meeting. "I think Frank was very threatened by what we offered."
Perhaps he was, but for various reasons Root's organizational efforts never

took hold in Dimock. She thinks this was because the Marcellus phenomenon was too new and too little was publicly known about the value of mineral rights prior to 2008. It's easy to point to the ignorance of others in hindsight, she explained, when you've already read in the papers and seen on television that Marcellus leases are going for $4,000 per acre. But it didn't happen that way for residents of Dimock, which was in effect the prospecting frontier of the shale gas age. "It wasn't because they were less intelligent than anybody else. It's because others were able to learn from their experiences."

On their own, some Dimock residents fared better than others in negotiating their leases in those early days, and each had his or her landman story. The time was ripe to be sold on anything billed as a new opportunity, and a lot of landowners in Susquehanna County were buying. Pat Farnelli, who lives in the Carters' old farmhouse with six of her eight children and her husband, remembers the day a stranger showed up in a large white sedan. He reminded her of the Marlboro Man. He didn't promise riches, exactly, but his very presence suggested the possibility. "It was like he was offering a lottery ticket," she said. "He made it sound like the chances weren't that great, but we had nothing to lose and everything to gain."

Pat, in her mid-thirties, is apple-cheeked and talkative, with spectacles and straight brown hair spilling about her shoulders. She and her husband Martin had bought the Carters' old farmhouse along with a 25-acre tract with the dream of farming it. Things were not going well when the landman appeared. Bills were mounting. Work to generate outside income to support the farm was scarce and growing scarcer. The family was surviving, sort of, on Pat's monthly Social Security checks, food stamps, and her husband's income as a part-time cook at the Flying J diner and truck stop by Interstate 81. The $25 per acre in lease money, to be paid up front, would just about cover a shortfall for the mortgage.

NO WAY OF KNOWING

Many of Pat's neighbors also could use the money, some desperately. Gossip about the landmen's visits spread. At lunch counters, grocers, churches, schools, the post office, and barber shops, the residents of Dimock swapped stories. Who signed and who didn't, and for how much—this all soon became public knowledge. "Basically, everybody was talking about this all the time," Pat said. "This became a constant conversation."

That conversation began every morning at 5 a.m. when regulars started arriving at Lockhart's lunch counter in South Montrose and shared news of their neighbors, their animals, and their neighbors' animals over coffee, hash browns, and eggs. Don Lockhart, a short, gregarious man with a barrel chest and thick glasses, has held court in the small market, gas station, and diner for twenty-five years, and the establishment's appearance has changed little in that time. Patrons sit on round stools with fluted chrome bases fixed to the linoleum-tiled floor in front of a faux wood–laminated counter. A hall opening near the far end of the lunch counter leads to the office of Shirley Lockhart, town notary (and Don's wife). She works from a large steel desk amid bureaucratic certificates, notices, stacks of licensing information, and photos of children and great-grandchildren. This is where landmen and landowners go to get their contracts notarized. There is little that happens in Dimock that the Lockharts don't know about.

Underwood and Fletcher began showing up with residents at Shirley's office in 2006. By the middle of 2008, the ability of these and other landmen to close deals was keeping Shirley very busy. "People were excited," Don recalled years later. "Everybody was very positive about it." He himself was among the first to sign—for $25 per acre. "I'm not bitter about it," he continued. "They were just good salesmen." He remained bullish on the gas rush—a boon for his and Shirley's business. Drilling rigs and equipment require plenty of fuel and licensing paperwork. The men who operate them are not delicate eaters, either.

The sheer number of landmen preceding these crews, and their doggedness in both New York and Pennsylvania, was a sign of something monumental. Everyone at Lockhart's sensed that, and excitement went beyond the lunch counter. The landmen also carried bits of news and gossip from one household to the next. They leveraged information and speculation—who had signed, what their neighbors thought, and what the future held—to suit their pitch. Not everybody was willing to accept Don's characterization that this was simply good salesmanship. Stories of questionable practices circulated. As events unfolded, some of these stories became allegations at the root of legal claims that the industry unfairly and illegally exploited residents on both sides of the state border.

In New York, the attorney general's office is free to pursue complaints of its choosing, based on their merits and, perhaps, the recognition of their potential for publicity. That may or may not have been the case when Andrew Cuomo, the New York attorney general when the land rush began, characterized industry tactics as misleading, bullying, and deceptive. "Many of these companies used their size and extensive resources to manipulate individual property owners," he said in November 2009. "This land grab must stop." His scolding followed an investigation by his office into the methods used by Fortuna Energy to claim property from more than three hundred landowners in southern New York. Fortuna (which later became Talisman) agreed to relinquish its claims to the property and pay the state $192,500. The problems were not isolated to Fortuna or any one company, Cuomo said. They were industry-wide, and his office was continuing to investigate problems and seek reforms.

In Pennsylvania, contract disputes between private parties, including those involving land leases, fall outside the jurisdiction of the attorney general's office. Landman disputes are thus left to civil courts, which tend to operate outside the political limelight. Residents from fifteen households in the Carter Road area ended up taking their landman complaints to Pennsylvania's U.S. middle district court in 2009, where they sued Cabot on several grounds, including "fraudulent misrepresentation" by landmen who "misstated material facts and omitted other material facts" to get residents to sign leases. The material facts in question related to the "amount, timing, and regularity" of royalties and "risks . . . to persons and property."

Cabot officials, in a prepared statement to the media after the suit was filed in November 2009, dismissed the claims as baseless. "While we respect the right of any resident to seek a judicial solution for a legitimate issue, we see no merit in these claims and are disappointed that these citizens felt it necessary to proceed in this fashion," stated Dan O. Dinges, chairman, president, and CEO of Cabot.

Industry representatives did not deny problems in the region, however. Some were related to training and experience, said John Holko, secretary for the Independent Oil and Gas Association (IOGA) of New York State. Staffing up to tackle a play the size of the Marcellus was no small matter. Even after experienced veterans were transferred from industry strongholds in Texas and Oklahoma, there were a lot of doors to knock on and relatively few available people with the necessary set of qualifications to solicit mineral deals. "The problem we have now is manpower," Holko said in 2008 when prospecting began accelerating in New York. "Some are uneducated. Some are less experienced. There is a learning curve." Robin Forte, executive vice president of the American Association of Professional Landmen, characterized questionable ethics as the exception rather than the rule among thousands of landmen working across Appalachia during the early stages of the Marcellus rush. As with any industry, some are better than others, and there may be occasional scoundrels, he told me. "I don't want to minimize that some of the people may have been subject to unethical treatment. We are concerned about it."

Residents and advocates turned to the legal system to settle claims; however, it didn't require adjudication to show that many property owners felt taken by the landmen. Based on residents' accounts, the pitch varied from household to household. It's hard to know exactly what was said without a tangible record. Yet, as I listened to story after story, themes emerged. A common thread involved pressure tactics. People who didn't sign were made to feel like they would be cut out of the deal or were holding up the good fortune of their neighbors. The time was always now, and the terms would never get better. "They made it sound like everybody else was against you," Pat Farnelli said. The other theme was betrayal. "We were all lied to and misled," said Julie Sautner, who signed with her husband Craig for $2,500 per acre. "I wish we knew then what I know now. But how could we?"

In 2007 they couldn't have known. But things might have turned out differently for them had the landman's pitch disclosed a bit of critical information known to everyone in the industry: the residents of Dimock were living on top of billions of dollars' worth of natural gas locked in a formation that posed a set of technical problems that the industry and regulators had little or no experience dealing with. Mistakes were inevitable, and they sometimes would come at the expense of landowners.

Forgoing this disclaimer, landmen successfully persuaded Dimock landowners to "partner" with Cabot to reap the good fortune they were sitting on. Still, some residents were less interested than others in what the landmen had to say, and others were downright skeptical about the whole venture. "I don't believe we want to do this. I don't believe we want to do this at all," Jeannie Carter insisted. She remembers uttering those words on the phone to Jim Underwood's wife, who had been relentlessly trying to convince Jean that she was sure to miss an opportunity if she and Ron didn't act soon. Ron deferred to his wife of fifty years, although he was more open to the idea. He knew his father had considered, and declined, leasing mineral rights to speculators never intending to drill but hoping to sell the leases at a huge profit in areas that might someday pan out. At $1 per acre, it was a cheap gamble. This was different. At $25 per acre, Cabot looked to be serious about drilling. The real money was in the royalties—residents were told this so many times that it became ingrained in their minds. Life on a fixed income would become a whole lot easier with monthly royalty payments. Real figures were elusive, but they always seemed to be enough for a new house and a comfortable retirement. Ron and Jeannie could take the trip to Vestal, New York, more frequently to eat at the Texas Road House, one of Ron's favorite places. When the royalty checks began flowing in, they might be able to afford to build a bigger home across the road, with a basement and room to accommodate their extended family of eighteen for family gatherings. In late 2006, the Carters finally agreed. "I asked the landman, if we didn't sign, what would happen," Ron said. "He told us, they would take the gas out anyway."

The Farnellis were told the same thing, Pat recalled. "It came at a time when we really needed it." The 12.5 percent royalty figure seemed reasonable. It was certainly tangible. It indicated a return directly proportional to the yield. Any farmer could appreciate that. Only with time and experience did they learn that it was, in fact, the minimum allowed by the state. Beyond that, prospective returns were an abstraction, subject to mathematical and legal interpretation and faith that companies would be able to accurately meter the flow of gas through rock fractures and the complex geology a mile below the surface.

A MODEL OF ENDURANCE

The landman who braved a wary Labrador retriever at a homestead near the Carters' land found a woman quite receptive to his proposal. Norma Fiorentino lived off and on with various children, step-children, grandchildren, and great-grandchildren on a modest homestead on State Route 2024, where it joins the north end of Carter Road. Her late husband had been a union plumber from New York City, and some of her boys are quarry workers. She worked for decades as an elder-care nurse and home health aide, even after disabling blows from a stroke, heart attack, and complications from diabetes. Physically demanding work seemed to be her calling. Some of her earliest memories are of helping stack rock fragments on pallets in the family quarry in Brooklyn, the town east of Dimock. "It kept us out of my dad's hair," she said, sharing details of her family history with me during a visit.

We sat in her family room, which was adorned with memorabilia, framed photos, and some of Norma's needlework. With limited mobility, she sat on the sofa next to her telephone and piles of bills, receipts, and various pieces of paperwork on a coffee table that doubles as a desk. The couch faces the side door, which opens on to the end of the driveway, offering a vantage point to track daily comings and goings—mostly of her kids and grandchildren and the occasional neighbor or in-law dropping by. Everybody knows Norma. She grew up in Susquehanna County, raised children with her contemporaries, and cared for many of their parents as they aged and their health failed. By the time I first visited in 2009, she had also become something of a public figure outside of Dimock, due to events she never could have anticipated.

The quarry of her youth was a place of hazards and great adventure, and the centerpiece of family life. On blistering July afternoons, Norma, her siblings, and their friends plunged into the cool green depths of quarry lakes. They climbed on the warm rocks to drip dry and then dig into paper bags to find peanut butter sandwiches. One day, her mother pulled up with a carload of kids and a picnic for her father and the workers. They were intercepted by her father, running and flailing his arms. They had pulled into a blasting zone. The charge didn't detonate, but the vision of her father rushing toward the family in desperation and urgency stayed with Norma.

On a winter day in 1970, she met her husband Joseph—a master plumber and handyman who had moved from New York City to manage a farm in Brooklyn Township—when he came to fix her furnace. They later married and settled on 3 acres of uneven pasture in Dimock. The centerpiece of the property is a doublewide that Joe artfully converted into a bungalow with a cinder-block chimney and pitched tin roof extending over a front porch. Two trailers on adjoining lots have housed various family members over the years. When Joe was alive, miniature ponies grazed in pastures between Norma's house and the trailers. By the time of my visits, the pasture had become mostly overgrown. Joe, Norma told me, was most happy when he was busy, making broken things useful again, tending the animals, or helping his boys in the quarry—where he held the positions of both "go-for" and senior advisor. He worked until he died of a heart attack in the kitchen one morning while getting ready to help his sons in the quarry. He was seventy-five. Younger than Joe by a decade, Norma continued her nursing work despite her grief and even as her own health began to fail.

When the landman came calling in late 2007, she had sold the five ponies and the house was falling into disrepair. A daughter was fighting terminal cancer. A son, struggling with drug addiction, would eventually land in prison. She was fresh with grief from the loss of her husband and coping with her own uncertain health. Yet she held the family together, running the homestead and often helping out with her six children's children and their children. Her dream was a new camper in which to spend summer vacations with her grandchildren or a small cottage on nearby Elk Lake. The man—who came to her door with good news and a raft of papers detailing lease money and royalty payments—could have been an agent for such dreams. Norma can't remember his name, and she has no record of his identity. He was a polite older man with a southern accent and cowboy boots, she told me. The lease money on 3 acres wouldn't amount to much, but the royalty payments, 12.5 percent, could change her life, she was led to believe.

"He told us money would not be a problem ever again," Norma recalled wistfully, as her daughter chatted with in-laws in the other room, an adolescent grandchild snoozed on the couch, and a great-grandson toddled about the house.

DREAM IN PROGRESS

Victoria Switzer could not remember the name of the landman she first spoke to either. But the "charming, nice old guy" who paid a visit to her trailer on a brilliant October day in 2007 found an altogether different set of circumstances and personalities. And the landman adjusted his pitch accordingly.

At the time, Victoria and her husband Jim were living in a trailer while they invested their life savings in their new home being constructed on 7 wooded acres off State Route 3023, which joins the south end of Carter Road. While Norma's idea of a cottage on the lake was little more than a distant vision, the Switzers' dream was in full swing, and the landman successfully presented his deal as not disrupting it. There was no promise of life-changing sums of money. In fact, Victoria remembers, he made it sound as if some "exploratory drilling" in the area would be incidental, at most—"a well or two" somewhere out of the way with few traces of drilling activity. It would not amount to much, but it just might provide some income to contribute to their nest egg as, right then, all their life savings were going into the house. Victoria routinely passed by gas utility lines running through parts of the countryside without giving much thought to where they went. Those lines, familiar and innocuous, reinforced the notion that, with the purchase of their new property and building a new home, they were in a serendipitous position to capitalize on development that was already an established feature of the landscape.

"He made it sound so benign," Victoria told me. "If he said we were going to get rich, flags would have gone up." Without much thought or negotiation, they agreed to the same terms as Norma and the Carters: $25 per acre and 12.5 percent royalties. Although Victoria doesn't remember the landman who made the deal with them, she does recall Frank Fletcher; he was there for the signing and he was the one who followed up with other papers that, as events began unfolding, did raise plenty of red flags. "I feel like we traded Manhattan for blankets and beads," Victoria said. She mimicked a patronizing drawl: " 'Oh, you're such a worrier, Mrs. Switzer,' he told me. 'Things are going to work out just fine.' "

Since she had been with Jim, things had indeed been working out better than she ever hoped, and perhaps she was taken off guard when the landman came. Victoria Heise, a school teacher for thirty years and single mom for more than twenty years, had begun dating Jim Switzer in 2003. They were a striking couple. Victoria, a brunette barely 5 feet 4 inches tall with penetrating blue eyes and high cheekbones, was a study of poise and focused intensity. Jim—known to his friend and Victoria as Jimmy—was an automotive technology teacher in the same school district. Jim has the physical charisma of an athlete on his game. After his first marriage ended in a divorce, he had become impassioned with bicycling and toured cross country. There seemed little he couldn't do. "If you can read a manual, you can do it," he was fond of saying.

Most of Victoria's adult life had been about working and saving for her daughter's college education, who ended up going to New York University. Victoria taught during the day and sold furniture at night, while rising to the demands of single parenthood. After her divorce as a young mother, there had been no room for a serious relationship—although not for lack of suitors— or even for the indulgence of a house of her own. She rented a place in Tunkhannock, where she taught, and even lived with her parents for a while.

Neither Jim nor Victoria had dated much after their divorces, and they were not looking for anyone during the years they were raising their kids. The two had, however, known each other a long time; their daughters played field hockey together. When they both became empty-nesters, romance happened quickly. He borrowed a friend's Corvette one day to take Victoria on a get-away to Rehoboth Beach, a resort in Delaware. The courtship was "fast and furious," and by the second date they were talking about their common desire for a house in the country. "That did it," Victoria said. "He was my man."

Settling down after all those years to build a house with Jim had a happily-ever-after appeal that became a catalyst for their courtship and a central component of their relationship. It became a multiyear undertaking involving all their free time, energy, and resources. They began drawing up plans. It was to be a post-and-beam beauty with huge windows and a vaulted ceiling rising 30 feet over a great room drawing to a point on the south end. They had beams and woodwork from native hemlock milled especially for the project and stacked under tarps to cure. They custom-ordered native bluestone, cut, ground, and polished as smooth as marble, for the counters, thresholds, and accents around the great hearth. They designed it all as a family destination-retreat, with rooms for children and grandchildren visiting for weekends and vacations.

The vision was not without sacrifices. Dimock did not strike her right off as the setting for happily-ever-after. "It was cow country and stone quarries," she said. "How am I going to coexist with that?" There was no nearby place for the sushi or martinis she enjoyed with her friends in Tunkhannock. She had a change of heart when Jim showed her the property, 7 acres of verdant hills with a creek coursing through the woods.

The property came with an old trailer, with moss on its roof and plastic on the windows, where they stayed while planning and building their dream. Victoria almost fell through the floor when she first walked in. It was cramped and uncomfortable but, in hindsight, at least, it possessed the kind of romance suited to newlywed life. "We figured together we could pull this off if we moved into that trailer and did it on our own." Moreover, her dog, a good-natured German shepherd named Bruin, grew especially attached to trailer life. Although the living quarters could stand improvement, the land was perfect.

Victoria told me that she is "a Jesus fan" and Jesus is central to her ethics and life. She was raised as a Methodist, but she is not a traditional church goer. The natural beauty she finds in daily life serves as her church. She and Bruin found plenty of that in their new woods, where they explored for hours at a stretch. One of her favorite spots was a secluded bank of a creek wandering under a canopy of hemlocks. It reminded her of the creek at her grandmother's house. The stream was connected to times and places far beyond that, however. As an earth science teacher and Susquehanna County native, Switzer was well aware of the geographical and ecological significance of the watershed and its lattice of waterways leading to the Chesapeake Bay.

The streams are part of a vast system flowing in all directions to the Susquehanna River, which coils in a semi-circle around the county that shares its name. It flows from its headwaters at Otsego Lake, on the southern banks of which lies the village of Cooperstown, New York, and dips into Pennsylvania before turning abruptly west at Great Bend, snaking along the Southern Tier of New York and then sweeping south again, passing back into Pennsylvania and working its way to the Chesapeake Bay at the end of its 444-mile course. With the Delaware River watershed to the east, the Susquehanna is one the most significant and productive freshwater resources in the country. The sliver of stream that Victoria mused over was a connection to the past, and although she was unaware of it at the time, it would become a powerful icon of the battle taking shape over Marcellus development in law and government offices from Albany, New York, to Washington, D.C.

Victoria, wandering along the stream one day on her new property, began exploring a new stretch of woods on the hillside when she encountered a stranger.

"Who might you be?" the man asked. His name was Ken Ely, and he owned the land on which Victoria was inadvertently trespassing. He had a thick build with strong, deeply tanned forearms and long gray sideburns. Victoria explained she was the new neighbor, and she just liked to hike. She was sorry if she was trespassing.

Switzer knew Ken owned a quarry at the top of the hill, and she fretted over the gash in the land and the dust and noise from trucks rambling over access roads. Ken didn't like the idea that Switzer had posted signs on her land to keep out hunters. "That's not very neighborly," he told her. Indeed, they were polar opposites in appearance and manner—Victoria, a picture of refinement in a color-coordinated hiking ensemble, and Ken looking as rough as unfinished quarry rock in a t-shirt and work boots. Victoria saw the land as a cathedral, a place of inspiration and reflection. Ken, too, was a devout Christian, but a churchgoer—a member of the Baptist Church with the Carters. He was also an avid outdoorsman and a hunter of bear, deer, and turkey. He knew the waterways and tributaries, the hills, and fields of the region well; he knew the fish and game they supported and their importance. He shot animals for sport and for food. Before he became a quarry man, he had owned an excavating business. He was not afraid to disturb the land. It was his livelihood.

At that time, before the landmen had made their mark on Dimock, neither Victoria nor Ken could have anticipated the events about to unfold that would make them close allies, along with the Carters, Norma Fiorentino, and the Farnellis, joined in a fight to salvage their respective dreams.

2. COMING TOGETHER

For Jackie Root, the Marcellus story began about 70 miles west of Dimock, near her home in Jackson Township, Pennsylvania. In fall 1999, years before the Marcellus rush drew landmen to the Endless Mountains, Jackie was driving her pickup truck on an errand at a neighboring farm when she noticed small blue and yellow flags lining Picnic Grove Road. Although there was something about them that looked peculiar, she dismissed them as survey work for road maintenance, or perhaps reference points for the 911 system the county was developing at the time. The next day, when she encountered a convoy of men and machinery advancing slowly along the marked road, she recognized the flags for what they were, and the implications they carried.

Earlier that year, Jackie had turned forty, and she was proceeding with her life ambition to be a farmer. After high school, she had earned a two-year agricultural degree from University of Delhi before buying a 400-acre dairy farm on Collum Road with her husband Cliff. When they moved into the three-story farmhouse in 1984, drilling rights to the land were held by a small independent company and they expired without incident two years later. A gas well had been drilled on their land in the 1940s when drillers were exploring the Oriskany Formation. The well never produced enough to justify building the infrastructure for transmission lines. All in all, mineral extraction was something the Roots thought of in terms of the land's history, not its future.

Jackie loved horses, and she married a man who loved cows. Owning a commercial dairy farm had always been an ambition for Cliff, and he began milking full-time as soon as he graduated from high school in the late 1970s. "They are not chores, they are things that just have to get done," he said about the dusk-to-dawn commitments of dairy farming. Horses were not a good fit with a serious dairy, so Jackie put her girlhood dream aside as she and Cliff settled into their new home. Over the years, she raised the couple's four children; helped out with many of those "things that needed doing" around the dairy; and fit in part-time jobs compiling information for the National Agricultural Statistics Service and preparing income tax returns, first through H&R Block and then for her own list of clients.

The colorful markers Jackie saw near her home in 1999 ushered in an altogether different life, and it began with a procession of ungainly vehicles approaching along Picnic Grove Road. They had articulating chassis and wheels as tall as Jackie's pickup. But the most unique feature was a broad piece of hardware attached to the undercarriage of each vehicle by a hydraulic system. After traveling a short distance between markers, the trucks stopped in unison. The hydraulic hardware descended from their bellies and pushed against the road, sometimes with force enough to lift the vehicles from their suspensions. Machinery throttled up with the sonic effect of a buzz saw meeting a log. The frequency of the noise rose and fell rhythmically for a minute or more. The hardware then receded into the undercarriages of the trucks, which rolled to the next spot a short distance down the road where the process was repeated. The operation was flanked by men traveling back and forth along the route in pickup trucks loaded with flashers, markers, signs, cables, and sensors.

The large vehicles were thumper trucks, a name reflecting the kind of equipment they carry. At one time, that was a heavy weight dropped from a crane to produce shock waves that rebounded off sub-surface structures and back to mobile electronic equipment that mapped them. A common contemporary method—the one that Jackie saw—is vibroseis. Using the same physical principle, vibroseis equipment creates sustained shock waves to produce three dimensional maps of geological formations. To sharpen the picture, crews sometimes use dynamite charges to produce additional waves from multiple directions and, where available, data from bore holes.

When Jackie sees strangers conducting business in her neighborhood, she told me, "I invite them to explain what they're doing." Jackie is not tall, but she is a substantial physical presence. She has a sturdy build, fair features, sandy brown hair, arched eyebrows, and dead-level eye contact that both projects and inspires confidence. She pulled her truck over and found the boss of the operation. Following a brief conversation, she learned the crew was "shooting seismic" for a client, and it had permits to use public roads.

She wasn't able to get much in the way of specifics from the crew chief, but her curiosity was whetted and her suspicions were raised. For one thing, she had gathered from her inquiry that the public rights of way gave the crews a vantage point to collect information on land extending under people's private property. Information is a form of equity. It has tangible value to the energy companies, which paid these independent geo-services contractors to harvest it. Like prospectors, Jackie had a knack for extracting opportunity from arcane places. When it came to tax work or crop yields, a skilled eye could discern critical trends and guide lucrative decisions; so she figured it was no different with respect to geology. Even without access to information that the companies were gathering from under the fields of Jackson Township, she knew ways to determine the value of whatever they were looking for. "Farmers are entrepreneurial by nature" was how she summed it up for me once during an interview at her home office, with a stack of files in front of her and reading glasses dangling from a chain around her neck. "They work for themselves and they take risks." The most successful entrepreneurs, she might have added, see things that others don't, things that seem obvious to everybody in retrospect.

OUTPLAYING THE LANDMAN

Jackie began nurturing the seed of an idea when, a few days after the seismic crews began appearing on Picnic Grove Road as well as on other local routes, a landman pulled into the driveway in front of the Root's three-story farmhouse. He climbed the steps of the spindled wrap-around porch and gave a prolonged series of raps on the storm door. After her conversations with neighbors and revelations about the seismic operations, Jackie was expecting this visit. She invited him in, and as he began his pitch, she paid him the attention of a proctor. She recalled that he was soft-spoken and that he told her and Cliff that he had just closed a mineral rights deal with Cliff's brother, who owns a nearby farm. After some discussion of the prospecting going on, the landman told the Roots he might be able to swing a deal for them that allowed for some exploration and maybe even drilling on their land because they were in good position. But please understand, he added, it was not something he could do for everybody. It would be best they keep it between themselves.

The lease he had brought for them to look over, he explained, was "a standard lease." If the operators struck gas, it would entitle the Roots to a royalty percentage governed by state law—12.5 percent. Aside from the lease payments, which he could offer at a premium of $5 per acre, all leases were pretty much the same. Jackie said they'd think on it. Before he left, she learned that until recently the visitor had driven a parcel-delivery truck and he was still learning the land trade. She didn't mention they shared that last point in common. Jackie was a fast learner, and she meant to stay a step ahead of landmen.

As soon as he was out the door, she called Cliff's brother to confirm what she strongly suspected: he had not signed anything. Rather than taking the inexperienced landman's lie as an affront, she analyzed it from a strategic perspective; and she saw how it fit neatly with a broader realization—landmen acquiring mineral rights from numerous parties benefited from a lack of reliable information and communication among the sellers. Jackie studied the lease left on the table. It would force the Roots to relinquish assets to their land while burdening them with liabilities. She didn't need a lawyer to see that. According to the landman, her neighbors were already on board with this lease, and she wondered just how many had actually signed. The "just-between-us" tactic especially made sense in certain rural communities, she later explained to me, where farms are run by men of a particular generation. "It's based on the belief that you probably won't speak to your neighbor about personal financial matters, and you might not even speak to your brother or brother-in-law, because you just don't butt into other people's business," she said. "Well, that's not me."

Understanding the pitch was one thing. Understanding the geology that was driving the pitch was something else. Often, rank-and-file landmen lacked geological specifics about the territory they were assigned to lease. They knew their employers were willing to pay a certain price, however, and the successful ones had an instinct for reading the market just as they read people. That was something Jackie was also good at. Charting who was buying mineral rights, where, and for how much would tell her what she needed to know about what the gas companies were finding with their seismic studies. To this end, Jackie set about building a parcel-by-parcel rendering of the entire township from pages she cut out and taped together from plat books of land surveys. On these maps, she outlined seismic routes with pins. Then she began color-coding the lots according to lease holdings of various companies. The work required cross-referencing the parcel numbers with deeds and title information filed at the Tioga County Courthouse in Wellsboro. It was tedious work but easy for her. She once shoed horses—that was work.

The courthouse in Wellsboro, about a twenty-minute drive from Jackie's farm, has broad granite steps leading to a structure with white pillars, tall lattice windows, and lofty ceilings. Jackie especially liked that it is filled with information that can be counted on, and in this regard, she wasn't alone. Upon stopping to chat with a clerk, Jackie learned the real property office had been busy with gas company functionaries and landmen investigating the very same records she was interested in.

When she wasn't at the courthouse, she worked from home, stealing time from the management of the farm, her other jobs, and family. She read about the culture and history of the petroleum business, including *Oil,* Upton Sinclair's influential novel about a self-made oil tycoon in the 1920s. (She also watched the movie inspired by the book, *There Will Be Blood,* which left her with the impression that the oil and gas culture has changed very little over the better part of a century.) In those early days of the Internet, she used a dial-up service that slowly produced bits of industry news and regulatory information, and crashed with maddening frequency.

Jackie lives within a fifteen-minute drive of the northern boundary of Pennsylvania. Locals call northern Pennsylvania the Northern Tier and southern New York the Southern Tier. Collectively, the area is called the Twin Tiers. Jackie's quest took her across the border into the Southern Tier to attend meetings held by the New York State Farm Bureau. It was at a meeting in Elmira that she learned the seismic surveys in Pennsylvania were extensions of prospecting that began in the Southern Tier, extended into parts of Pennsylvania off and on, and followed the Appalachian Range all the way to West Virginia. At the time, prospecting in the Twin Tiers generally traversed the eastern boundary of what was proving to be a lucrative series of pay zones in the Trenton Black River Formation, an elusive target of limestone 2 miles underground. Because gas reservoirs were sporadic within this formation, it was a "seismic-dependent" play, meaning drillers needed a good amount of seismic research to justify pursuing it; and even then, it was a hit-and-miss pursuit. But when operators hit one of those reservoirs, they often hit it big. Profits from a good Trenton Black River well justified the effort for those who could afford the cash outlay and the financial risks.

In 1999, when Jackie began her research related to Trenton Black River development, there was little discussion of the Marcellus or shale gas in the Northeast. Those who followed the industry closely, however, would have known that operators were experimenting with new drilling and stimulation techniques with the Barnett Shale in northern Texas. Up until then, neither horizontal drilling nor hydraulic fracturing was considered an effective tool for shale gas. By 1999, experimentation on using them together set the stage for a transformative effect on the industry. Because the gas of the Barnett proved accessible, the going rate for leasing land in northern Texas climbed into thousands and sometimes tens of thousands of dollars per acre, with handsome royalties. Those paying close attention would have also known the Twin Tiers covered some of the largest shale gas reserves in the world—many of them unexplored. Since the time of Drake's 1859 discovery, this area had produced oil and gas from many horizons, and it was rich with possibilities for new ones.

At first, Jackie's homemade map of leasing trends was a hodgepodge of colors and blank spots. Patterns took years to develop. Prospecting activity fluctuated with the price of gas and the accumulation of seismic information, with an upward trend over time as buyers entered the market. Sometimes, a party would lease rights to a parcel and then sell the lease to another party for a higher price. Lease prices began rising from $2 per acre to $5–$10 as more land was leased. The big properties were most coveted by companies seeking contiguous tracks on which they could map out "drilling units," surveyed parcels which they would use to model the flow of gas and corresponding revenue from a given well. Slowly, Jackie's map reflected lease holdings consolidated among several big players including Fortuna Energy (then a U.S. subsidiary of the Canadian firm Talisman) and East Resources (an Appalachian operator based in Pennsylvania). Although shale gas was not (at first) their target of record, these companies continued to expand their landholdings in the Twin Tiers as a decade of exploration in the Barnett was reaching fruition.

In addition to the blocks leased by the big companies, Jackie's map showed uncolored parcels that included the Root's 400 acres, and 17,000 acres owned by neighbors who cared to listen to her advice to hold out and work together. Not everybody did. Some of the old-timers, she told me, "weren't going to take advice from a neighbor, let alone a neighbor's wife. A lot of people thought I was the village idiot for talking about gas all the time." As she was compiling her research, she was visited by numerous landmen of various styles and dispositions. Some of their advances, she suspected, were coordinated. "If the young innocent guy fails, call in the old guy with the southern accent," she said, speculating on their strategies. A few individuals stand out in her memory. There was the lean man with a cowboy hat and shiny boots who drove a black Mercedes, accompanied by a young, fidgety woman. Then there were the two men who made up what she and Cliff came to call the "Ed and Bill show," a calculated pairing of alternating styles and approaches. She also remembered the one landman who stepped over the seismic cables, which by that time had been strung along the road in front of her house, and insisted there was little in the way of active exploration happening.

She allowed each landman his say before turning him away with her standard "We'll think on it." In 2005, after a prospecting lull of about six months, Jackie decided it was time to deal, and she did so with a neighboring farmer who had become a landman for East Resources. He approached Jackie knowing many of the neighbors wouldn't sign before she did. With that leverage, she settled on a deal that paid a total of $190 per acre and 12.5 percent royalties for a five-year lease, and a separate agreement that allowed the company to take seismic readings—or "shoot seismic"—on the leased property for one year. She added an addendum that increased setbacks from structures from 200 feet to 500 feet. She also insisted on something called a "Pugh clause." Without it, the company was entitled to extend its claim indefinitely over the entire leased acreage once production began, even if only a tiny fraction of that acreage was included in a production unit. The Pugh clause that Jackie demanded allowed the lease to expire on nonproducing land.

The deal became an early benchmark for Jackie and those who held out with her because it established the effectiveness of a collective approach to leasing. As it turned out, the Trenton Black River formation ended up being only marginally productive in that part of the country. After Jackie and her neighbors collected the lease payments, prospecting grew quiet again. By that time, however, Jackie had honed her skills, and she was well prepared when the mineral rights game took a turn and the stakes rose dramatically.

THE FARMER'S LESSON

Even though Trenton Black River production never met expectations in Pennsylvania, it flourished just 25 miles north in western New York. There, in a Chemung County Farm Bureau meeting in Elmira, Jackie gleaned insights from a self-described "hillbilly farmer" by the name of Ashur Terwilliger and an attorney with an Ivy League education named Chris Denton.

Chris, who owns a law practice in Elmira, first heard the news of a "Texas-sized well" coming on line in nearby Steuben County in spring 1999. A client who worked in the land business called Denton to tell him that the Jimerson well was coming in at 3,500 pounds per square inch bottom-hole pressure. There was a pause, and his client asked, "You don't know what that means, do you, Chris?" Another pause: "That means that's a world class well, and you better brush up on your mineral leasing law if you want my business."

The competitive nature of a gas rush suited Chris's nature. At Dartmouth, he had lettered in lacrosse and hockey before he went on to earn his law degree at Syracuse University. He played and coached hockey throughout his adult life, and the passion with which he pursued the puck and the law outlasted his first marriage. Now, after verifying the Jimerson well was indeed a breakthrough, he began another quest. His mind was filled with an estimation of the number of people on the verge of signing an industry lease without legal counsel—a concern akin to a doctor imagining a convention of diabetics being served sugared punch. It was a Saturday afternoon when he thumbed through the yellow pages to the church listings and began calling pastors in rural communities. He left a message that he hoped could be delivered at the next day's services: members of your congregation must not sign gas leases without consulting an attorney, no matter what landmen might tell them; call me back and I will explain.

No one called back.

By Monday, he came up with another approach—government meetings. The schedule of civic events happened to be light that Monday, but he noted in the newspaper that the Chemung County Water and Soil Conservation Board was convening. That would be a start. He showed up and listened to several hours of talk about erosion control, manure run-off, and flood protection plans before getting a chance to speak about gas leasing. The members of the board were unimpressed, Chris remembered, except for one farmer, who walked up and introduced himself after the meeting. His name was Ashur Terwilliger. He knew about the Jimerson gas well, and he was interested in what the attorney had to say.

Their shared interest in gas development turned out to be well-founded. Led by the western New York Trenton Black River wells, the gas production in New York State would more than triple between 1998 and 2005, with output reaching more than 55 billion cubic feet by 2005. Along with this success came more landmen competing for mineral rights, more seismic crews gathering data, and more and more questions from landowners. As president of the Chemung County Farm Bureau, Ashur felt responsible for providing answers to those questions.

Ashur lived with his wife, Margaret, on a 152-acre family farm about 15 miles east of Horseheads. He had broad features, a booming voice and a fondness for caps, which he collected by the dozens. A favorite was a red-and-white one on which was printed "8th Annual Finger Lakes Equipment Show, 1992." He also kept a Fortuna Energy cap in his collection. It had a faux bullet hole in it and red-marker-simulated blood stains. "Confiscated this one from a landman," he told me. Like most farmers, Ashur did a lot of other work, and he had a special knack and fondness for buying and fixing up old homes, which he then rented out or sold at a profit. When he was not involved in building projects, public meetings, fishing with his grandkids, tending the few animals remaining on his farm, or at the supper table enjoying Margaret's cooking, he spent much of his time in his den amid shelves of petroleum industry information, geological maps, books, trade publications, academic journals, and files of clippings from magazines and newspapers. He spent many hours studying these files in his easy chair, with the television tuned to the cable station that played vintage country and western movies.

Like Root, Terwilliger was self-educated in gas industry matters. In addition to his leadership role in the farm bureau, he had served as town highway superintendent, and was still active on various community boards and planning groups. With this experience, he understood the sometimes-messy interface between institutional motives and public interest. As Trenton production picked up, gas development proved to have a gravitational pull on all his public and personal affairs. He knew when and where landmen were pursuing leases, and which wells were being drilled and under what terms. He consulted with people who were informed or seeking information about such things, and he felt that it was important to keep residents apprised of developments with regular meetings organized through the farm bureau.

In addition to recruiting Chris Denton to speak about legal matters, Ashur was joined at these meetings by Lindsay Wickham, a newly hired field representative for the New York Farm Bureau. As natural gas development began affecting more rural property owners in Chemung and Steuben counties, and eventually throughout the state, the issue became a focus for the state Farm Bureau. Wickham had grown up on a farm in Seneca County and earned a degree in agriculture from Cornell University. He was in his early thirties when he got the job in summer 2000; and, as things developed, he would become the primary source of advice related to gas-related issues for the state agency. But, first, he would get another form of schooling by Ashur at these grassroots affairs held in cafeterias and auditoriums across western New York.

One of the first challenges Lindsay had to master was conveying a coherent agency position as it applied to a volunteer leadership base of farmers—like Ashur—who were mostly his elders and often made a show of being independent, blunt, and ornery. Lindsay happened to have a knack for it. Although he had learned a lot about agriculture through his formal education, he gained credibility with farmers from shared experiences growing up on the family's orchard and vineyard farm. He knew of the expectations that old farmers placed on their children to follow their career paths, and the hard labor and low income that often prevented those expectations from being filled. He had experience with this, as a son who had opted not to return to the family farm and as the brother of a farmer who did. Wickham was good-natured and could take a lot of ribbing; he could also dish it out at the right moments. He tended to like what farmers liked: race cars, machinery, and tipping a few glasses at a local establishment after meetings or conventions. He also knew a thing or two about planting, growing, and harvesting. But perhaps the key to getting along with these farmers was his ability to listen. The older farmers, he found, had plenty to say.

At these early meetings organized by Denton and Terwilliger, Wickham mostly listened and learned as Ashur extolled his audience to resist institutional exploitation much in the way an old-style labor leader stirred up workers. Jackie Root remembers attending one of these meetings in fall 2000. Several hundred farmers had gathered in the auditorium of an old school building in Elmira. Ashur's capped head was hard to miss, even from Jackie's seat in the back of the auditorium, as he strolled back and forth in front of the stage. Jackie remembered that he wore a t-shirt stretched across a well-fed physique and spoke into a microphone that provided unnecessary amplification. He told the crowd about the coming of a world-class gas bonanza. It started with the Trenton Black River Formation, but there was more. Once companies made the capital investment in a well, the exploitation of one horizon would lead to another. Farmers had to wake up to the "things that was happening on their own land, under their own feet." They were "settin' on

18 different layers of gas," he said; and he warned them not to be sold-short by landmen who downplayed the value of mineral rights and tried to rush property owners into leasing cheap. "You can tell a landman's lying by when his lips are moving," Ashur said. He likened them to used car salesmen. "They're telling you, if you buy the car tomorrow, they'll put the tires on it. Oh, by the way, there's no motor in it." Then he talked about what land-men called a "standard lease," which typically included a 12.5 percent roy-alty. "The landmen will tell you that it's fixed by state law. Yes it is: 12 and a half percent is the legal *minimum*, and it's an insult. What's the maximum? That's open to *negotiation*."

"Negotiation . . . ," he repeated, letting the word hang over the audience for a moment. Before Ashur introduced Chris, he explained the kind of lease deals property owners got in Texas, where farmers are savvy about such things and command $5,000 or more per acre and royalties of 15, 20, and sometimes 25 percent.

Chris, who was customarily dressed in a tweed sport coat and gray wool trousers, took the microphone and began with lesson number one: "A land lease is a complex business transaction masquerading as a lottery ticket." He portrayed landmen as predatory and scheming, and he listed their tricks: pro-viding fake maps colored in to suggest falsely that everybody has signed but you; offering to pay your attorney's fees, providing that you chose from their list of attorneys; refusing to deal with you if you hire an attorney the company didn't like; threatening that you will lose the deal and ruin it for others by not signing; portraying the deal as found money and that the lease will have little or no consequences for the landowner; using multiple landmen to tag-team landowners, especially the elderly; portraying state law as limiting royalties to 12.5 percent; and tampering with contracts after they have been signed.

There were other legal snares, apart from the landmen's tricks. "Compulsory integration" allowed companies to take gas from under un-leased property once they controlled at least 60 percent of the adjacent prop-erty within a unit. "Force majeure" enabled companies to extend the time frame of leases against the will of landowners if lessees could prove drilling had been delayed by events beyond their control. The surest way to prevent problems was through informed legal council, and many of the holdouts would hire Denton for just that.

Chris, Ashur, and Lindsay refined and developed their presentations as local interest in the Trenton Black River gave way to regional interest in the Marcellus Shale, but the core message was always the same: Beware of the landman. I interviewed landowners after some of these meetings and found that they generally left with the realization that they were the gatekeepers to mineral riches coveted by influential companies. These mineral riches could do a lot of good for the people and the country, they believed, and it was up to landowners to ensure they were developed in a responsible way. Landowners listened to Ashur when he warned them of the consequences of being unprepared; and they agreed with him when he said landowners should not deny gas companies the fundamental tools of doing their job: access for seismic testing, pipelines, roads, and well sites. Without the proper development of these elements, no good would come of it for anyone.

At some of these presentations, Ashur told a story of when a landman by the name of J. B. Brotherton called on him. Brotherton represented Chesapeake Energy, and the company was interested in collecting seismic information on Ashur's property. "I'm all for it," Ashur told the landman, "but not with that piece of paper you brought me to sign. If you want to come on my land, I'll draw up the papers and you can come on my land." Ashur's lease set various conditions that allowed Chesapeake to use dynamite charges to map the seismic landscape. They could talk mineral rights after he saw the data. And one more thing, he added, "I want to be there when you shoot it, so I can tell people what I seen and what I heard." Not long after signing the agreement, he saw workers running wires across his property to a distant field. He called Brotherton. "I told him: 'Looks like something is happening. Come and pick me up so we can go out and have a look.' He told me he couldn't get there in time. I said 'That's OK. I got my cable cutters. They won't be blasting for a while.' There was a car at my door within a few minutes, and they took me right out to where I had a good view. I got along fine with those boys after that."

THE NEW MILLENNIUM

While oil and gas development in northern Pennsylvania was shaped by the legacy of Drake, western New York had a natural gas history of its own. In 1825, a gunsmith named William Aaron Hart drilled the country's first viable gas well in Fredonia. It provided gas, transported through wooden pipes with tar-laden cloth wrapped over their joints, to light shops, stores, and a mill in the village. Since that time, natural gas has been tapped with varying success throughout western New York in both deep and shallow formations. According to an estimate from the New York State Department of Environmental Conservation (DEC), there have been about 75,000 wells drilled in New York from the earliest days until the present (most of them now abandoned). Yet gas production has remained no more than a small niche in the state economy, lacking both the large-scale development of infrastructure and institutional buy-in that match investments in Pennsylvania and other gas- and oil-producing states.

Still, as many upstate farmers know, there is plenty of gas in the ground, and some landowners can relate stories of finding naturally occurring methane in their wells. The notion that the countryside had an unseen dimension, with hidden reservoirs of water and subterranean fields of natural gas, had intrigued Ashur since he was a boy. One of his earliest memories is of a drive with his father up Hardscrabble Road. He can't remember the circumstances that brought them to a spot along the side of the country lane not far from the family homestead, but he vividly recalls his father showing him an open pipe sticking up from the ground. Ashur stuck his face in it and inhaled a musty odor of rust, earth, and gas. He cupped his hands and hollered down the pipe, and enjoyed the echo of his voice magnified through unknown depths. It was the sort of thing that would inspire wonder and imagination in any boy's mind. Did it lead to China? Hell? Was it a bottomless pit? His father produced a box of wooden matches from his pocket and told his son to stand back as he lit one and dropped it into the pipe. "It went WHOOOF," Ashur remembered. He told me the story of the discovery of that abandoned gas well more than once, and each time his face lit up with the memory. Since then, he has paid attention to anything related to gas development in the area.

There were a few wells—some successful and some not—drilled throughout the western New York countryside in the more recent past. Development of the Oriskany Sandstone Formation led to a spike in gas prospecting and production in the 1930s and 1940s, followed by a lull until advances in seismic techniques led to the development of the Onondaga Reef Fields in the 1960s. Along with these came a modest network of pipelines, the biggest of which was a 20-inch line, operated by Columbia Gas, running west to east across the state. These developments were often followed by lulls in exploration and extraction, and interest and competition never sustained levels that had a significant impact on land value.

As a life-long Chemung County resident with family ties that go back to a trading post on Watercure Hill Road, Ashur knew well the features of the land, including the small network of pipelines and feeder lines produced in the early days of gas exploration. But it was in 1998 that he came across some information that, like the flags in Jackie's neighborhood, signaled bigger events to come. Columbia Gas (a pipeline company) had filed an application with the Federal Energy Regulatory Commission to build a new pipeline directly through the Southern Tier, parallel with the Pennsylvania border. The 30-inch-diameter line represented a major upgrade to the existing infrastructure, and it would be named the Millennium Pipeline. On the western end of the state, the Millennium would tie into a pipeline system from Canada and points north. In the eastern part, it would connect with lines that ran south to New York City. The story of record, reported in the newspapers and on company press releases, was that it would be an artery for the system that transmitted Canadian gas to major metropolitan markets in and adjacent to downstate New York. I reported on the development of the Millennium in the early 2000s as plans were drawn and redrawn to overcome regulatory hurdles and local opposition. At the time, pipeline representatives denied that the project was related to gas production and storage infrastructure developing in the Southern Tier.

With news of the Jimerson well in 1999, Ashur recognized the Millennium project as being something more than the press releases made it out to be. As its name suggested, the pipeline would link the past and future interests of the industry within New York State. Its path eastward ran directly through the place where the Trenton Black River wells were being proven and then bisected a line of depleted Oriskany wells. These exhausted reservoirs were deep, porous sandstone cavities ideally suited for gas storage, allowing companies to warehouse gross quantities of product when prices were low and supplies were high. From there, the Millennium would cross directly over the northern portion of the Marcellus, extending from the Pennsylvania border into the Southern Tier and central New York—an area that would prove to include the thickest and most ideally positioned sections of the shale formation. Ashur never believed that it was simply a coincidence that, in connecting points north with points south, the Millennium was routed directly through the largest swath of gas land in the Northeast. "They kept that real quiet," he told me one day while reflecting on these events. "How are they going to lease land if everybody is on to it? Even a hillbilly farmer knows that."

The timing, location, and capacity of the Millennium Pipeline supported Ashur's message about a forthcoming gas bonanza. As construction on the project began in summer 2007, the price of gas continued to climb, prospecting intensified, and so did talk of Marcellus exploration. The "hillbilly farmer's" audiences and influence grew accordingly at town hall meetings throughout upstate New York. Chris, whose free advice was a draw for auditoriums full of prospective clients, also became a significant figure in shaping awareness of Marcellus development in New York. So did Lindsay, who traveled with Chris and Ashur and began participating as a moderator or panelist at these meetings. Lindsay's perspective tended to serve as a leavening agent for the heavier approach of his two colleagues. (After Ashur told his audiences not to believe anything the landmen said, Lindsay would tell the same audience not to believe 50 percent of what the landmen said.) By the time the Millennium Pipeline began service in late December 2008, the warnings of the farmer, the lawyer, and the bureau representative were known in hundreds of farming communities across the state.

Communities in Pennsylvania were also organizing town meetings. But across the state line from New York, Marcellus Shale prospecting was greeted with a different attitude. The words "jackpot" and "lottery" were routinely tossed around by economic developers, elected officials, industry groups, and even agricultural agents and advisors as they anticipated Marcellus development. The idea of found money was also an irresistible angle for news crews; and newspapers, radio broadcasts, and Web publications amplified the optimism in the Northern Tier and beyond. "You have whole communities essentially that have won the lottery because of these lease payments," Tom Murphy, an educator at Penn State Cooperative Extension in Lycoming County, told the Associated Press. "This is a huge mineral resource that is coming to market at a time when the market is asking for more and more gas." Governor Ed Rendell likened the industry to "a golden goose." Seizing on an angle that served the industry, Energy in Depth, the Washington, D.C., industry group, compiled these proclamations and others in a press release called "What They're Saying: Marcellus Shale Creating 'A Modern-Day Gold Rush.'"

After her success in negotiating the deal with East Resources, Jackie Root was recruited to speak at town meetings organized through Penn State Cooperative Extension. Jackie made her presentations with extension agent Earl Robbins, and the pair often teamed up with Murphy, the extension associate in Lycoming County, and Kenneth Balliet, who represented the extension in Snyder County. The standard presentation—first offered to Dimock residents in 2006—was based loosely on the format provided by the Cornell University Cooperative Extension and the New York Farm Bureau, but there were significant differences. The Pennsylvania group conveyed a buyer-beware theme, but the tone and message was less adversarial than that conveyed by Terwilliger and Denton. Root and Robbins focused on the advantages of forming landowner organizations to attract major suitors and prevent speculators from working one neighbor against the other—an approach Chris and the New York Farm Bureau would later adopt. Eventually, Robbins and Root resigned from their roles with the extension and formed R&R Energy Consulting to work with landowner groups for a profit. They charged by the acre for successful deals—first $1, then $2, and later a percentage of the overall transaction. Jackie became a president of the Pennsylvania Chapter of Royalty Owners and a board member of the national affiliate of the agency, which represents landowners' interests in mineral leasing. Her critics—and there were several on the New York side of the border, at least—saw her role as a professional broker as a breach of the grassroots ethics that are the underpinnings of farm bureau activism. "Jackie learned well," said Ashur, who takes no money as Farm Bureau president or when he advises landowners. "She learned to help the gas companies."

INDUSTRY PLAYERS

Mineral prospecting tends to follow patterns. Large, well-capitalized companies are inclined to acquire land with long-term development potential, with an eye on risks, infrastructure, and markets. These companies often are publicly owned, and they employ economies of scale that enable them to afford paying a premium for vast contiguous tracts that are proven and well positioned for long-term gains. They tend not to stake their future on single horizons. When drilling into one formation, they anticipate acquiring not just natural gas but new data that will lead to other pay zones. Small and medium-size independent operators and speculators, by comparison, are more inclined to lease unproven land on the cheap in the hope of making it big. With comparatively limited capital, their common strategy involves trying to "lock up" land that will later prove valuable to large companies. The intention of these smaller operations is to sell rights to the land—"flipping it"—at a higher price as the play develops and bigger companies become involved. Some independent speculators, wildcatters, might drill wells to prove the land's viability as a gas resource and further boost its worth. Others might simply make a business out of brokering mineral leases as investments.

So it was in Chemung and Steuben counties in western New York, where prospecting picked up as plans for the Millennium developed. Independent land agents and drilling operators had been working on and off around Ashur's Farm Bureau region for generations. Fair Drilling, Abarta, and Pennsylvania General Oil and Gas were among those that put together landholdings and drilled wells that proved the Trenton Black River play in the late 1990s. By 2002, these original prospectors were being bought out by larger companies with more capital and resources to develop deeper and more challenging resources. By the time Marcellus prospecting took off in 2007, the biggest players in western New York were Talisman, the Canadian company that had been prospecting in Jackie's neighborhood, and Chesapeake, a company based in Oklahoma City with a history of exploration in the area (including comparatively modest operations in the Medina and Queenston formations in northwestern New York).

Eastern and central New York had remained relatively undeveloped throughout the twentieth century. Then, after news of the Millennium Pipeline became public and before the Marcellus gained widespread attention, a Norwegian drilling company quietly acquired mineral rights to 130,000 acres of land in the middle of the state. That company was Norse Energy, and its leasehold was over the Herkimer Sandstone, a geological feature extending in a broad, comma-shaped band approximately from Syracuse in Onondaga County to Middletown in Orange County. Using a U.S. affiliate called Nornew, Norse secured rights to acreage in the middle of this band, close to the Millennium route. Here, it began developing Herkimer wells with enough success to justify building a midstream system in Chenango County and a compressor station to pump gas into the Dominion Pipeline running through Madison County to the north. As the decade progressed, and companies realized that shale gas production was profitable in Texas, Norse began developing Herkimer wells south toward Broome County, where gathering systems could feed into the Millennium. It was no accident, company officials told me, that Norse was positioning itself to explore the Marcellus while developing the Herkimer and building infrastructure designed for full-scale shale gas development.

The Millennium route through the Southern Tier runs parallel to another burgeoning gas distribution system in northern Pennsylvania, the heart of which is the Tennessee Pipeline. As Norse positioned itself in New York, Cabot, Chesapeake, Chief, East Resources, and Fortuna began to develop midstream lines and compressors to gather and push gas into the Tennessee artery to reach markets in New Jersey and the East Coast. Schlumberger, a $27 billion well development firm, set up a regional office in Horseheads, New York. The spot was well situated on thoroughfares extending through the Southern Tier and northern Pennsylvania, giving the company fleet of trucks access to development on both sides of the border.

There was synergy between the development of pipelines and wells. In New York, the success of the Herkimer and Trenton Black River led to more pipeline development, and pipeline development led to more prospecting. Other, more wide-ranging factors also played a part in setting the stage for bigger things to come: the price of gas, wars and conflicts in the Middle East, advances in global warming research, and the state of the national economy. Much of the time, these things were characterized by confusion, lack of consensus, and ideological pandering by politicians and interest groups; but in spring and summer 2008, events seemed to bring clarity to the minds of many investors and land speculators: tolerance for foreign oil was reaching a new low with the prolonged Iraq War, natural gas prices rode an economic bubble to a new high, and global warming was front and center as a political issue. A line of thinking developed—supported by science, common sense, and financial markets—promoting natural gas as a clean-burning (and profitable) alternative to oil and coal. Natural gas could be adapted to either electric power plants or automobiles, and therefore it was seen as a response to climate change. Banks began tightening financing on coal plants while investing more heavily in those powered by natural gas; and the natural gas industry began designing advertising campaigns to build its reputation as a clean-burning alternative.

Shale gas development was also tied to a more enduring influence in the national energy policy debate—the concept of peak oil. The theory, advanced in 1956 by M. King Hubbert, a geophysicist, is based on the maxim that there is a finite amount of petroleum reserves in the world and that supply will begin to decline as the easy-to-reach reserves are developed. Therefore, the ever-rising energy demands nationally and globally will have to be met by petroleum reserves that are growing harder to reach. Eventually, as the peak arrives and production wanes, all energy consumers will be at the mercy of economic circumstances, including high demand and correspondingly high prices. Energy producers, motivated by high prices driven by high demand, will pioneer new technologies to locate and produce less accessible resources. There is little dispute at the foundation of this theory; fossil fuels are a finite resource and the easiest to reach are the first to be extracted. Knowing when we have arrived at the peak, however, is tricky. A reliable answer requires an ability to foresee advances in the drilling know-how necessary to keep petroleum flowing from harder-to-reach places. It also hinges on accurately gauging the world tolerance for bearing the financial and ecological costs of proving new extraction technologies, be they off-shore rigs in the Gulf of Mexico or high-volume hydraulic fracturing of horizontal wells in the Twin Tiers.

HIGH STAKES IN DEPOSIT

The shale gas rush spilled over from Pennsylvania into upstate New York in late 2007, beginning in a rural community flanking the western foothills of the Catskill mountains. At the center of this community is the village of Deposit, on the west branch of the Delaware River. The village, partly in the town of Sanford in Broome County and in the town of Deposit in Delaware County, was named in the late eighteenth century, the story goes, because it was here that loggers brought their timber for transport south via the Delaware River. As the area was logged out, it survived as an agricultural hub for the surrounding countryside.

This is where Dewey Decker was born on the family homestead in the town of Sanford in 1935. One of his birthrights was land. At that time, the family owned 2,000 acres, which included two farms and 400 head of Holstein cattle. Some, with prize-winning lineage, found buyers in Japan, Saudi Arabia, and Chile. Dewey's life is woven into the community. He is a lifelong member and benefactor of the McClure Methodist Church. He had served on the school board and town board for a decade, and after that was elected as town supervisor. To many of his neighbors who elected and re-elected him term after term, he is a role model for how the affairs of farm, family, government, and faith are managed. They saw his values reflected in his clean, orderly farm, where he lived with Dawn, his wife of more than fifty years. His steadfast belief is that no problem exists that cannot be solved by hard work.

In Dewey and Dawn's hometown, the prospects of pumping billions of dollars of wealth from the ground remained largely an abstraction—something that happened in Texas or Saudi Arabia. Compared to Pennsylvania and western New York, the Catskills and adjoining areas had little or no petroleum legacy and not much in the way of midstream pipelines or the accrual of geological information that might encourage gas development. That all changed in fall 2007. After five years of developing the Herkimer, Norse geologists had collected data on other formations, and they liked what they saw.

The company hired landmen to secure the rights to unleased territory to the south, where the Herkimer reservoirs were situated under a promising section of the Marcellus and over a promising section of the Utica Shale. The size, shape, and location of these shale formations were generally known, but their potential as gas providers was uncharted and had been of little commercial interest before horizontal drilling and high-volume hydraulic fracturing began changing the industry.

Other companies began following Norse's prospecting venture into the town of Sanford. These were nothing like the wildcatters who had drilled an inconsequential well somewhere in Deposit long before anybody could remember. The landmen who arrived in the town of Sanford represented multinationals, including Norse and Chesapeake. Dewey and most of his neighbors could tell right away that they meant business. At that time, the price of gas was on the rise and so was the bidding for mineral rights. Offers rose from $25 per acre, to $50, and then to $100. When they continued to $150, to $200, and then hit $350, residents didn't need a geology lesson to understand the significance of their mineral rights. Unlike the type of uncompetitive and unchallenged prospecting in Dimock a year earlier, landmen working in the town of Sanford upped their bids before landowners finalized a deal with a competitor.

Once the rush was on, Dewey dug into his file cabinet to find an old pamphlet from the New York Farm Bureau about mineral leasing. He called the field advisor, Lindsay Wickham, who recommended Chris Denton for legal advice. Dewey called Chris and explained that the advice he needed wasn't just for himself—the entire town needed council. Landmen were everywhere, pushing contracts. Chris and Dewey arranged a town meeting at the community theater. As Dewey was placing flyers for the meeting at the feed store and diners, he bumped into a town worker who owned land just across the border in Pennsylvania. A woman by the name of Jackie Root had organized a group of residents in Susquehanna County, he told Dewey. They had just signed a deal with Chesapeake for $750 per acre and 12.5 percent royalties. Jackie, who by that time was identified on her website as "the lease guru," was having a meeting that night in a nearby town to see if adjacent property owners might be interested in signing on. Maybe they could get some useful information if they attended, Dewey's friend suggested.

Dewey invited himself to that meeting, held in a Sons of Norway lodge in Sherman, Pennsylvania, 8 miles south of Deposit. He sat in the back of the room and listened to Jackie explain the terms of the deal. She told the several

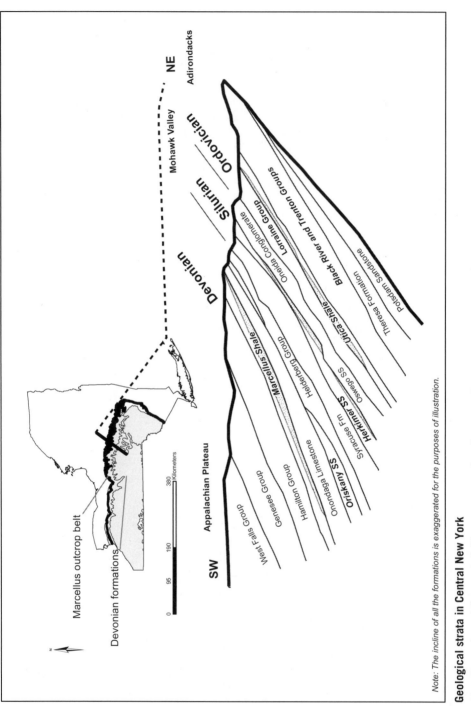

Note: The incline of all the formations is exaggerated for the purposes of illustration.

Geological strata in Central New York

Note: Formations with names in bold type are natural gas-bearing rock. Fm, Formation; SS, Sandstone

Source: Bruce Selleck (Colgate University), "Geological Framework for Natural Gas Development in Central New York," August 2009 Source: Bruce Selleck (Colgate University), "Geological Framework for Natural Gas Development in Central New York," August 2009 hundred assembled landowners that Chesapeake was still looking to add to its leasehold and was considering rights to bordering properties. If they wanted to be part of the group, they could be included as "add-ons" under the same terms, but there would be no tailoring the leases to suit special needs. If a person wanted to negotiate specific surface rights, where pipelines or well pads would go, for example, or they wanted to preserve certain tracts for hunting, he or she would be better off negotiating an individual deal. Dewey told me that Root impressed him with "the way she handled herself." She had a no-nonsense style, and her address conveyed confidence without hype. He asked her if she would be interested in working with Deposit residents if they were to organize into a group. Though Dewey had sold off land over the years, he still had more than 1,200 acres—enough to command special attention from prospectors. He also had influence over a close-knit community that controlled tens of thousands of contiguous acres. "Dewey could have negotiated a deal on his own," Jackie later told me. "You get to drive the bus when you have that kind of acreage. He wanted to make sure his neighbors got a good deal, too."

In early 2008, after sorting through some red tape that came with doing business in New York, Root met with Dewey and a core group of several dozen landowners in the warehouse of a local stone and forest products business in Deposit that had a stake in leasing its land. She was right at home, standing in the middle of a group of farmers sitting on pallets and stacks of lumber. Dealing as a group takes a sense of loyalty, she told them. There was no binding contract that required them to stay with the group after the offers were on the table. There was no obligation to sign on with the group and nothing legally to prevent people from using the group to test their leveraging power, drop out, and then try for an even better price as individuals. In practice, Jackie added, those strategies lead to ill will with the energy companies and sometimes bad consequences among neighbors. "If you can't play in the sandbox with other people, this group thing is not for you," she concluded.

As Jackie began doing research for the group in the winter of 2008, Chris gave the first of two presentations at the Deposit town theater. The seats were filled with representatives of several hundred families. The crux of his two-hour presentation was that gas companies must be held accountable and that the only way to do this was with sound leases. He highlighted the many considerations that sound leases should include. Chief among them was protecting the landowners' surface rights and making sure the lease specifically addressed how operations were conducted to minimize disruption and unexpected impacts. There were basic things, such as forbidding open pits and waste disposal on site, and more complicated things, such as ensuring landowners had control over the placement of access roads, pipelines, and well pads. The way the lease was worded could matter significantly in the development of the site. Wording was also critical in establishing royalty payments. It was common for industry leases (in the fine print) to deduct certain expenses from the landowners' share of royalties, and a sound lease would prevent that. Chris also warned of the "flashbulb effect"—when everybody is blinded by the dazzle of large sums of money. "When that money reaches a certain point, they stop reading the lease."

In the Deposit presentation, as with dozens of farm bureau meetings in years prior, Chris painted a bleak picture of drilling operators left to their own devices and warned landowners of consequences amounting to nothing less than the complete loss of control over their land. Although Chris felt Jackie's proposal—a "company lease with some addendums"—lacked safeguards to prevent this, he avoided criticizing it outright at the meeting. Instead, he emphasized his own approach, which involved a comprehensive contract, drafted from scratch with the landowners' interests in mind, rather than an industry-drafted lease with modifications and stipulations. Chris's standards and expectations, some believed, were too high to attract development. "Chris is a good man to have on your side if you are fighting the gas companies," Dewey told me later. "We weren't doing that." Jackie also saw Chris's presentation at the theater. She recalled that he "made it sound like drilling would be a disaster."

Years later, Chris remained conflicted about the outcome of the meeting. "I saw Jackie Root in the crowd," he told me. "Somebody had pointed her out. Afterwards she came up to me and said 'thanks for not trashing me.'" Jackie said she had the same recollection. Whether due to Chris's uncharacteristic restraint, Jackie's resourcefulness, or possibly owing to the flashbulb effect, the farm girl with a two-year agricultural degree landed the deal of a lifetime that got away from the lawyer.

As word got out about the prospecting, Root and Denton suddenly faced competition from large firms and independent landmen looking to put together career-making deals of their own. Shortly after Jackie and Chris made their pitches to residents in the Deposit area, Representatives of Hinman Howard and Kattel (HH&K), a Binghamton firm, rented the Binghamton Regency Grand Ball Room as a venue for a free seminar on mineral rights issues. The firm had ample means, with more than ninety lawyers in eight offices in New York and Pennsylvania. The event, in late May, 2008, drew more than three hundred people, mostly farmers and rural landowners, who were greeted by tables laden with expensive cheese, crackers, fresh fruit, bottled water and juices, and glossy HH&K folders and pens. "They are treating me quite well," one farmer told me, referring not only to the HH&K event but to previous overtures by energy companies and estate planners and lawyers. He was getting schooled in the business of mineral rights, and after helping himself to the food table at the HH&K event, he tucked his folder under his arm and took a seat in the ballroom to hear Rob Wedlake, a partner with the firm, explain the basics of the Marcellus rush and the leasing prospects that came with it.

By that time, the 300 landowners with 37,000 acres in the Deposit area led by Dewey Decker had cast their lots with Jackie, believing she had the savvy to capitalize on their position. Events were lining up in Jackie's favor as negotiations got underway during an unprecedented surge in natural gas prices. It didn't take long for the companies to respond to her proposal. The community theater was not immediately available, so Dewey arranged a meeting at the Deposit school, where the group gathered to hear Jackie's status report. Norse and Chesapeake, along with some smaller firms and brokers, were strong prospects. Bidding was at $750 per acre and 12.5 percent. Then something unexpected happened. The night before, Jackie received a fax from a representative of Whitmar, a gas exploration and prospecting firm in Colorado interested in the deal. Whitmar was working with a company from Fort Worth, Texas, called XTO. It was one of the largest gas drillers in the country, and it already had significant land holdings in the Barnett Shale. Now it wanted to enter the Marcellus play. Unlike Fortuna, Chesapeake, and Norse, XTO was a newcomer to the Northeast.

Company executives clearly liked what they saw in the Deposit land—the acreage was close to a major pipeline, near major markets, unleased, and big enough to develop. All of these factors improved the chances of a profitable return on investment, even if gas prices fell. XTO would pay $2,411 per acre in cash, up front, and 15 percent royalties for a five-year lease, with an option for the same terms for the next five years on undeveloped land. For many in the group, the upfront money—totaling $90 million split among the three hundred people based on property size—was more than they had made in a lifetime. For landowners with 210 acres, it meant nearly $506,310 in cash. Landowners with 420 acres would receive more than $1 million. Dewey, with more than 1,200 acres, would be a millionaire several times over—before the drilling even began. The legalities would have to be finalized, and for that the coalition hired an attorney to execute the transaction engineered by Jackie and overwhelmingly accepted by its members.

The next meeting was held at a banquet room at the Regency Hotel in Binghamton. Rather than stacks of lumber and sawdust, the farmers were greeted with tables covered with white linens and china, and a buffet of tenderloin tips, chocolate mousse, and other dishes provided by XTO. Along the wall of the dining room, notaries and clerks sat behind a bank of tables with files bearing contracts, deeds, and titles, all sorted and organized by landowners' names. The group of three hundred was broken into smaller groups, each of which was appointed a time for the signing. The process was planned to span two days in late June, and the whole event was marked by punctuality and efficiency. The landowners filed into the room at their appointed times. Some partook in the banquet before and some after their turn at the signing table, working the pen under the notary's watch and receiving vouchers for lump-sum payments that were to come due within ninety days.

Clusters of landowners, some of them newly minted millionaires, lingered at the banquet table. There were plenty of smiles and handshakes. There was also some talk, facilitated by questions from reporters, of big-screen televisions, all-terrain vehicles, and mortgage payments. Jackie sat at a separate table near a drilling display, taking interviews and chatting with landowners between phone calls. (She would soon have her horses—Morgans, in fact—and a magnificent new barn and stables where she could raise and breed them.) Many of the landowners saw the XTO deal as the first chapter of a new American Dream that would play out over the countryside. I reported this for the paper, and while editing my coverage, Ed Christine, the metro editor, sang a few lines from the theme song from *The Beverly Hillbillies*—referencing the 1970s TV series about the fictional Jed Clampett and his kin, who struck it rich after oil was discovered on their homestead. Rags-to-riches stories resonate with readers. It was one of the first things that came to Christine's mind, and that angle, which was on my mind as well, could not be ignored.

I wrote about Kathi Albrecht, a fifty-four-year-old family matriarch, who lived on the farm in the town of Sanford where her mother was born. By day, she worked for a records-processing firm. She also kept busy hosting and feeding her immediate and extended family, doing chores, volunteering for community dinners for the McClure United Methodist Church, and attending Deposit school meetings, where she had served on the board for twenty years. Her husband, Gary, was an industrial painter at CH Thompson in Binghamton. Their situation, along with their hopes and expectations, were not so different from that of the Carters and others I had met across the border in Pennsylvania. The Albrecht homestead, once encompassing 1,200 acres, had been sold off, piece by piece, to pay for expenses as the agriculture economy gave way to industry and then to a service economy. In 2008, Kathi and Gary were left with 225 acres and a $10,000 tax bill when the landman called. The XTO deal, Kathi said, was "a miracle." Suddenly, family vacations were being planned, and the crumbling farmhouse chimney would finally get fixed.

As inviting as *The Beverly Hillbillies* comparison was, it fell well short of the complete story. The money would allow the Deposit landowners to stay on the farm, not move to Beverly Hills or New York City, I was told time and again. (Although I also knew of instances in which farmers sold their herds after receiving gas money.) With the XTO deal and others expected to come, the economic dynamics of rural living in eastern Broome and western Delaware counties would shift, and many believed the money from natural gas extraction would reverse the fate of the small family farmers who had been facing declining profitability for generations.

Another theme soon emerged, beginning with people who did not own vast tracts, yet were swept up by the forces of gas development apparently beyond their control. I met some of these landowners at a second signing, also at the Regency Hotel, several months after the first group signed. There were 200 of them, and they collectively owned about 7,000 acres. Many expressed a join-them-if-you-can't-beat-them mindset and agreed to sign their rights over to the gas companies in the belief that gas development was inevitable. Jackie arranged for this group of voluntary "add-ons" to sign, just as she had done for others in Pennsylvania. For these people who split $16.9 million, I found the glamour of the money was a little tarnished, but it was still enough to bring them to the table. For this group of latecomers, the company served bagels and donuts rather than tenderloin and chocolate mousse. My interviews with them reflected a sense of resignation rather than celebration.

"It's going to be all around us," said Barbara Lester, who owned a little more than 2 acres with her husband, Greg. "We hope it's the right thing. We pray it's the right thing." Ed and Sylvia Theis, from Long Island, were both losing and gaining peace of mind as they signed the lease for their 10 acres. It was the place where they had raised their two boys, who had caught tadpoles and fished away the long summer days of their youth. The boys were grown, and Sylvia had been recently diagnosed with breast cancer. The money would help with medical expenses. "I'm afraid it's going to change the whole character of the place," she said.

There were others who didn't sign at all. Some of them would hold out and hire Chris Denton or another attorney for a compulsory integration hearing, or perhaps go unrepresented, as their subsurface land was apportioned to drilling units locked up by prospectors.

"It's like a giant coming over the mountain," Dewey said. "Even though it's a friendly giant and a good giant, it makes people afraid, and you wonder how you're going to handle him when he gets here."

THE GIANT'S PATH

XTO was big and growing bigger when it acquired rights to the Deposit land. With nearly $2 billion in profits, $12.9 billion in assets, and a 20 percent return on invested capital for 2007, it was listed as the seventh fastest growing American energy company by Platts, the investment research firm. This kind of success was built on sound business decisions, and the Deposit deal looked like another one of those. It gave the company access to two of the world's largest shale gas reserves—the Marcellus and the Utica—at a point where available research suggested that their depth, thickness, carbon content, maturity, and structure were suited for production.

These and other factors created what geologists call a "high probability fairway" for drilling through the heart of northern Appalachia. The fairway reached to the edge of the Allegheny region to the west and encompassed the Endless Mountains to the south, the Catskills to the east, and part of the Finger Lakes region to the northwest. The natural history and features—geology, climate, and biology—of these regions in the northern part of the shale gas fairway are not so different one to the next. Each sits over the same ancient sedimentary seas that formed the Utica and the Marcellus. The visible contours and features of these regions are products of the same glacial history; they are seasonally transformed by snow-swept winters and temperate summers. Each area has endured logging and development that further altered its natural features. By the late twentieth century, scrub brush and trees began reclaiming tracts left fallow by farmers as agricultural development waned and then declined in all these areas. Today, the hills and valleys, woodlands and fields, lakes, rivers, and streams are habitats for coyotes, black bear, deer, fox, bobcat, beavers, bald eagles, and other wildlife making a comeback in the postindustrial age.

The northeastern United States and the Marcellus Shale formation

Sources: United States Geological Survey; Bruce Selleck (Colgate University); El Paso Corporation; Millennium Pipeline Company, LLC

As similar as the natural features are, the New York communities that call this prospecting fairway home are dissimilar and often contrary in social and political matters. The drilling culture migrating from Texas and Oklahoma with its celebrated roughnecks and cowboys would be welcomed by people in some parts and shunned by those in others. The Catskills to the northeast, although part of Appalachia by virtue of geology and terrain, are shaped by diverse populations and cosmopolitan influences. A century and a half after authors Washington Irving and James Fenimore Cooper contributed to their mystique, the mountains east of Dewey Decker's farm and west of the Hudson River became the site for the "Borscht Belt," a collection of resorts run by and for New York City Jews. More broadly, since the middle of the nineteenth century the 6,000-square-mile Catskill region has been the site of a heterogeneous collection of farming communities, vacation homes, sanitariums, sleep-away camps, spas, ski resorts, guided adventure attractions, and hunting and fishing lodges, all of which have been populated and visited by local folk and residents of downstate areas and the Hudson Valley. The influence of downstate New York and New York City is a defining component of the history of the Catskills, with emphasis on natural preservation.

Since declaring parts of the region a preserve in 1885, New York State bought close to 300,000 acres within the park. New York City purchased more than 40,000 additional acres of the Delaware watershed to preserve its drinking water supply. This purchase included Cannonsville, a small farming community sacrificed to create the Cannonsville Reservoir in the early 1960s. The impoundment on the unspoiled west branch of the Delaware River flowing past Deposit now serves as both the drinking supply for New York City and a rich habitat for trout in what is one of the premier fisheries in the Northeast. The water flows from upland sources in the Catskills to the New York City waterworks more than 160 miles away, requiring no filtration and little treatment.

Opposite the Catskills, the Finger Lakes occupy the northwestern flank of the fairway. The largest of these lakes, Seneca and Cayuga, are deep enough to resist freezing and provide a sustaining buffer against the winter cold for the commercial vineyards and orchards on the slopes rising from their shores. Tompkins County, encompassing the southern part of Cayuga Lake, is more diverse, better educated, younger, and wealthier than any of the Twin Tier communities. Here wineries, yacht clubs, parks, and summer homes provide visual reference points to the identity of the region. In Tompkins County, this identity is also heavily tied to colleges and universities. The largest and most influential of these are Cornell University and Ithaca College, both located in Ithaca at the southern tip of the lake. With a collective body of students and faculty larger than the city population, and no shortage of workshops and lectures on topics such as social justice and sustainable growth, these institutions are a substantial part of the regional legacy of activism and reformist politics stretching back to the mid-nineteenth century. Citizens of Ithaca elected a socialist mayor for three terms in the 1980s and 1990s, and cast more votes for Ralph Nader than George W. Bush in the 2000 presidential campaign. Ithaca is a few hours drive from Montrose. Ideologically, they represent places separated by light years.

A large part of the fairway in the area between the Finger Lakes, the Endless Mountains, and the Catskills is occupied by farmers or the children of farmers, land rich and cash poor. The Marcellus boom was an extension of trends unfolding through the late twentieth century: small family farms in the Northeast on the decline while farms in the heartland adapted to economies of scale that lowered prices and fueled demand for agricultural products. If the fields of the Midwest were producing ethanol to relieve dependence on foreign oil, there was no reason the fields in the east shouldn't be producing natural gas. At least, that was the thinking of many area farmers and the energy companies vying for rights to their land. Talisman and Chesapeake, also among the top Platt global energy companies, joined Norse and XTO in the northern section of the fairway. In Broome, Tioga, and Chemung counties in New York, they found groups of owners of large fallow tracts who were willing and eager to begin dealing.

It's a safe bet that many landowners around Deposit had never seen a working drilling rig when they signed over their rights to XTO. Those who had were probably familiar with conventional wells—sunk vertically, as wells had been for generations in western New York, into traditional gas reservoirs. These conventional operations tended to affect relatively small populations in rural areas. Their cost-effectiveness was influenced heavily by the difficulty of finding gas and the risks of drilling a dry hole.

The shale gas boom progressing up the fairway through Pennsylvania and toward the Finger Lakes and Catskills was altogether different. The Marcellus fairway covers tens of thousands of square miles; and while some spots within the fairway are more promising than others, the risks of drilling dry holes in the vast shale mantle are comparatively low. Extracting the gas, however, is difficult. This is known in the industry as an "unconventional" play, and it requires unconventional methods and infrastructure. Shale gas pads tend to be four to six times larger than conventional single-acre pads. The bigger size is required for the greater number of tank farms, reservoirs, mixing areas, waste pits, pipes, equipment, and trucks needed to drill and stimulate a horizontal well. Unconventional wells are drilled vertically, from the center of the pad to the shale, and then steered horizontally through the "pay zone." The horizontal pieces (called laterals) are developed one at a time, each extending from the vertical axis. Draining a shale gas reserve requires a grid of wells and related infrastructure covering entire regions, unit by unit, town by town, county by county, and state by state.

The biggest difference between traditional and unconventional gas development is found by assessing the impact of the respective technologies over a broad area—a factor referred to by regulators as cumulative impact. Plumbing an unconventional resource takes decades. Land must be secured; equipment and workforce mobilized; and wells, pipelines, and generators spread uniformly over every square mile or two. On each pad, multiple wells are individually developed and redeveloped to create laterals reaching in various directions. It took twenty years for the Barnett Shale development to advance across 5,000 square miles of northern Texas, eventually encompassing the Dallas–Fort Worth metropolitan area. From 2000 through 2009, more than 2,400 wells were drilled within a single 1,000-square-mile area of the Barnett. Development was still going strong when XTO and others, confident in the experience they had gained there, began staking out the Marcellus, covering an area more than ten times larger. Expectations for economic returns in Broome County were based on 4,000 wells distributed over the landscape. Tens of thousands more wells would be developed in the counties to the north, south, east, and west if the play reached its maximum potential within the fairway. These wells would be clustered in groups of six to eight, extending laterally from the center of a single pad. The collection of wells at each pad would be interconnected with gathering lines and compressor stations spaced throughout to push the gas into the Millennium Pipeline.

For all this to happen in the most efficient way, laws would have to change to help operators calculate whose gas was being extracted from where and to prevent wells from encroaching on unassigned property. To do this, leased land is divided into "drilling units." Royalty payments are appropriated to a landowner depending on how much of the landowner's property falls within a given unit. The size and shape of drilling units allowed by laws that had been designed for conventional wells, however, did not neatly fit the foot-

0

-1,000

-2,000

-3,000

-4,000

-5,000

Sandstones & Shale of the Gensee,
Sonyea & West Falls Groups

Surface Casing

Platteskill, Ashokan, Mahantanga,
& Mt. Marlon Shales & Sandstone

Intermediate Casing

Drill Pipe

MARCELLUS SHALE

Onondaga Limestone

Tristates Group

Helderberg Group

0   1,000   2,000   3,000   4000

A horizontal well in the Marcellus Shale formation in the Catskill region

*Source: New York City, Department of Environmental Protection, "Final Impact Assessment
Report: The Impact Assessment of Natural Gas Production in the New York City Water Supply
Watershed," December 2009*

print of unconventional development that extend outward rather than downward through the pay zone.

To grant permits for unconventional wells, variances would have to be submitted and approved unit by unit; and this meant a bottleneck of paperwork, public hearings, and administrative work that would slow development and, according to industry supporters, discourage interest in development. As prospecting heated up in 2008, elected officials, guided by the New York DEC and industry advisors, set about to eliminate this encumbrance. With technical input from industry attorneys, the regulatory agency drafted a bill to amend spacing laws and streamline the permitting process. Among other things, the changes standardized units of 640 acres—1 square mile—to more closely correspond to an area that could be reached from a single Marcellus pad.

The amendment, S-8169-A, was sponsored in New York State Senate by Thomas Libous, a Republican whose district contained some of the most lucrative shale gas prospects in Broome County. A week after the Deposit deal became known, S-8169-A sailed through the senate by a vote of 45 to 16 and then awaited approval from the New York State Assembly and the governor's signature.

After the Deposit deal, anybody in the Marcellus fairway who hadn't been paying attention to Marcellus prospecting now was. Awareness raised by Terwilliger and Denton at farm bureau meetings extended to municipal and planning meetings throughout the region. Local and regional governments were left to figure out public policy regarding the handling of matters such as aesthetics, noise, and traffic.

The overriding issue, however, had to do with water. The Finger Lakes and tributaries provide drinking water to the Syracuse and Finger Lakes communities. The Delaware provides water to New York City and Philadelphia. The Susquehanna provides water to Binghamton. Out of view but no less important was an extensive aquifer that 125,000 residents in Broome and Tioga counties depend on for drinking water. The Clinton Street Ball Park Aquifer—a system of glacial deposits underlying the Chenango and Susquehanna rivers—is recharged by a network of streams and valleys within the watershed. In 1984, the U.S. Environmental Protection Agency (EPA) designated it as a "sole source aquifer," in recognition of its hydrological connection to all municipal and private wells in the area.

Before it can begin producing, each horizontal Marcellus well needs between 2.4 million and 7.8 million gallons of freshwater. This water, pumped from streams, rivers, aquifers, or municipal supplies, is mixed with chemicals and sand and then injected into the well bore under high pressure. Sometimes exceeding 10,000 pounds per square inch, the force of the liquid is enough to fracture the shale and stimulate the flow of gas. As gas begins flowing, the chemical solution is regurgitated along with other substances from deep in the ground: heavy metals, brine, and some naturally occurring radioactivity. Some of the water and chemicals stay in the ground. What comes out, "flowback," varies from well to well; and it all must be contained, treated, and disposed of. At 4 million gallons per well, 4,000 wells in Broome County (over ten years) would require 1.6 billion gallons of freshwater annually—about 64 percent of the total volume produced by the waterworks of Binghamton, the largest city in the area. The wells would also produce waste that—in its composition, quantity, and concentration—was generally unknown to municipal planners at the time. In more rural places such as Delaware County, the water needs of drilling, critics argued, could easily exceed limits of the water supply and treatment infrastructure.

In addition to sand and water, hydraulic fracturing fluids contain various combinations of chemical additives. They are used to kill bacteria, prevent corrosion, and inhibit the build up of scale and deposits inside pipes. Additionally, friction reducers and surfactants are added to make the solution easier to pump over long distances through small spaces. These chemicals, in simple terms, make the frack solution go further by making it slippery—a process known in the industry as a "slickwater frack."

Some fracking additives are caustic, poisonous, or explosive. Although the exact chemical recipes are publicly unavailable, they are broken down for New York State regulatory purposes into a dozen broad chemical classifications, each with a dozen or more subgroups. They include alcohols and glycol ethers, petroleum distillates, aromatic hydrocarbons, and microbiocides. Some of these agents, such as the glycol ethers, are common components of antifreeze. Microbiocides include Dazomet, commonly used as a soil fumigation pesticide; and aromatic hydrocarbons contain benzene, toluene, and xylene, which are toxic and carcinogenic.

Sewage treatment plants, struggling to fund upgrades to meet federal discharge requirements that were growing more stringent by the year, were not equipped to test for, much less handle, the kind of waste coming from these new wells. The metals and solvents were one issue. Another was the salts, sands, and other suspended and dissolved matter that could overload municipal or natural systems. Wells also produced radioactive material, liquid and solid, for which the treatment plants had no testing protocol.

Where would the water come from, and how would the waste be handled and treated? As a politician who built her career on an environmental platform, New York State Assemblywoman Donna Lupardo heard these types of questions and concerns from her constituents. She served as a member of the Assembly's Committee on Environmental Conservation, which put her in a good position to find the answers. Issues related to the Marcellus Shale development, she told me, represented "by far the most contentious issue that I've worked on." As emails, letters, and phone calls poured into her office in Binghamton in summer 2008, she arranged a public meeting with the New York Farm Bureau and the DEC to address them.

"NO PROBLEM"

The meeting, on July 16 in the municipal building of the town of Chenango, was part of a series of forums featuring staff from the permitting arm of the DEC to explain how the impacts from shale gas development would be managed in New York. The Deposit deal, signed just weeks prior, had triggered a full-blown prospecting rush in the Southern Tier. Landowners, following the Deposit model, were organizing into groups to capitalize on the sudden competition for their mineral rights as the price of gas spiked. The DEC revision of the spacing regulations to streamline the permitting process—already approved overwhelmingly by the New York State Senate—had passed the Assembly 135 to 7 a few days earlier. The beginning of Marcellus production in New York now awaited Governor David Paterson's signature.

At the meeting, the DEC was represented by Linda Collart, regional supervisor with the Division of Mineral Resources. The panel also included Wickham, representing the farm bureau, and Mike Brownell of the Susquehanna River Basin Commission (a regulatory body that oversees regional water issues). They sat in front of several hundred residents—suburbanites, farmers, and officials from the town, county, and state governments—who packed the public hall beyond capacity. The meeting was also streamed online. Collart ran through a PowerPoint presentation of how Marcellus development would proceed, using information and photographs from conventional plays in upstate New York. She paused on a slide showing a lush meadow of wildflowers and grasses with a bank of trees in the back-ground. This was a reclaimed natural gas site, she said, and an example of the expected long-term impact from Marcellus development. I recognized the slide from a DEC display two years earlier, when I was covering public hear-ings on a proposal by the Office of Mineral Resources to lease the mineral rights of state forests in the Southern Tier and Central New York. Marcellus prospecting had yet to become a public phenomenon, and there had been no mention of shale gas at those meetings. Nonetheless, prospects of any kind of drilling on state land drew fierce opposition from outdoor enthusi-asts ranging from hunters to hikers. At these hearings, Jack Dahl, director of the DEC Bureau of Oil and Gas Regulation, set up a display with his col-leagues pitching clean, well-organized drill sites, including before-and-after landscape depictions. This display had included the photograph that Collart now showed; and Dahl had given a similar assessment that the impacts from gas development, due to the "well established regulatory program" and "rig-orous permitting process" of the agency, would be minimal or even environ-mentally beneficial over the long term.

The crowd at the Chenango Town Hall was skeptical. People fired ques-tions at the panel before Collart's presentation was over. A person in back stood up and asked how local emergency responders could prepare for a spill, fire, or explosion when the industry did not fully disclose the complete chemical content and concentrations of fracking fluids. Collart, who projects the disarming manner of a grade school teacher in her presentations, looked at other members of the panel to see if anybody "might want to take a crack at that one." They looked back at her expectantly. "We don't anticipate any sig-nificant emergencies," she said after a pause. "These things are rare." Another person stood up and asked how regulators were preparing for an influx of drilling that would exceed any historical comparison. Collart responded, "We have been doing fine so far. . . . No problems." She returned to her PowerPoint and was interrupted again by a person who noted that incentives for roadside dumping would go up as waste increased faster than the options for treatment. How would the agency handle that? "You have landowners out there. You have neighbors out there. We would hear about it," Collart said. "Hopefully, the operators will be responsible." More questioning along the same lines followed her presentation, and she delivered the same answers. Flowback is classified as an industrial waste and, therefore, requires a permit for transport, she said again, and again came a question: Where does it go? "I can't answer that," she said. "It's all regulated," she added.

Brownell, representing the commission that oversees the wellbeing of the Susquehanna River, addressed from where the water would come. Many wells were planned in headwater areas, away from the main river stems. Drawing from these streams in the volumes necessary for horizontal drilling is not allowed, he explained, without mentioning that some companies in Pennsylvania had been caught doing it anyway. Water would have to be trucked by tankers to the well sites, significantly increasing truck traffic in rural areas. "The water requirements are so much different [for horizontal fracking], it has caught us a bit off guard," he said.

After the meeting, I interviewed several county and state elected officials. All were unsatisfied with the DEC responses, including Lupardo, who told me, "It's clear these officials will need a lot more information and the DEC will need a lot more people." Steve Herz, a Broome County legislator, property owner, and proponent of gas development agreed. "The answers could have been more to the point, especially from the DEC. There are still a lot of unanswered questions."

The governor's office was immediately tuned into these and other complaints coming from local and state representatives in response to the DEC presentations in Broome County and elsewhere. The prospecting rush was drawing more and more media coverage, and reporters were focused on public responses to the issues arising at these forums. A day after the meeting in Broome County, Collart was scheduled to give her presentation at a meeting in Greene, a town about twenty minutes north of the Town of Chenango. That morning, I got calls from press offices of both the DEC and the governor informing me of last-minute changes to the agenda. Attending the meeting would be Judith Enck, the governor's top advisor on environmental issues; Stuart Gruskin, executive deputy DEC commissioner; and Jack Dahl, Collart's boss—all dispatched from Albany to appear on the panel in this rural town of fewer than 6,000 people in Chenango County.

Enck had risen to a position of influence from an altogether different background than the civil service rank and file in the DEC Mineral Resources Division. Staff members of this part of the DEC typically worked within or alongside industry for much of their careers; their jobs involved overseeing operations and initiatives related to the interests of the state in exploiting rather than preserving natural resources. Enck had gotten her start outside government as an environmental advocate, first advancing to a senior staff position with the New York Public Interest Research Group (NYPIRG) and later becoming head of Environmental Advocates of New York (both organizations are liberal nonprofit agencies dedicated to environmental interests). I remembered Enck as an advocate against lawn pesticides in the late 1990s. She later became policy adviser to Elliot Spitzer when he was attorney general of New York State. In that role, much of her focus was on Spitzer's campaign against pollution from antiquated coal-fired power plants. When Spitzer became governor, he appointed Enck as deputy secretary of the environment. It was the top state environmental post, and she kept her position when Spitzer resigned amid a scandal and Paterson took over as governor. Enck, with short, sandy brown hair with an outdoorsy, tousled look, tends toward an approachable casualness in manner and dress that served her well as a community organizer for environmental causes. Her fondness for grassroots activism is evident when she mingles at public forums. When I interviewed her before the meeting in Greene, she was guarded about discrediting the thrust of Collart's and the DEC's message to date, but her talking points were altogether different from those previously followed by agency staffers. She told me the state would need resources to provide oversight of shale gas development. She also suggested that the DEC might need more regulatory power, possibly through legislative changes. "The thing that stands out the most is the need to have a comprehensive mechanism to protect groundwater and surface water," she said.

The high school auditorium in Greene, which held about six hundred people, was filling up as Enck and others took their places behind microphones at a table on the stage. Enck began by thanking the public for the many good questions (raised at previous forums and in general) about how shale gas development would be handled; and she emphasized the importance of this type of public input. Governor Paterson intended that all these questions be answered before drilling proceeded. "The DEC is going back and doing its homework," she said. "I'm sure you will hold our feet to the fire and make sure it gets done." The crowd didn't need encouragement. It pressed the panel with questions about the contents and handling of fracking fluid and wastewater, and about radioactivity picked up from brine and fluids passing through elements in the shale. The answers from Dahl and his mineral resources staff tended to be general, pointing to regulations that ensure things are handled safely and enforced by DEC staff. Through the course of questioning, the audience learned there were nineteen inspectors to oversee tens of thousands of wells.

"The DEC's history does not instill much confidence," one man said. "There are problems with local quarries and loggers. How does the DEC expect to monitor the oil and gas industry, especially when you're not planning on expanding staff?" In response to this and other comments, Dahl began explaining the role of industry logs and reports to help regulators keep tabs on operations. He was interrupted by shouts: "Do it yourself! Take your own samples!"

The questions, comments, and occasional outbursts went on for hours. One of the DEC mineral resources staff who was not on the panel walked up and down the aisles of the auditorium with a microphone to field questions. One person asked for a show of hands of people in favor of gas development. A smattering of hands went up, but the audience on the whole seemed uncommitted about this. The DEC representative holding the microphone was not—he shot his free hand in the air. Up on the stage, Enck rubbed her forehead.

After consulting with his administration, Paterson announced a decision six days after the Greene meeting: there were too many uncertainties about high-volume hydraulic fracturing to let Marcellus development proceed under the current regulatory framework. Although Paterson signed the spacing bill, in a separate action he directed the DEC to review the impact that fracking horizontal wells would have on the environment and communities. In the meantime, the agency would not issue permits for these wells until new guidelines could be established, based on the outcome of the review.

Looking back on these events, Gruskin characterized the spacing bill as "a fork in the road" for shale gas development in New York and the meeting in Greene as an eye-opener regarding what was at stake. "If there was ever any doubt about the significance of all this, going to that meeting made it clear that it was going to be a really big issue in New York," Gruskin told me. "We had to make a decision as to how we were going to approach it." That decision—to establish the spacing rules while buying time to update permitting regulations—evolved from discussions among agency administrators that began before the meeting in Greene, according to Gruskin. Another account I received points to Enck as the impetus behind the moratorium. According to this account, she engineered the plan with DEC Commissioner Grannis, Gruskin, and Val Washington, head of the Mineral Resources Division, after the Greene meeting.

Regardless of its genesis, the plan placed a de facto moratorium on shale gas drilling until the agency made the necessary updates to guidelines designed in 1992 for conventional drilling. These were spelled out in a document called the Generic Environmental Impact Statement (GEIS), which guided DEC staff on matters related to granting permits. Before industry could apply the technology needed to develop the Marcellus in New York—high-volume hydraulic fracturing coupled with horizontal drilling—the rules would have to be amended through a Supplemental GEIS (SGEIS). The review was to focus on the impact from unconventional well development on groundwater, surface water, wetlands, air quality, aesthetics, noise, traffic, and community character, including an assessment of the cumulative impact on all these things. Before it could be finalized, the assessment would have to be publicly critiqued both through hearings and a process to address written comments. The state was "committed to working with the public and local governments to ensure that if [Marcellus] drilling does go forward, it takes place in the most environmentally responsible way possible," Paterson declared in a statement accompanying the order. The review was initially expected to take less than a year, and it drew little criticism. For those who were opposed to drilling, it was a clear and early victory. Most drilling proponents, meanwhile, saw it as a temporary delay in return for safeguards that would make critics happy. Libous, the sponsor of the spacing bill in the state senate, summed it up: "It's a process where we will be learning as we move forward. We have to make sure the environment is protected. I am pleased so far." Lupardo, one of the few in the assembly to vote against it, said she was "relieved" by Paterson's decision and "satisfied that the right questions are being asked."

When Paterson signed the bill, checks—some in excess of $1 million—were arriving by certified mail in the Deposit area. A surveying crew from XTO came out and placed a stake in the ground in a field of lush wildflowers and hay at the Albrecht Farm, which would be one of the first Marcellus sites to be developed once regulations were squared away in 2009. There was no reason to believe otherwise.

3. GAS RUSH

The caravans rolled into Dimock, Pennsylvania, two years after the landmen first began knocking on doors. Tankers and flatbeds with oversized loads navigated lanes threading through the Endless Mountains. One convoy advanced along Meshoppen Creek Road, a tar and stone route wrapping around Ken Ely's mountain. With transmissions whining and blasts of compression from Jake brakes, the rigs decelerated for a hard left and then throttled up a steep unpaved access road. After bouncing along for several minutes, the trucks neared the top and slowed as they passed a small cabin hidden in a stand of trees near a glassy pond. From there, they negotiated a switchback to a final pitch opening to a broad clearing filled with men and machinery.

It was September 2008, two months after the moratorium was declared in New York, and I was in Dimock to see the budding gas development first hand and report on the early assessments of residents. Coalitions forming in New York had big expectations, and Dimock was the place where those expectations would be nurtured or dashed. I drove up Ely's mountain in a small press car, careful to avoid damaging the undercarriage. Near the switchback at the top the crushed stone path passed through an open gate posted with white signs with the words "WARNING" and "RESTRICTED ACCESS" spelled out in red letters. From here, I followed a dirt lane that branched off the access road and led to a freshly mown field of clover at the end of the pond that was Ken's front yard.

Ken walked over from the pond to greet me. He was wearing a flannel shirt over a green t-shirt, along with work boots, jeans, and a cap with a logo featuring a red, white, and blue derrick. He had seen the press car and guessed why I was there, and the conversation turned immediately to the venture at the top of his land. Ken was enthusiastic about it. It was providing work for locals in the construction business, he told me. Quarry businesses were selling stone for access roads and well pads. Even though most of the skilled jobs on drilling crews were filled by industry veterans from West Virginia and Texas, locals experienced with heavy equipment, welding, and construction were finding work on pipelines, well pads, and roads. Ken's son, Scott, was part of that group. He had taken a job with Cabot as an equipment operator while becoming an apprentice of the gas and oil business, and he had ambitions of advancing up the ladder.

Like many of his neighbors, Ken had signed for $25 per acre, before the potential of the Marcellus became known. He was confident he would be well compensated from royalties of 12.5 percent. "They seem like pretty good people to me," Ken said. "I can't knock Cabot Oil. They seem like one of the better companies, I guess, in the area, so. . . ." He left the sentence unfinished and looked toward the activity on the hill above his cabin.

The plans for his land had changed, he told me. They were drilling a horizontal rather than a vertical well. That meant the well would draw gas from a larger unit—extending beyond his property lines—and more people would share the revenue from its production. It would also produce more gas. He couldn't say whether the expansion would strengthen or weaken his returns.

"We don't know exactly how much money we're going to make," he said. "I assume we'll do well, but I guess you have to wait and see. I'm a millionaire without any money." He gave what passed for a chuckle, but the smile didn't last but a moment.

I asked about the traffic. The rigs I had encountered on the back roads of Dimock proceeded slowly and drivers, perhaps schooled on the importance of leaving a good first impression, often waved cordially to oncoming traffic forced to pull off the road to make room. "More than what you might anticipate," he said. "They could do a little better with dust control."

Was he worried about pollution? Ken nodded toward the pond, which reflected cumulus clouds against blue sky and leaves touched with the first hints of autumn. "It's beautiful," he said. "It's beautiful so long as it's not polluted." He looked at me expectantly. I nodded.

The pond was a favorite spot for him and his grandchildren. They fished for bass and crappies with night crawlers and bobbers. Ken oversaw the gathering of bait, the cleaning of fish, and lessons about living off the land that he had learned from his parents and grandparents. "I assume it's not going to really affect the land much," he said, looking again toward the open gate and warning signs on the road leading to the pad above his cabin.

Ken Ely's 200-acre mountain, as he liked to point out, is flanked by tributaries of Meshoppen Creek that flow in valleys to the east and the west and that eventually join to the south on their way to the Susquehanna River. The tributary on the west side, Burdick Creek, is named after Ely's maternal grandfather, and Ken's younger brother Bill lives in the family farmhouse that his ancestors built more than a century and a half ago on the west bank of the creek. Scott, Ken's son, lives with his wife and three young children on the east side of the creek, at the foot of Ely's mountain. Less than a half mile upstream, the creek bisects the Switzer property, also at the base of the west flank of the mountain.

The September that I met Ken at his pond, the frame of Victoria and Jim Switzer's home had been erected with the help of Richard and Wendy Seymour, artisans who ran an organic garlic farm next door. It was three years after Victoria had first encountered Ken while exploring the woods. As neighbors, Ken and Victoria had reached an unspoken truce. He tacitly allowed her to continue her hikes on his land after he had made his point about the posted signs forbidding hunting on her land. She didn't take them down, but she didn't make any effort to keep them up either; most eventually fell either by the forces of nature or the hands of her neighbors. Anyway, hunting had become a secondary concern for Victoria. The Ely woods were becoming a less peaceful place with the construction of well pads and gathering lines. As she was taking stock of this intrusion, Harold Lewis, her neighbor to the west, told Victoria that Cabot had chosen his land for a well also.

"Harold loves to freak me out," Victoria told me, and he had succeeded with this bit of news. She recalled the landman's words: "Maybe a well or two out there, Ma'am. . . ." First, there was the test well on the Teel farm to the south, then the Ely well on the hill to the east, followed by plans for wells on Carter Road to the north. Now she was learning that caravans would be streaming to and from the Lewis farm to the west, directly across from her property.

Lewis, a tall, good-natured man with a shaved head, had more news. He stood in her driveway, admiring the frame of the magnificent house towering

The Cabot well field in Dimock, Pennsylvania

Source: Consent Order and Agreement between the Commonwealth of Pennsylvania Department of Environmental Protection and Cabot Oil and Gas Corporation, November 4, 2009 over the moss-covered trailer where the Switzers lived during construction. Then he handed her a map—it was a detailed topographical survey representing operational plans of Cabot Oil and Gas covering 5 square miles. Victoria quickly identified the points of reference: Burdick Creek, Meshoppen Creek, and Carter Road. She noted the contour lines indicating Ken's mountain that rose between the creeks and the other hills and valleys where she hiked with her dog, Bruin. And there was her property, near the middle of the map. Inky dots were evenly distributed across the page, likes bugs on a windshield. Each one represented a drill site. There were thirty of them, all within hiking distance.

CABOT'S STRATEGY

Plans for the Dimock well field were inspired by the results of seismic surveys, followed by a test well drilled on the Teel property in 2006. Back then, much of Susquehanna County was still unleased, and landmen had secured enough acreage on a few large tracts—including the Teels—to drill two secluded test wells. They were vertical wells, big enough to test the slickwater fracking techniques used in other shale plays, but small by Marcellus standards. They lacked the long horizontal legs spidering out from the center that would have required land units involving leases from multiple property owners. Even with these limits, the results were prolific, each producing between 800,000 and 1 million cubic feet per day.

The crews capped the wells and packed up and left with no indication of what they had found or their future plans. The drilling was a private affair as far as the Teels were concerned, and they didn't share the details with neighbors, much less talk with the press. Soon after the crews left, however, prospecting efforts intensified, and rumors circulated among hopeful residents not wanting to be left out of whatever windfall they imagined was being discovered in those woods.

In summer 2008, Dan Dinges, president and CEO of Cabot, revealed much broader plans for Susquehanna County in a conference call to investors from his headquarters office in Houston, Texas. By then, Cabot had secured the rights to 120,000 acres—87 square miles—all around the Teel property where the first well was tested. "We have been working on this project since 2006 and have leased what we believe to be the core of the Marcellus play," he explained. "The Susquehanna area has the thickest and richest Marcellus we are aware of, and we believe our initial wells bear this out."

The capacity of the Marcellus remained unknown when drilling ramped up in Dimock, but data from academics and industry suggested it was vastly larger than once thought. In January 2008, Terry Engelder (the Penn State University geologist who took me on the field trip along Cayuga Lake) announced the Marcellus held between 168 trillion cubic feet and as much as 516 trillion cubic feet of gas—two orders of magnitude larger than the estimate of 1.9 trillion cubic feet by the USGS that had been the benchmark until then.

Some were skeptical. John A. Harper, a geologist with Bureau of Topographic and Geologic Survey, wrote in the spring 2008 issue of *Pennsylvania Geology* that the Marcellus potential was far from a sure thing: "One would think, from all the fuss about the Marcellus, that it was a newly discovered gas reservoir containing enough gas to sustain America's needs for decades. In reality, the Marcellus has been a known gas reservoir for more than 75 years. What has made it newsworthy, besides much hyperbole, is that the oil and gas industry has both new technology and price incentives that make this otherwise difficult gas play economical." Still, Harper concluded, results from the Cabot test wells in 2007 were noteworthy. More data would soon determine the "true value" of the Marcellus. "It is possible," Harper wrote, "that the Marcellus will ultimately turn out to be the great gas reservoir everyone is fussing about."

When Harper wrote this, a number of factors were adding to a prevailing bullishness in shale gas investment on Wall Street: rising natural gas prices; breakthroughs in horizontal drilling and fracking in shale gas plays in Texas, Arkansas, Louisiana, and other states; and concern about global warming and dirty coal supporting the industry message that it was providing clean fuel. By the second quarter of 2008, Cabot stock—reflecting a rise in energy stocks in general—peaked at $67.73, nearly double its value from the same time a year earlier.

By that fall, noise from well pad construction and drilling echoed through the hills and hollows of Dimock. In specific places and for certain periods, the din was loud and sustained. In other spots, it amounted to little more than persistent background noise, occasionally punctuated by a boom or a sustained jetlike release. Flatbeds hauled rigs, generators, tanks, and massive pieces of plumbing to fields and woodlots cleared and leveled with crushed stone. Heavy excavation equipment cut a network of pipelines through fields and woods.

The activity, the part landowners could see, at least, had a degree of familiarity—young men and heavy equipment working the land. "It's a hodgepodge of noise," Pat Farnelli explained. "There's screeching, banging, mechanical noise—a lot like quarry noise, but more varied. Nothing we can't stand," she added, "as long as it doesn't affect the water quality." At night, from her gabled bedroom window in the old Carter farmhouse she could see the glow of floodlights in the woods and the tip of a derrick poking through the treetops on the far side of the Burdick Creek hollow. To Pat, it was more a beacon of hope than an irritation. Crews were preparing to drill another well right next to her house, and the Farnellis were expecting royalties from that. The family income was less than $25,000 that year. They were overextended after buying fuel oil for the heating season just ahead, so they were eager for the work to begin.

Dinges was also counting on production from those wells. He made $4.5 million that year, a class of executive compensation based on leadership that had gained the confidence of investors, whom he briefed via quarterly teleconference calls as the company poured more resources into leasing and drilling in Susquehanna County. "We are in the largest land grab in decades," Dinges said in a July 24 update. There was no more cheap land in the northern Marcellus fairway after the Deposit deal became public in May 2008.

The company responded by upping the ante. In October, Dinges announced a $600 million capital plan for 2009, 90 percent of which was allocated for the Marcellus and a shale play in east Texas. In addition to focusing on production and infrastructure, the company planned to lease more land to bring its total holdings in northern Pennsylvania to 190,000 acres, mostly in Susquehanna County.

By the end of 2008, nascent Cabot operations in northern Pennsylvania included twenty wells concentrated in Dimock, producing 13 million cubic feet per day—more than twice original projections—flowing through 10 miles of gathering lines.

Even after Wall Street crashed and energy prices tumbled with the recession in late 2008, Cabot's faith in the Marcellus remained firm. In 2009, the company added more than 60 miles of gathering lines to connect 100 wells in Susquehanna County. Daily production—despite being limited by pipeline capacity—accelerated steadily, reaching 72 million cubic feet by the end of 2009 and more than 100 million cubic feet by February 2010.

Cabot's venture in northeastern Pennsylvania drew the media spotlight as the first prospecting wave crested in 2008, but it was not the first big strike in the Marcellus. Range Resources, a company with a long history of exploration in Appalachia, had begun testing Marcellus wells in southern and central Pennsylvania and West Virginia in 2004. As with Cabot, Range crews applied the slickwater fracking techniques pioneered on the Barnett Shale in Texas and liked the results. Range kept the venture out of the public eye while locking up more than 1 million acres of the Marcellus before its potential became common industry knowledge. From 2005 to 2010, burgeoning Marcellus development led to a doubling of the number of Range shares.

"The bloom hasn't even opened yet," Range Chief Executive John Pinkerton said in an article that appeared in *Forbes Magazine* in August 2010. By then, several international companies, including Chesapeake and Fortuna, had entered the competition. During the last six months of 2010, Marcellus production in Pennsylvania surged to more than 256 billion cubic feet from 1,147 wells—a volume greater than the annual production from wells tapping all other horizons combined.

"NOTHING TO SNEEZE AT"

Lacking a crystal ball, Ron and Jeannie Carter signed over rights to their land for $25 per acre. Still, as drilling ramped up in 2008, they expected the royalties to make things right.

That summer, a survey crew placed a stake with a red flag not far from the hickory nut trees—where Ron and Jeannie had met on church picnics in their youth—in the field between the Carters' trailer and the family's old farmhouse, where the Farnellis lived. Wells, like some of the old roads, took the names of property owners. The one built almost within casting distance of the Carters' porch was called Gesford 7, after Karl and Colleen Gesford, Dimock farmers and the current owners of the land.

On a late-summer morning, Jeannie was drawn to the kitchen window by the noise of heavy machinery. She watched as a bulldozer, taller than their trailer and almost as long, backed off a flatbed and advanced across the field toward the hickory trees. The machine dropped its blade and plowed into the trunks. The trees resisted for a moment, then tilted and fell.

Ron was unhappy to see the trees go but less bothered than Jeannie. There were plenty of other things occupying his attention. The excavating equipment that graded the field was followed by more flatbeds with more machinery, pipes, and tanks, all of which soon filled a 10-acre staging area bordering their property. To him, it was a carnival of activity and characters unfolding on the field that had been fallow for decades, and it was exciting.

Not long after the trees fell, the Carters returned home after a supper at the Green Gable in New Milford and found a derrick towering over the field. A fresh crew arrived before dark and continued working through the night under floodlights. The next morning, the day shift returned, marking the beginning of a cycle that continued without a break through the fall. On hazy September mornings, Ron continued reading his spy novels and newspapers on the porch. His once commanding view of countryside, framed by birdfeeders and hanging plants, was now filled with machinery and men in hard hats.

"Actually, it's been good entertainment," he said. "Now we're hoping for some money out of it." The work crews were approachable and polite, and he became a student of gas field operations. "Yeah, there's dust," he said when I asked him about it. "You live on a dirt road, you have to expect dust."

Soon after Gesford 7 was underway, another derrick appeared over the trees across the Carters' northern property line. This, the Baker well, shared its northern border with the Fiorentino's homestead.

Norma Fiorentino developed a tolerance for the trucks rumbling to and from the Baker well on the state road in front of her house and the private access road circling behind it. Occasionally, blasts shook the ground and made her heart skip. Those she could do without, she told me. She also noticed, from time to time, a foreign smell behind her house that she linked to the venture. It was unlike the familiar smells of grease, quarry dust, and heavy machinery that her boys routinely brought into the house on their work clothes.

Norma didn't dwell on any of this, which was merely background for far more pressing issues. Her daughter's cancer was terminal. Her son was beginning drug rehab. She had to keep track of the comings and goings of her grandchildren and great-grandchildren, all of whom needed to be fed, clothed, watched, and disciplined.

Maybe the gas rush was helping business at some of the big stone yards in the area, but it wasn't doing anything for the small family operations where her boys worked. What she knew for certain about the drilling was that, as 2008 drew to a close, it had yet to produce a royalty check. Despite that, news of success at neighboring wells buoyed her hopes.

The Carters' first monthly check, $4,400, was in Ron's words "nothing to sneeze at." The $2,000 check that arrived by certified mail to Victoria and Jim wasn't life changing, but it gave them hope that, after the commotion settled and the wells were in place, royalties, even from their small tract, might lighten the financial burden of the house, which was turning into more of a project than they had anticipated. "I don't thumb my nose at the money," Victoria said. "I don't pretend that I'm above that."

Ken Ely had never aspired to become a millionaire. He never even played the lottery; he wasn't a gambling man. He was happy working in the quarry, his brother told me, on land he had helped pay for with logging money. Bluestone, although he didn't sell a lot of it, was like a fine cherry timber, a brightly speckled pan fish, or a majestic buck—something beautiful from the land, and he took great pleasure in harvesting it.

The well at the top of his hill, and others on his land, soon produced millions of cubic feet of natural gas per day. Like many of his neighbors, he had signed for royalties of 12.5 percent, the minimum allowed by law. Even so, the sheer volume of gas coming from beneath his land provided income beyond his imagination. The influx of money, as far as those who knew him could tell, had little discernable impact on his lifestyle. He still lived in the two-room cabin on top of the hill with his third wife, Emma, and he kept a critical eye on drilling operations. He did buy a four-wheeler and a new skid steer for his quarry with the check that came in, but his most prized possession remained his fishing pond and the land on which he hunted.

The Monday after Thanksgiving, the start of buck season, has long been an important date on the Ely calendar. Ken's younger brother Bill and his son Scott would meet at Ken's cabin before dawn, along with various others from the Ely clan, and then break off into various quarters of the mountain. There they would spend the morning in tree stands, rifles at the ready, searching the shadows cast by the autumn sunrise. By late morning, they would return to Ken's cabin, where they would dish up steaming bowls of potato soup and chili, drink strong coffee, and swap stories of the trophies that had wandered into view but maybe not quite close or clear enough for a shot. Occasionally, one of them would have more than a story to show for the day.

Trophies were fine, but it was the stories and the family ritual that sustained Ken. He was leading the good life; and sometimes, at family gatherings, he liked to daydream out loud about "the better life." When they were done drilling, that would be the better life, he told Scott. He would plant game crops in the clearings created by the well infrastructure. The deer in his woods, wandering into view, would be like fish in his pond—well-fed and there for the taking. This hunter's paradise in itself was a fine story; the story was made even better because it was now within financial reach.

The better life for Ken would have to wait. After one well was completed on his property, another was begun. Vertical wells were redeveloped into horizontal wells. Large clearings became staging areas for frack tanks, and some of them leaked.

A SYSTEM UNDER PRESSURE

In September 2009, Cabot closed its regional offices in Charleston, West Virginia, and Denver and opened a new headquarters in Pittsburgh to manage its workers and resources moving from other plays to the Marcellus. In his year-end conference call to investors, Dinges characterized the venture in northern Pennsylvania as "the centerpiece for Cabot's future strategy." He continued, "It is developing into a true company-maker and it's a world class resource. Early on, we were conservative with our judgment regarding the potential of this play. Now it has met and exceeded all our expectations and we know it will be the driving force for Cabot for years to come."

Victoria began to call Dinges's venture "The Occupation" when she realized that, contrary to her original expectations, it would not end in the foreseeable future. By 2010, with one hundred wells drilled in Susquehanna County, Cabot was just getting started. In time, old wells would be revisited by crews drilling new laterals or stimulating production with new fracks. Crews expanded the network of gathering lines and compressor stations to push the gas from the growing connection of wells through the Tennessee Pipeline, the major gas artery running through Susquehanna County on its way from the Gulf of Mexico to Canada and a supply line to major metropolitan markets in the mid-Atlantic and Northeast.

Activity in Susquehanna County reflected statewide trends. An acceleration in Marcellus permit approvals—478 in 2008 to 1,984 in 2009 to 3,314 in 2010—drew warnings from officials and politicians that development was outpacing the ability of regulators to oversee it. Craig Lobins, the Pennsylvania Department of Environmental Protection (DEP) regional oil and gas manager, told an advisory board in October 2008 that the department was understaffed, even if the annual number of new permits held steady. Lacking enough staff, DEP inspectors typically go to a well site only when looking into a complaint. For completed wells (where problems arise from deterioration and abandonment), they could manage no more than one routine inspection every ten years.

The message of stress and overextension became thematic as DEP officials appealed to lawmakers for an industry severance tax to pay for more staff and resources to manage the workload. Testifying before the Pennsylvania House Appropriations Committee, J. Scott Roberts, deputy secretary for mineral resources management, referenced the land rush that followed Colonel Edward L. Drake's first oil well in Titusville in 1859. A similar revolution was now unfolding in the Marcellus with "large, often multi-national corporations" placing "added burdens on local and state government."

There were monetary incentives to keep permits flowing. The agency had forty-five days to process applications or return the application fee, which ranged between $900 and $3,000 for each well, depending on the depth of the well bore. Reviews of applications typically lasted thirty-five minutes or less, and they were approved more than 99.5 percent of the time, according to a deposition of regulators in front of the Pennsylvania Environmental Hearing Board. Wells near "high-quality waterways" protected by state and federal law received no additional scrutiny because permitting officials lacked a working definition of what the designation meant as it applied to their review process. These and other revelations came to light with a lawsuit filed by residents and environmental groups contesting a permit that the DEP issued to Newfield Appalachia in the protected Delaware River watershed, about 300 feet from a pristine stream. "What these depositions reveal is that the state is doing next to nothing in approving permits, even in the Delaware River basin, even in high quality watersheds, even in the wild and scenic river corridor," said Jordan Yeager, an attorney representing Damascus Citizens, a group challenging the permit. "They've got limited time to do a massive job. . . . If we're getting it wrong in this case, we're getting it wrong for every well site that's being developed."

In West Virginia, where Marcellus permits tripled between 2007 and 2009, the number of inspectors increased by only one, to a total of eighteen. These eighteen inspectors were charged with overseeing more than one thousand new Marcellus wells, with thousands more, recently permitted, coming on line, plus tens of thousands of older, traditional wells. "We simply do not have the number of people necessary to do the job," West Virginia Department of Environmental Protection Secretary Randy Huffman said in an interview with the Associated Press. "It's easy to issue a permit. What I think we're doing is issuing permits faster than we have the ability to keep up with them on the ground."

Although officials had suspended permitting for high-volume hydraulic fracturing in New York pending the environmental review that was part of the SGEIS, there was renewed interest from prospectors in conventional formations below the Marcellus not subject to the moratorium. Regardless of their intrinsic value, wells into these horizons provided a way for prospectors to stake their claims on pieces of the Marcellus, and get a head start on the infrastructure and lease holdings to produce it.

With permits outpacing regulatory manpower across the region, regulators called on the public to help manage the industry. Bryan Swistock, a senior water resources associate with Penn State Cooperative Extension, urged landowners at informational workshops around the state to "do their best to keep an eye on things," while cautioning them to avoid trespassing. (Companies have the right to restrict access to well sites, even by landowners.) "It's uncanny how good people are at just using their instincts and just understanding things aren't right." Landowners sent him photographs of torn or degraded waste pit liners, leaking tanks, and other potential problems that should be "immediately" reported to the DEP, Swistock said.

EPA officials, while lacking the jurisdiction to oversee hydraulic fracturing, were "very concerned about the proper disposal of waste products, and protecting air and water resources," according to a 2010 press release that announced a citizen watch program called "Eyes on Drilling." Through the program, which included a tip line, the agency was seeking "a better understanding of what people are experiencing and observing as a result of these drilling activities" and "counting on concerned citizens to report unusual or suspicious activity related to drilling operations." That included, in addition to a time and place, details on "materials, equipment and vehicles involved and any observable environmental impacts." The EPA program did not mention the issue of trespassing.

"AN INDUSTRIAL PROCESS"

The acute disruption from the construction of infrastructure and wells—"plumbing the oil patch," in industry parlance—was easy enough to see and often impossible to ignore. Trucks, some hauling 100,000 pounds or more, were the first and most visible sign. A single well requires between 900 and 1,300 round trips by trucks hauling equipment, water, sand, chemicals, and flowback to and from the site before production can begin.

In northern Appalachia, the rugged, wooded hills—and changeable weather—were causing problems for crews used to the open, arid landscapes of Texas and Oklahoma. Although the Marcellus is the second largest contiguous expanse of shale gas known in the world, the terrain that covers it affords little open space to work in, plenty of obstacles, and close proximity to sensitive water supplies. Drilling and support crews had little margin for error.

Problems in Dimock first appeared as work-a-day construction accidents. In early July 2008, a truck knocked over a storage tank, spilling between 600 and 800 gallons of diesel on the Teel property—the second fuel spill on that location in two months. Several weeks later, a truck rumbling down Ken Ely's mountain reportedly ran over his hound dog, Crybaby. In February 2009, a cement truck slid off an icy access road leading to a well on the Lewis property and crashed in the family's yard. Any one of the accidents, to the south, east, and west of the Switzer's dream home, might be forgettable, at least to those not personally involved, if not for the growing level of public scrutiny on the Marcellus. The social, economic, and political dynamics of the coalition movement, dire warnings from the environmental movement, and vast sums of capital on the line had drawn the attention of media, critics, and advocates; and now all eyes were on Dimock to see how the rush would pan out.

Information about these incidents was not always easily available, and accounts varied depending on the sources. Neither the company nor the DEP issued a public notification of the second diesel spill on the Teel property (which I came across by chance in a records search). The driver of the truck that ran over Crybaby disposed of the dog's body without notifying Ken, according to Ken's son Scott, who at the time worked as a Cabot contractor. The truck that crashed on the Lewis property caused a leak in a line or hose under the truck that left "a spot" on the ground, a Cabot spokesman told me—a characterization that differed from the DEP record indicating 100 gallons of diesel had spilled from the truck's ruptured saddle tanks.

"These kinds of things happen when you are moving a lot of equipment," Ken Komoroski, the Cabot spokesman, told me. "The weather conditions are adding to the difficulties."

Regulators offered a similar assessment of these and more serious problems to follow. Roberts, deputy secretary for mineral resources management, told the Pennsylvania House Republican Policy Committee that "even with strong regulations, impacts will not be zero. Drilling is an industrial process. . . . Spills and accidents will occur."

While lumping gas extraction with industry in general, Roberts, along with other regulators and industry officials, overlooked an important distinction—most industry tends to operate in a fixed location, zoned for certain uses based on its expected impact on the environment and communities. Drilling involves an industry—exempt from federal regulations—that works on other people's land, over widespread, often environmentally sensitive areas. Although the surface operations are visible, most of its functioning is hidden under the surface, where risks are complicated by geologic uncertainties.

It was events under the surface that changed the lives of Julie and Craig Sautner, beginning in September 2008. Julie Sautner pulled laundry from the rinse cycle of her washing machine and found it dirtier than when she had put it in. The new clothes she had bought her children for the school year were ruined with stains. The water in the toilet was brown, and so was the tap water. The dramatic change in the quality of their water well, which had no history of problems, came as Cabot crews completed a gas well less than 1,000 feet away. The Sautners saw this as more than a coincidence, and they notified the company of the problem.

When this kind of situation arose, the person who represented Cabot—as liaison to both the regional media and local residents—was Ken Komoroski, a Pittsburgh attorney with an undergraduate engineering degree from Penn State. With a large, broad physique, he looked like a crew foreman when he donned a hard hat and gave rig tours to elected officials, media, and landowners. He was an avid bass fisherman and an organizer of the $2 million Forrest Wood Cup professional bass tournament in Pittsburgh's three rivers. In 2008, he served as Cabot's environmental attorney and media spokesman. For Dimock residents, he became the face of Cabot.

The company denied that the drilling had degraded the Sautner's water, but it supplied them with bottled water and installed a filtration system in an attempt to fix the problem. This was not an admission of guilt but a response to regulations that hold companies responsible when water well problems crop up within 1,000 feet of a gas well. Wells go bad due to naturally occurring conditions, from time to time, Komoroski said, and this was one of those times. "We want a good relationship with landowners and we are happy to help them out," he added.

There was no doubt in Julie's mind that Cabot's drilling had ruined her well, but—at that point—she had faith the company would make things right. "Everybody is aware that there could be this kind of potential problem," she told me. "They are nice people, and they are cooperating." She hesitated, "Our water was crystal clear."

Ron Carter heard about the Sautner's well, and he was skeptical. The drilling crews had told him, "There is absolutely no way this can affect your water," he recounted, and he trusted them. His confidence dimmed one October morning when Jeannie drew a glass of water. It stank of mildew. After she dumped it, the smell hung in the air. They soon learned their water was also useless for showering or laundry because the smell clung to their clothing and skin. Cabot sent a technician to collect samples at the Carter well, and a few days later Ron got a call from a company official in West Virginia. The water tested positive for coliform bacteria, and Ron could fix that by dumping chlorine down the well, the man on the phone said. He didn't know how much chlorine to use, and it didn't matter, exactly, he told Ron, as long as he didn't overdo it. "I wondered why he would have results on his desk in West Virginia, and I didn't have them," Ron told me. "He said, 'I'll send a landman over and he'll show you how to do it.' I said, 'No, you won't. I'm not going to dump chlorine down my well.'"

The well, which also provided water to his family in the trailer next door—including Ron's pregnant granddaughter—was drilled into 500 feet of bedrock in 1970 and, until now, there had not been the slightest problem with it, Ron said. Rather than carrying out the plan with the chlorine, Ron spent his royalty check on a built-in water treatment system.

A few months later, the water well of Ron and Anne Teel—within 1,000 feet of another Cabot well—suddenly went bad. The Teels would not talk to the press. A letter they sent to the DEP states the well became "clogged with some type of fine black, greasy feeling sediment." They were "cooperating with Cabot in an attempt to remedy the problem."

The Lewises also reported problems to regulators that they didn't share with neighbors or the media. A strong odor of diesel lingered for months after the cement truck crash, even after a Cabot contractor excavated contaminated soil as ordered by the DEP, according to a DEP memo. In response to their complaints, the DEP returned to the site and "detected strong diesel fuel odors in the soil at several locations," prompting it to order a second, then a third excavation, this last one completed with "on-site supervision" from DEP staff.

Ken Ely kept a critical eye on drilling crews and wasn't shy about confronting them when he was unhappy about the way things were going. The surface disruption didn't bother him—it was what he expected. Yet he grew irritated at some crew members, whom he was sure were helping themselves to choice pieces of bluestone when they ended their two-week shifts and loaded up their pickups before heading back home to West Virginia. He set up his shooting range at a conspicuous spot near the quarry, and his armed presence could be seen and heard when he squeezed off rounds at targets and, occasionally, into the air. The accident with Crybaby also nagged at Ken. The dog couldn't be buried because, as he learned from Scott, the driver had disposed of the carcass.

There was something else that bothered Ken. He hadn't anticipated the wells on his land would produce so much waste. This included not only the spent chemical solution used to stimulate each well, but tens of millions of gallons of brine and whatever else came up from the holes. The flowback was supposed to be treated and disposed of at plants equipped to handle it, yet nobody could tell Ken exactly where those plants were.

In June 2008, a few months before the water problems on Carter Road came to light, the DEP ordered municipal sewage treatment plants to stop accepting drilling wastewater without knowing its composition and how to treat it. The suite of tests required to comprehensively screen the material was expensive and, for many treatment plants, prohibitive. The problem was complicated by the fact that drilling companies did not have to report what they used and their characterization of the waste tended to be too broad to establish parameters for testing.

One of the treatment destinations that began refusing flowback was in Sayre, Pennsylvania, about 30 miles northwest of Dimock. Dave Allis, manager of the plant, told me they stopped taking drilling fluids after the DEP warning. Lacking the wherewithal to test for metals and other contamination in a timely and cost-efficient way, they decided "it was more headache than it's worth."

When I asked regulators where the flowback was being treated, they told me it was a question for the companies. When I asked company representatives, they told me, "It's all regulated." In 2008, I learned of two plants in Pennsylvania: one about 50 miles north of Pittsburgh and another about 40 miles east of Pittsburgh. Collectively, they handled about 360,000 gallons per day. A single Marcellus well could produce several million gallons of waste, and the DEP had permitted more than 2,500 of them in 2008 and 2009. In addition, the industry had to do something with the waste from tens of thousands of conventional wells, which also produce brine and other constituents.

Ken was by then formulating a view of the industry from his own observations and from questions he put to Scott about what he saw on the job; and he was convinced the wells in Dimock were producing waste faster than the company could legally get rid of it. He was now mostly worried about his pond, the watershed, and the collection of 20,000-gallon frack tanks taking up more and more space on his land, where a second and third well had been drilled, with plans for more. According to Komoroski, the tanks contained freshwater. The flowback was piped directly into tanker trucks and taken to designated treatment facilities. "The entire process is regulated," he added.

When Ely became sure his complaints were going unheeded by Cabot and DEP officials, he called Laura Legere, a reporter covering Marcellus development for the *Times-Tribune*, a Scranton paper with readership concentrated in northeastern Pennsylvania. He brought her to the tanks and showed her a steady trickle of fluid coming out from the bottoms. Regulators had told him it was freshwater, but he was sure it was flowback, and he wanted it off his property.

"We'd like to have them not wait two, three months to take this stuff out and take care of it," he said. "People get lax and I hate to say it, maybe we need a little more regulation. . . . The landowners want the money, and I understand that. We all want the money. But gee, we'd like to have fish in our pond."

The interview was interrupted by something in the distance that sounded like a jet engine. "You hear that?" he asked her. "That's another well going off, the pressure. Awesome isn't it? There's so much gas up here. These are world-class wells."

GESFORD 3

From the laundry line in her side yard, Pat Farnelli had a sweeping view of a scene that would determine her future—men and equipment had moved from the hill across the Burdick Creek hollow to a fallow pasture directly below her house. In late September 2008, bulldozers cut through a field ablaze with goldenrod to level a pad for Gesford 3. The well would draw gas from under the adjoining Farnelli land, and the family's mortgage depended on corresponding royalties.

The derrick went up in early October, and soon the platform straddling the hole was busy with men in coveralls and hard hats who were lifting, swinging, and lowering pipes with hydraulics and heavy hardware hanging from chains. Shouts of men over machinery and generators carried up the valley, sometimes with the smell of heavy machinery and diesel exhaust. As work progressed, it was easy for Pat to believe destiny was working in her favor. At schools, churches, and the Lockhart lunch counter and gas mart, the news of the landmen and their promises was giving way to the excitement of wells producing millions of cubic feet per day from the Teel and Ely properties. There was also news of corresponding royalties. Crews were now busy building compressor stations and pipelines on the Teel land—at fair compensation, she heard—and drilling two other wells along Carter Road.

This is what Cabot's $600 million in shale gas development looked like. The size and intensity of the operation, its manic focus and energy, were all directed at producing wealth, and Pat took comfort in knowing her 20 acres were locked into the equation. Even a drop of that wealth—$3,000 or $4,000 per month—would make things good for them. They could pay their mortgage, buy horses and other animals, and make a go at farming. Her oldest daughter was getting married that spring, and they would have enough for a nice wedding. Pat would no longer have to worry about making ends meet for six dependent children with Social Security, food stamps, and the lousy hours and low pay Martin was getting at the Flying J.

The shouts from men over the drone of diesel motors and generators on October 8, 2008, a bright Wednesday morning, didn't sound peculiar to Pat. The crew had drilled to 2,000 feet—still several thousand feet above the Marcellus pay zone—when they encountered a problem that brought the multimillion-dollar operation to an inglorious stop. Debris from upper layers had fallen into the hole and jammed the drill.

The Devonian bedrock under Gesford 3 is covered by 400 feet of glacial till—unconsolidated stone and gravel known in the industry as "overburden." Drilling through the till is like trying to bore through a gravel pile; and although there are techniques to deal with it, they are not foolproof. A drill that jammed somewhere in the overburden might have been less of a problem, but the Gesford crews had already worked through the till and were well into a gas-bearing zone of bedrock above the Marcellus. Gas had begun flowing, and the crews, left with an open, uncased hole, had no way to control it.

The problem persisted throughout the fall. To Pat, it looked pretty much the same from day to day, with the round-the-clock procession of equipment and men. The yelling at the site might have been laced with a little more profanity than usual, she reflected in hindsight, but really there was no way for her to know the problems would soon amount to more than a lost piece of hardware.

DEP inspectors were also ignorant of complications at Gesford 3 until an event on New Year's Day 2009 put them on notice. A blast echoed through the hills. A mile north of the Gesford well, concrete dust billowed from the ground and hung in the frigid air over the ruins of Norma Fiorentino's well.

Norma was having supper at her daughter Brenda's home with her daughter, grandchildren, and a great-grandchild. She was feeling optimistic that this year would be better than the last. Brenda's chemo treatments were keeping the cancer in check, and her granddaughter was expecting a second child. With luck, she could buy some nice baby things with royalty payments like the ones her neighbors had begun receiving. Now, upon her return home, she stood trying to fathom the gaping hole and the shattered remains of a massive concrete slab once covering her well. A random act of violence? Who would want to blow up her well? She called 911.

As the Springville Volunteer Fire Department cordoned off Norma's yard with bright yellow caution tape, Cabot representatives arrived on the scene. They took some tests around her house and determined that any gas, if it was there, was not lingering.

INFORMATION VOIDS

The explosion at Norma Fiorentino's property was caused by methane migration, according to a subsequent DEP investigation, a phenomenon that occurs when natural gas—forced by pressure naturally occurring in the earth or introduced by crews trying to extract it—ends up in places where it's unintended. Unlike a blowout, when gas overcomes a pressure barrier and erupts sensationally from a vertical hole, methane migration involves gas forced through the ground along the path of least resistance. That path, often beginning with a faulty well casing designed to seal off the aquifer from the well bore, leads to basements, well heads, crawl spaces, and aquifers. Methane migration is often imperceptible, at first—a fact that makes it deadly.

Although Norma's well soon became famous, it was neither the first nor the worst case of methane migration related to gas drilling in Pennsylvania and elsewhere. Much of what officials knew about the dangers of methane migration they had learned years before from more costly instances. The Pennsylvania DEP Bureau of Oil and Gas Management had files on more than fifty other cases dating from the beginning of 2004 to the time Norma's well exploded. All involved dangerous and sometime fatal accumulations of gas migrating from new or abandoned wells into enclosed spaces. They had happened before the shale gas rush became big news, and they had gained relatively little public attention.

In 2004, DEP records documented the collection of gas in the basement of the Harper residence, near several wells being drilled by Snyder Brothers in Jefferson County about 80 miles northeast of Pittsburgh. On March 5, the furnace kicked on, and an explosion leveled the house and killed Charles Harper, his wife Dorothy, and their grandson, Baelee. A report by the Pittsburgh Geological Society includes a photograph of the scene: a foundation, covered by charred rubble, and the shells of burnt-out automobiles in the driveway. "Although it rarely makes headlines, damage or threats caused by gas migration is a common problem in Western Pennsylvania," states the document.

In July 2008, an explosion killed a resident of Marion Township who tried to light a candle in the bathroom. The DEP record of the event—one paragraph long—states that the agency "became aware" of the problem after the fatality, which it linked to gas migrating into the septic system from an old gas well with deteriorated casing. The DEP files also contained some cases noteworthy for what was unknown or, at least, undocumented: "Unknown name, Armstrong County—SWRO [South West Regional Office]—1999: House explosion, resulting in destruction of residence and one fatality. Investigation is not well documented. Origin/mechanism of migration is an operating gas well. Pressurization of casing. Status: Resolved." For every fatality in the DEP files, there are dozens of close calls, evacuations, injuries, and water problems.

Shortly after the Dimock problems came to light, the DEP began receiving a rash of complaints involving murky and foul-smelling tap water in Bradford Township, a municipality 200 miles west of Dimock along the northern border of Pennsylvania. Investigators found methane in two wells and levels of iron and manganese above safe drinking water standards in five others on Hedgehog Lane. They determined the problem was caused by overpressurized oil and gas wells operated by Schreiner Oil and Gas.

There is potential for methane migration wherever wells are drilled. The problem over the Marcellus, however, is a function of its geographical size and the volume of gas it holds, both unprecedented. In western New York and Pennsylvania, abandoned wells in other formations create voids in the ground that act as methane seeps, further complicating the picture. Many of the disused wells that fall within the Marcellus footprint are poorly documented, creating a wild card for drillers who might not know they even exist. The same applies to abandoned coal mines and seams in eastern Pennsylvania and in West Virginia.

The study of methane migration is complicated by the fact that it can be caused by conditions unrelated to gas drilling. Biogenic, or microbial, gas is the result of bacteria acting on organic material. It occurs in garbage dumps, landfills, and, to a lesser extent, compost piles, manure storage tanks, swamps, and septic systems. The decomposition produces methane that naturally pools or moves through the earth, sometimes mixing with water. When gas contaminates a water supply in areas near drilling, industry representatives often blame biogenic gas.

Shale gas, like that targeted by energy companies in the Marcellus, is thermogenic. It has the same combustible properties of biogenic methane, but it is formed over millions of years from organic matter deep within the earth, under intense heat and pressure. Extracting this gas requires drilling through the water table, and that's where problems often begin. Casings are designed to seal off the water-bearing zone from gas that comes up the well bore under high pressure from deep gas-bearing zones. Crews insert steel pipes into the well bore and fill the void between the outside of the pipes and the ground, called the "annular space," with cement. It's a simple concept that becomes complicated in practice. The void must be accurately calculated in order to know the volume of cement required. The cement, ideally, must have properties that make it indefinitely durable, which is also technically unfeasible. It must also set quickly enough to withstand pressure soon after it is poured. The practice is vulnerable to unexpected geological features and human error. The features include undetected faults and fissures that divert the cement from the well bore and create imperfections and weak spots. Human errors can include miscalculation of the amount of cement needed to fill the annular space; failure to keep the pipes centered in the bore hole, which raises the likelihood of uneven distribution of cement around them; and continuing production operations before the cement has cured.

Lacking staff to oversee every site, inspectors must trust that companies will comply with measures to properly case, monitor, and control wells. To check this, they rely on a paper trail of well logs and completion reports. When problems arise, they face extensive field work to pinpoint the source and, when necessary, hold companies accountable. Cement casings are the first place inspectors look when confronted with methane migration and water-well problems. It wasn't until after Norma's well exploded that the DEP began a comprehensive field evaluation of Cabot's operation. With problems developing in Bradford County and elsewhere, a torrent of permits pouring in across Pennsylvania, and an absence of paperwork in Dimock, what inspectors didn't know was as significant as what they did.

As the extent of water problems in Dimock gradually became apparent in 2009, Ken Komoroski visited homes to assure residents that the company was taking problems seriously. He sat in the Switzers' small trailer, listening to their complaints about being without potable water, and he reassured them the company was looking into the problems. Victoria's white Angora cat, Lucy, perhaps fond of the expensive fabric of Ken's suit, jumped up on his lap and began kneading. "Oh he's fine," Ken said. "I love cats."

When Ken finished what he had to say—that Cabot was sparing no expense to act as a responsible neighbor—Victoria closed the door behind him and looked at Jimmy. "He loves cats," she said. "If we had an iguana, he would have loved that."

Other residents told me their stories about Komoroski's visits in 2009, and they became less cordial as the year progressed. "They say they don't know where it came from," said Jean Carter, talking about the migrating methane while standing on her porch with Brandee jumping at her feet. "We know where it came from. It came from the gas well. We don't drink the water, we don't cook with it, and I don't give it to my doggy."

Cabot denied that it was responsible for Norma's well explosion and at first refused her request for deliveries of potable water. "I was told 'we aren't in the water business,'" Norma told me. "'No,' I told him. 'You're in the business of ruining it.'"

I asked Ken about that. "I like Mrs. Fiorentino," he said. "She's a good, honest person. We are going to re-evaluate this and work with her and the residents to make things right."

4. FIGURES, FACTS, AND INFORMATION

Several years before Norma's well exploded, Terry Engelder, the shale expert from Penn State, began paying attention to early attempts to test unconventional drilling in the Marcellus. As production from the Barnett Shale flourished in Texas in the first decade of the millennium, companies began testing slickwater fracking on the Marcellus in Pennsylvania. Prior to tests by Cabot on the Teel property in Dimock in 2006, Range Resources had been working on a project in southwest Pennsylvania for two years without revealing its progress.

That all changed on Monday, December 10, 2007. Range reported bringing five horizontal Marcellus wells on line in Washington County, each ranging in immediate production of between 1.4 million and 4.7 million cubic feet per day. These results helped the company exceed third-quarter projections "while making solid progress in refining . . . drilling and completion technologies and reducing well costs." This report piqued competitive interest in the industry; and competitors, and the people who financed them, turned to Engelder for information.

Engelder's influence is based on his extensive and distinguished vitae. He is a Fulbright senior fellow with more than 150 published research papers to his credit, a frequent international traveler on lecture tours and teaching assignments, a consultant to staff of multinational energy companies, and an advisor to capital market interests. He is not shy about his influence. As events unfolded with early Marcellus exploration, he told me, "It became well known in industry, including capital markets, that I'm the go-to guy." His research put him at the center of corporate discussions and media accounts of the commercial value of the Marcellus. His calculation of the Marcellus as one of the world's largest energy resources, announced in a university press release in January 2008, was quickly picked up by international news agencies. Engelder was later featured in a cover story about shale gas in *Time Magazine* as a person who "played a key role in the discovery of the Marcellus." He sees this work as a landmark in energy development that will interest historians and, accordingly, he keeps detailed journals and desk logs—filling hundreds of notebooks—chronicling everything from phone conversations to meetings with reporters. He shared a collection of these materials with me when I asked him to recount the discovery of the Marcellus as a world-class energy reserve.

Within hours after the Range press release was issued, Subash Chandra, an analyst at the global investment firm Jefferies and Company, emailed Engelder to ask him to lead a conference call on shale prospects for Jefferies' clients, a group that included investors tracking Cabot Oil and Gas, Range Resources, and other operators exploring Pennsylvania shale gas. "The holy grail for a scientist is to participate in research that has an impact on the public," Engelder told me. At this conference call, scheduled for the Friday after Range issued its report, he would have a virtual classroom full of interested investors.

Providing a single value for the capacity of the Marcellus had never been part of Engelder's research goals. At about this time, however, he saw that it would be useful to derive such a number. While preparing for the Jefferies call, he did a "back of the envelope" calculation based on what he knew offhand about the formation's expanse, depth, thickness, thermal maturity, and organic content (known in the industry as total organic carbon, TOC). The number seemed large, but lacking context and concrete data it had little meaning, so he discarded the paper. Still, he was curious, and he did some checking to see what was documented about this. He found only an estimate by USGS that the Marcellus held slightly under 2 trillion cubic feet of gas. His rough calculation, he remembered, was two orders of magnitude larger. He contacted Gary Lash, a research partner at the State University of New York (SUNY) Fredonia, and they began working on a formal calculation.

Meanwhile, Range officials were guarding their Marcellus secrets while releasing only enough information to entice investors—a reticence that strengthened Engelder's position as the "go-to guy." He and Lash had a consulting company—Appalachian Fracture Systems—that specialized in providing the kind of information necessary to capitalize on shale gas development. The heart of this information was Engelder's research documenting the orientation and physical characteristics of natural fractures in Devonian shales (work built on the research conducted by Pearl Sheldon along Cayuga Lake). Understanding the pattern and orientation of the natural fractures (which Sheldon had called J-1 joints) would enable operators to drill well bores that intersected the greatest number of cracks in the shortest distance, thereby compounding the effects of hydraulic fracturing.

Shortly after news broke about Range's success, Terry received an email from a former student now working for Range. He told Terry that wells with laterals drilled along a northwestern to southeastern axis—perpendicular to the J-1 joints—were indeed the biggest producers. These Marcellus gushers were validating the research. To underscore the classified nature of the operations that produced those results, Terry's informant added that Range had armed guards at the Marcellus well sites.

Email reflecting excitement over the Marcellus poured into Engelder's inbox, including another note from a friend in the oil and gas industry. This friend had seen a press release about Engelder's upcoming call with Jefferies and recognized the impact that the shale gas rush would have on Engelder's career: "At certain points in some people's lives, events occur that are fortuitous, and when the Brass Ring is presented, it must be grasped. The Marcellus play in NY, PA & WV is now the hottest gas play in North America and maybe the world. The Energy Forum Seminar in Pittsburgh last week was standing room only with LOTS of new companies from outside the basin. I'm not sure how you can position yourself to benefit from this, but opportunity is there."

Two days later, Engelder spent close to two hours on the Jefferies conference call educating about seventy-five Wall Street investors on the basics of Marcellus geology. His work with Lash to quantify the volume of gas was still unfinished, so he focused on how his research of the J-1 fractures pertained to financial returns.

Industry is often circumspect in revealing the science behind their discovery. Now what this means often is that the majors—Exxon Mobil, for example—can afford to do the research themselves and consider it proprietary. On the other hand, independents—these are the people that operate

in the Appalachian Basin—often don't have the resources to fully understand the science behind the play. And in a sense then, they're poking
around almost in the dark. And I think in part it's fair to say this concerning the J1 play, the Devonian black shale play right here.

He then connected his research to the Range wells:

> I think it's fair to say that about a day after this Range Resources web cast
> [announcing production figures of the first Marcellus wells], a Range geol-
> ogist looked at this particular paper submitted to the AAPG [American
> Association of Petroleum Geologists] Bulletin and sent his congratulations
> back to us. And implicitly we—at least we take that to mean that what we
> said in this particular paper was quite in line with their thinking concern-
> ing this J-1 play.

After the conference call, Engelder informed Lash in an email that
"Appalachian Fracture Systems is in business, and I PREDICT it will be the
Wall Street investors who will see that AFS is a success."

As the year drew to a close, Engelder worked late nights in his office on
"an adrenalin high," he told me, to complete the calculation of the volume of
gas in the Marcellus. He thought of the Marcellus discovery as a Christmas
present for America; and on his drives home, after days full of thinking about
the possibilities, he was given to playing the Radetzky March by Johann
Strauss. In a celebratory mood on one of those nights before leaving his
office, he wrote Lash:

> What a week!
>
> Anyway, we have to be very very careful not to over commit until we
> really understand the magnitude of the prize that we might have come
> upon. We could still screw up early in the game by underestimating what is
> really at stake.
>
> Whatever is out there (and I can't see it) it is bigger than a whale . . . and
> I want to know more about it before getting too close. Chandra from Jefferies
> said that the money listening in on my talk this morning was "Billions and
> Billions of Dollars." . . . I don't think Chandra is a man to exaggerate.

A few days later they finished their estimate. The Marcellus held
between 168 trillion cubic feet to 516 trillion cubic feet of gas. In geology

parlance, a reservoir of more than 30 trillion cubic feet is classified as "a supergiant," which is the largest classification and tends to draw a proportional amount of commercial attention. Bigger prizes draw more brainpower, capital, and tolerance for risks. Lash and Engelder had just shown the Marcellus prize to be 100–250 times bigger than previously thought; it was a supergiant of titanic proportions. Not all the gas was recoverable, of course. But a better understanding of natural fractures, combined with evolving hydraulic fracturing technology, would maximize the yield. Engelder originally put that number at 50 trillion cubic feet, or about 10 percent of the whole reservoir.

The dean of the College of Earth and Mineral Sciences at Penn State, William Easterling, notified the university president, Graham Spanier, of the calculation. Spanier, according to Engelder's notes, asked Engelder to "hold off" on announcing the figure until Spanier could make the news public at a Board of Trustees meeting the following month on January 17. On that day, the university released the news through a press conference after the trustees meeting.

Engelder characterized the events uncovering the Marcellus potential as an epiphany for himself and the country. "The value of this science could increment the net worth of U.S. energy resources by a trillion dollars, plus or minus billions," he was quoted as saying in the university press release on January 17. His schedule became booked with engagements to explain Marcellus opportunities to investors, landowners, the press, and communities. It was a story that started big and continued to build for months and then years. After analyzing production data in 2009, Engelder revised his calculation of recoverable gas from 50 trillion to 489 trillion cubic feet—enough to meet U.S. natural gas demands for decades. By the middle of 2009, he had, by his own count, spoken with 139 reporters, with 194 natural gas companies, and at 150 community presentations on the Marcellus. Engelder's message was overwhelmingly positive. The Marcellus was a gift—a fix for federal debts and state deficits, the unfavorable imbalance in international trade, high unemployment rates, global wars driven in part by perceived threats to supplies of foreign oil, and uncertainties of global climate change. Engelder's view meshed perfectly with the industry's pitch to sell the world on the benefits of natural gas. "It's almost divine intervention," said Aubrey K. McClendon, chairman and chief executive of the Chesapeake Energy Corporation, which began aggressively pursuing Marcellus drilling in 2009. "Right at the time oil prices are skyrocketing, we're struggling with the economy, we're concerned about global warming, and national security threats remain intense, we wake up and we've got this abundance of natural gas around us."

THE "PENN STATE REPORT"

The careers of many industry supporters were tied, by way of reputation or investment, to the trillion-dollar Marcellus find. For these stakeholders, Engelder's projection provided a starting point for building a case that Marcellus exploitation would benefit not just industry chiefs, money managers, and researchers but also patriots, environmentalists, and the working class. Sure, there would be those, such as Ken Ely, who could expect to make a lifetime income from royalties, but there would be many more like Scott Ely, Ken's son, who could expect a good job where there once was none. This expectation was advanced through the work of two of Engelder's colleagues at Penn State: Timothy Considine, professor of natural resource economics, and Robert Watson, associate professor of petroleum and natural gas engineering. Both academics, like Engelder, served as advisors to industry and investors. As chairman of the Oil and Gas Technical Advisory Board, Watson also headed the panel that guided Pennsylvania DEP regulators.

Watson and Considine authored *An Emerging Giant: Prospects and Economic Impacts of Developing the Marcellus Shale Natural Gas Play,* published in July 2009 through the Penn State College of Earth and Mineral Science. The report began with a reference to Engelder's work showing "estimates of recoverable reserves of at least 489 trillion cubic feet seem increasingly reasonable." From there, it expanded on themes Engelder had struck during his public appearances. "The market and strategic value of the Marcellus Shale will no doubt grow as conventional natural gas reserves are depleted and our economy adjusts to a path with lower greenhouse gas emissions." Natural gas has carbon emissions 60 percent lower than coal and 30 percent lower than oil and "is widely viewed as a bridge between the age of oil and the next energy paradigm" of sustainable energy sources, according to the assessment. The authors assigned numbers to the economics of developing the reserve in Pennsylvania; the industry would generate more than 175,000 (mostly blue-collar) jobs over the course of a decade (ten times the number projected by the Pennsylvania Department of Labor). Marcellus development would also produce $13.5 billion in value-added revenue to communities and $12 billion in state and local taxes.

As robust as these prospects were, Considine and Watson warned, they were vulnerable to the dampening effect of regulations and taxes. In summer 2009, when the report appeared, a proposal to tax gas as it was drawn from the ground (called a severance tax) was under consideration by Pennsylvania lawmakers. (Pennsylvania was the only significant gas-producing state without such a tax, which was commonly used to fund oversight of the industry in places such as Texas, Oklahoma, and West Virginia.) If such a tax passed, the authors cautioned, "drilling activity would decline by more than 30 percent and result in an estimated $880 million net loss in the present value of tax revenue between now and 2020." Natural gas extraction, they continued, "is a very competitive business prone to sharp contractions in drilling activity from adverse swings in costs, prices and taxes. As a result, many states have adopted policies that promote development. As the Pennsylvania Marcellus shale industry develops, policymakers should keep in mind the trade-offs between any short-term gains from taxation or regulation with the long-term effects on industry development." All in all, an industry encouraged to grow with favorable policies would generate more tax revenues than an industry "stunted by high taxes and costly regulations," the report advised.

The credentials of the authors of this report, and the university seal printed on the cover and each of its thirty-two pages, lent credibility to its contents, which were routinely cited in press reports and community presentations. The Marcellus Shale Coalition, an industry group that had funded the report and touted its findings, referred to it as the "Penn State report" in press releases. The reference was adopted by national media and trade groups; and the report became a common instrument used by industry supporters to lobby against taxes and regulation.

"The report makes plain that hydraulic fracturing is a non-negotiable tool for making these resources and revenues possible," stated Lee Fuller, policy director for Energy in Depth. Fuller reiterated the findings of the report that stated federal legislation to regulate hydraulic fracturing "would be disastrous in terms of the domestic oil and gas industry, raise prices for gasoline and natural gas, and ultimately derail any efforts to address the need to reduce carbon emissions." These types of claims eventually drew counterattacks by industry opponents. A year after the report was released, university president Spanier received a letter from an advocacy group complaining that the so-called Penn State report confused academic research with propaganda by its industry sponsors. At the base of the complaint was the lack of transparency about its funding and origins. "It is completely inappropriate, Dr. Spanier, that Pennsylvania's flagship university would allow its reputation to be used to support the gas industry's positions by dressing them up as independent research," wrote Drake Saxton, president of the board of Responsible Drilling Alliance. The report, and a subsequent update (with similar numbers) released in May 2010, "greatly exaggerate economic potential of gas exploration and are distorting the legislative process." Saxton concluded, "This is dirty business, Dr. Spanier, and it is time for Pennsylvania State University to cleanse itself."

Spanier ordered William Easterling, dean of the college that published the report, to address the complaint. A few weeks later, Saxton received a reply from Easterling conceding that an internal review "found flaws in the way the report was written and presented to the public." The fact that the report failed to identify its sponsor was a "clear error." Further, "The authors could and probably should have been more circumspect in connecting their findings to policy implications for Pennsylvania, and may have well crossed the line between policy analysis and policy advocacy." A disclaimer was added to subsequent versions of the report, noting the role of the Marcellus Shale Coalition and stating that no parties from the university warranted its accuracy, completeness, or usefulness. The Penn State shield remained on the cover but was taken off the other pages. The university also asked the Marcellus Shale Coalition to stop referring to the report as the Penn State report, Easterling explained in his reply to Saxton. He added, "The Committee is under no obligation to accept our recommendation."

As a debate over the severance tax—supported by Governor Rendell and Democrats and opposed by Republicans—reached a head in the Pennsylvania state legislature in fall 2010, references by industry representatives to the findings in the Penn State report continued to find their way into news accounts, sometimes with more conservative estimates of Marcellus-related economic potential from other sources. The Pennsylvania Department of Labor and Industry projected that Marcellus-related jobs would grow statewide, rising from about 8,000 in 2006 to about 12,400 in 2016. A report by the Marcellus Shale Education and Training Center at the Pennsylvania College of Technology (affiliated with Penn State and located in Williamsport) projected between 8,000 and 9,000 predominately blue-collar Marcellus jobs would be created by 2013 in a fourteen-county area in central and northern Pennsylvania. Most of the direct gains in employment would come from the 11.5 jobs that economists found were created by the average drilling rig, said Larry Michael, executive director for Workforce and Economic Development at the Pennsylvania College of Technology. These were considered short-term jobs, created during the boom cycle of development; and there was uncertainty about how long they would last. "We've heard estimates as low as 15 to 20 years and as high as 50 to 100 years," he said. "The bottom line is nobody knows for sure how long the boom cycle will be."

Projections pertaining to New York State were clouded by similar unknowns and marked by even greater extremes of opinion. A study commissioned by the Broome County government found full-scale Marcellus production would generate between $7 billion and $14 billion in local spending that would support between 810 and 1,600 new jobs for a decade. A paper by J. M. Barth and Associates Inc., an economic research and consulting firm in Croton-on-Hudson, New York, concluded that any projected gains from Marcellus development in New York had been based on assumptions that were "either biased, dated, seriously flawed or simply not applicable to the region that would be affected." Jannette Barth, the author of this paper, concluded that gains in Pennsylvania, in addition to being irrelevant to New York, had been overstated. All extraction industries combined employed only about 21,000 workers in Pennsylvania. Tourism jobs, by comparison, numbered 400,000. More than 677,000 were employed in the retail sector, with more than 48,000 jobs produced by Walmart. Barth, who holds a doctorate in economics from the University of Maryland, dismissed the work by Considine and Watson as "an exercise commissioned by the natural gas industry to try to prevent the State of Pennsylvania from imposing a severance tax on natural gas. Any intelligent lawmaker should not take this study seriously."

Christian Harris, a senior economist for the New York State Department of Labor, told me as I was reporting these developments that it was hard to project Marcellus jobs in New York because the gas industry was such a small part of the overall economy there and because there were so many uncertainties about its future, including unresolved regulatory issues and market factors. In general, he said, gas development could help buttress the weak economy, making up for some jobs lost in manufacturing. He also pointed out unresolved questions about the economic downside of the industry, including the effect of mineral leasing on property values.

TOO TOUGH A JOB?

Even as future employment projections remained a point of contention and speculation, industry proponents and critics both noted a trend as drilling ramped up in Pennsylvania in 2008: the majority of workers on drilling rigs were from out of state. Larry Michael, of the Pennsylvania College of Technology, put the figure at between 70 and 80 percent. Local workers lacked the knowledge and experience to work the rigs, he said, and his agency began training programs aimed to fix this. The jobs that were most demanding were the hardest to fill. These included roustabouts, entry-level workers who, like their namesakes in the carnival business, set up equipment before operations began and broke it down before moving it all to a new site; and roughnecks, who operated the rigs. Both jobs require physical endurance and strength, and long stretches away from family and friends. Twelve-hour shifts in all kinds of weather are typical. Roustabouts and roughnecks often work fourteen days straight before leaving the area for a break of one or two weeks. They are stationed for months or years in one region—some housed temporarily in "man camps" or sharing rental properties or hotel rooms—before moving on. Entry-level wages for roughnecks and roustabouts generally ranged between $15 and $20 per hour.

Larry Milliken, director of energy programs at Lackawanna College's Towanda center, told a Scranton *Times-Tribune* reporter that companies couldn't find workers willing to stay on the job once they found out how difficult it was; "I think there's a disconnect for people going into the business about what the demands are," he said. "This business is very intensive and very demanding." Tracy Brundage, a director of workforce development at Pennsylvania College of Technology, characterized it as "a whole different type of work culture and work schedule than what we're used to here. . . . We try to train them in that, what the industry expects." This disconnect was suggested by statistics kept by the Pennsylvania Department of Labor and Industry. Marcellus development would produce 72,000 "new hires," for a net increase of 9,300 "new jobs" related to the industry between the fourth quarter of 2009 and the first quarter of 2011. In other words, due to high turnover rates, more than seven people on average would be hired to fill one Marcellus position over the course of a year and a half.

Engelder was more blunt in his assessment of the situation, which he expressed in an email to a critic who asked him how an out-of-town workforce serves Pennsylvanian economic development interests: "The drilling rigs are run by people who can stand to labor under very, very difficult conditions. Unfortunately, there are not enough people like this in the state of Pennsylvania, sad to say, hence people who are willing to work very hard for a wage come to Pennsylvania from other parts of the country. Sounds a little like the flood across the Mexican border doesn't it? This does not reflect well on the work ethic and the blue-collar labor force indigenous to Pennsylvania." This conviction, he told me, was reinforced by his conversations with job-training specialists.

Engelder's response left Janine Dymond—the activist who had queried Engelder—wondering why Marcellus drilling would not be well served by the unionized labor force that had built Pennsylvania's steel industries or by the work ethic embodied in its farming, quarrying, and mineral extraction sectors. Perhaps it was less expensive to hire an itinerant work force that would endure exhausting shifts under dangerous conditions for low hourly pay. State Representative David Levdansky had heard the claim that there were too few drilling hands in Pennsylvania, and he found it suspect. "That's the excuse—Pennsylvania doesn't have qualified workers," he told a reporter for the Pittsburgh *Post-Gazette*. Levdansky supported an extraction tax, along with a bill that would create a tax credit for each Pennsylvania job created (unionized or not).

Whether by design of the industry or by the choice of the blue-collar workforce in Pennsylvania, more than two years after drilling began the high number of out-of-town workers persisted despite training programs. Labor officials complained to the Pennsylvania House Labor Relations Committee in September 2010 that qualified local workers were being shut out of contracting work. "We have tried every conceivable way to get our signatory contractors work with every gas company currently present in the Northern Tier counties of our state with little or no success at all," said Vern Johnson, council representative for the Greater Pennsylvania Regional Council of Carpenters. "Most of these companies will make us chase our tail until we're right back where we started." Union leaders said the industry lacked a system for unions to bid on jobs and that sometimes they were invited to bid on projects already underway by another contractor. Richard Bloomingdale, president of the Pennsylvania American Federation of Labor and Congress of Industrial Organizations (AFL-CIO) representing 900,000 Pennsylvanian workers, urged the committee to strengthen incentives for local employment and voiced support of the severance tax (introduced by Levdansky) as a way to provide money for occupational safety and health training for workers and emergency responders.

The severance tax, resisted by Republicans, died a slow death, and a compromise version never made it to the floor before the 2010 session came to a close. As events developed, the debate about jobs and energy extraction in general became more divisive. Those at one extreme embraced the industry as an expression of old-fashioned free enterprise. It offered work that built character and brought deserving rewards for those with initiative, whether they be roughnecks working twelve-hour shifts, investors staking their capital, or researchers staking their reputation on the next big discovery. At the other end of the spectrum were those who saw the industry as a relic of grandfather clauses and cronyism that dated to a period of predatory exploitation, when fantastical deals were pitched by door-to-door peddlers, manufacturing waste was buried in lagoons on private property, and unions were nonexistent. The middle ground was occupied by an untold number of consumers used to cheap plentiful energy, and property owners, who had their worries but also were able to calculate how much a mineral rights lease might be worth.

LANDOWNERS AND LAWYERS

Jim Worden, a dairy farmer from Windsor, New York, began making those calculations in December 2007. Jim was born and raised on a dairy farm, and perhaps by natural selection, was equipped with the kind of resilient temperament farmers often rely on to endure events beyond their control—bad weather, weak markets, and the occasional natural disaster. Jim faced all of these. His youngest daughter, Madison, was born with mitochondrial disease, a chronic disorder of the cells that gradually robbed her of strength and eventually, her ability to walk. In 2006, Worden lost much of his feed crop and a barn to flooding on the Susquehanna River, and he faced foreclosure. Jim and his wife, Bobbie, made up for the financial shortfalls just as many other farmers had—selling off his land bit by bit. By December 2007, Jim, in his late thirties, had about 250 acres left of the family farm and was resigned to selling the rest when the landman showed up and offered $60 per acre for their mineral rights. That would mean a $15,000 check upon signing—enough to take some stress off the family and make repairs to the van they used to take Madison to specialists in Boston and Syracuse. "Trying times come before good times," Jim told me on several occasions; and a seriously ill child kept him trained on the belief that every day was a gift.

Jim's farm, in eastern Broome County, was about one hundred miles east of the conventional drilling unfolding in the Trenton Black River in the western part of the state; and Jim had not yet heard the advice that Ashur Terwilliger, Chris Denton, and Lindsay Wickham were dispensing at farm bureau meetings. Instead, he heard the advice of Bobbie, whose intuition prevented them from dealing with the landman. "My wife just didn't like the man," Jim recalled. "She said he was peddling snake oil. She won that fight, thank goodness." They both felt something better was on its way, and they were now ready to take the remaining farmland off the market and wait for it.

Worden believes in fate, to which he attributes the difficult sequences of events that came with the New Year in 2008. In January, eight weeks after the landman's visit, he lost two fingers in the radiator fan of a tractor. He got his fingers reattached, but he couldn't milk for a while, so he spent his convalescence researching gas prospects. He began attending town meetings that he otherwise would have had no time for. Windsor was next to the Town of Sanford, and Jim happened to meet Jackie Root at one of the community meetings Dewey Decker had organized in the winter of 2008. Jim also attended the meeting with Chris Denton at the Deposit theater that winter.

Jim and Dewey had been talking about the possibility of Windsor landowners joining up with the Deposit coalition to negotiate a gas deal together, but both farmers concluded that a group that big would be unmanageable. Windsor—with more than 1,000 landowners and 70,000 acres—was better off forming a group on its own. That spring (as Root was shopping out the Deposit acreage to bidders) Jim, still recovering from his accident, began putting flyers in the windows of area businesses advertising a meeting at the Windsor Elementary School featuring Chris Denton on May 8. With this meeting, Chris had another shot at landing the business of a coalition, this one even bigger than the one he had lost to Jackie.

Worden credited fate, then, when news of the mega-deal between the Deposit Coalition and XTO broke on the front page of the *Press & Sun-Bulletin* the same day Denton was to give his presentation in Windsor. That night, more than a thousand residents—all living in the heart of the prospecting fairway and suddenly filled with dreams of becoming rich—showed up to Jim Worden's meeting to hear Chris Denton explain all the aspects of mineral rights leases, including the "flashbulb effect."

The promise of the Marcellus shone even brighter after a cascade of economic disasters in late 2008. The stock market crashed. Oil and gas prices plummeted with the implosion of financing and of housing sectors inflated by cheap money and extravagant homes sold to unqualified buyers. The American automobile industry, heavily vested in the production and marketing of luxury SUVs, soon followed. Unemployment rates climbed as both private and public sectors cut expenses.

Despite all this, prospecting continued to flourish in the center of the shale gas fairway in the Twin Tiers. Tens of thousands of landowners began organizing into dozens of groups to lease their land. Following Jackie Root's model, they were informally organized, with a nucleus of land owners (often those with the largest tracts) forming a steering committee that chose attorneys or lease brokers to draft a proposal, shop it out to energy companies, and represent the coalition in negotiations. Every resident, government agency, school, church, and business with land over the Marcellus became a prospective client to anybody who could answer their questions and showcase knowledge about mineral leasing.

People tended to join these groups based on their expectations, desire or need for money, geographic location, and their tolerance for risk. The leadership and steering committees provided the foundation of a grassroots democratic process that guided respective leasing strategies, starting with their choice of attorneys. Through his farm bureau networking, Chris gained the business of several of these groups, Windsor included, representing more than 110,000 acres in the prime part of the fairway, immediately west of Deposit. Chris worked without a retainer, with the understanding he would be paid a flat per-acre fee when the deals were signed. Of all the coalition attorneys, he was the most aggressive in arming his proposal with stipulations on land use, an approach that sometimes cost him business from group leaders who felt too much lawyering encumbered the process. Chris's contracts, including a separate surface rights agreement, were sixty to seventy pages long. In addition, he set about planning a way to ensure that the terms of the leases were policed and enforced well into the future. He proposed dedicating a substantial sum—$10 million or more—to be included in the leasing terms with gas companies to endow an independent team of experts, including a lawyer, accountant, geologist, pipeline specialist, and field operations manager. Under the terms of the lease, they would work for the coalition to ensure all aspects of the contracts—including gas metering, royalty payments, pipeline development, drilling, and legal considerations—were abided by over the long term, rather than leaving this task to the industry and state regulators.

"Once you sign, it's not over," Chris said, as he explained the concept to me. "The drilling and operations of these wells and pipelines and compressors will go on for generations on your property. There is so much institutional knowledge that is lost from one generation to the next, and this is a way to preserve that knowledge." There was little for landowners to dislike about the principle, but its execution became a point of contention for some coalition leaders, who saw it as idealistic and unworkable. Some were also skeptical about how well it would serve them. Time was not an issue for Chris, however. "This does not have to be done quickly," he told me at the time; and as months passed, he did not waiver on this matter. But it did become an issue for some of the landowners, including Dan Fitzsimmons, who headed a coalition that controlled 18,000 acres in the towns of Conklin and Binghamton. Dan was a former farmer and roofer in his early fifties, slowed by an aggressive and disabling form of arthritis. He and his wife Ava had a disabled adult son who required full-time care, and the family was chronically pressed by financial worries. His 185-acre farm was mostly fallow, and his dream was to use gas money to start a winery. Dan, an articulate, forceful man who once unsuccessfully ran for county office, had the temperament and convictions to govern. He was an astute student of the industry who often called me with informed tips during the early days of the rush and also to bend my ear about what he saw as negative press about spills and mishaps. He feared New York would lose drilling opportunities to Pennsylvania. Dan headed the Conklin group that hired Chris Denton, and he grew impatient with the contingencies and big plans with indefinite time periods. One night, he approached Chris before a coalition meeting about these matters. The encounter escalated into a confrontation reflecting the frustration that had been building in both men over their different philosophies. "Some bad words were said," Dan recalled. Chris was fired.

"Chris has a way of educating," Dan later reflected. "Everybody needs to learn about this and go through it, and he was a big part of that. . . . As far as his negotiating, I don't think it was a good thing. If you are litigating, it's one thing. But we were trying to negotiate." Chris said Dan had a mind-set that ran counter to the counsel Chris was offering to the group. "There was the sense with Dan that they had won the lottery and they wanted the dough. The thinking was: 'Let's get it done. Let's get a lease. Let's get the money in our hands, I have things to do with that money.' . . . Dan is very much a drill-baby-drill kind of guy. There are many people out there who want to drill-safely-drill."

Shortly after firing Chris in early 2009, Fitzsimmon's group hired Levene Gouldin and Thompson (LG&T), a Vestal firm that got Dan's attention when it teamed up with Peter Hosey, a Texas attorney who had put together deals in Texas and was well versed in the drilling culture. Drawing on Hosey's experience from afar, Scott Kurkoski, a partner with LG&T, represented the group locally.

Other attorneys and landmen made business of the rush, but collectively Kurkoski, Denton, and Rob Wedlake of HH&K represented the concentration of landowner power through their work with the coalitions. With control over 28 percent of the total 458,000 acres of Broome County, along with large portions of neighboring Chenango and Tioga counties, they were in the position to influence how the Marcellus would develop in New York. But they had more to contend with than reconciling the interests of landowners and gas companies. Forces would soon organize that would take matters out of their hands, at least for the time being.

Public meetings once held in town halls now routinely filled gymnasiums and conference centers. Some landowners welcomed the attention of landmen and lawyers; others did not. There were people who wanted nothing to do with gas drilling; were uncomfortable about the all-for-one, one-for-all approach of a coalition; or were still trying to sort through intimidating amounts of information, much of which was colored by politics and ideology.

In September 2008, about four hundred people attended a "landowners' rights" conference organized by Donna Lupardo in the Broome Community College Gymnasium. It featured presentations from representatives from the state attorney general's office, who warned of pressure tactics from landmen and urged residents to take care when negotiating leases. Michael J. Danaher, the featured speaker, explained a lease in terms antithetical to the landman's pitch. He explained a sweeping phrase giving drillers unconditional control over the property: "The right of access for the purpose of extracting oil and gas and other constituents—that's a phrase that companies use—and all the exclusive rights needed to explore, develop, produce, measure, and market production." This or a similarly worded phrase, he explained, gives companies "unlimited interest" to property that includes not only drilling but seismic testing and the creation of roads and pipelines. Once signed, he went on, the agreement is filed in the county clerk's office as a lien that impacts the property title as "an encumbrance," or a right that somebody else has to that property. "You are bound," he said. "We can't get people out of a lease. . . . If you can walk away with one iota of information from this meeting, you should and must consult with your own attorney before you sign the lease."

Ashur Terwilliger was in attendance. During the question and answer part of the program he stood up in the audience with his own warning—gas companies could store what they wanted on your property, including natural gas or wastewater. He waved a paper. It was a permit application to the New York DEC to inject drilling waste into an empty Trenton Black River well in Van Etten, Chemung County. If landowners don't include lease provisions that specifically prohibit storage wells, Ashur warned, they might get some surprises like this. The issue could ensnare landowners, he said, because New York State had yet to develop a policy to deal with wastewater. "Had they done, what they should of, nine, ten years ago, we wouldn't be in the position we are now," he said. "Our wastewater out of our wells goes where? To Pennsylvania to be treated. Sometimes the trucks are back too quick to have taken it to Pennsylvania. I know because they are up and down my road. I'm convinced some of that went into a dry hole."

The questions and comments continued for hours, often digressing from the legal particulars of land leasing to opinions about fracturing, truck traffic, water use, and other environmental issues. A middle-aged woman in a lavender blouse stood up and spoke into the microphone. "There's been a great deal of attention given to protecting people making their contracts. But very, very little attention given to people who don't want to participate in these contracts. . . . We have very few rights if we don't want to have a contract with one of these companies, and money is not our concern. Nobody is looking out for us." In response, Danaher explained the basics of "compulsory integration," a New York statute that allows companies to extract gas under land after 60 percent of it has been leased. Those without a lease are required to join the pool of landowners that make up the unit and are entitled to the minimum compensation negotiated within the pool. Although drilling companies have no surface rights to the unleased land, they can drain gas from under it. "There is nothing we can do about that part of it," he added. "That's the way the law's set up. The legislature determined many years ago that these resources are valuable resources that need to be extracted to support all of our lives."

Lindsay Wickham, who was also on the panel, followed on Danaher's account and elaborated on horizontal drilling and related changes in the spacing bill. Rather than having dozens of vertical wells, he said, one well pad could accommodate six or more laterals to extract gas from the same area. The new spacing law "will make it so we won't have a patchwork of what we call non-conforming units. It's much more efficient."

"A word like efficient is the language used to express the side of the drilling companies," the woman replied. "It's not efficient to the people who don't want the drilling."

RECORDS FROM THE PAST

Walter Hang had heard the promise that gas production in New York was well regulated and problem-free, and he intended to put it to the test. Hang headed a firm called Toxics Targeting, based in Ithaca, which compiled and packaged records and reports on polluted sites using files and databases kept by the New York DEC, the federal EPA, and other regulatory bodies. His clients, many from the New York City area, included engineers, consultants, and municipalities conducting due diligence required for construction permits and financing. In a nutshell, Hang made his living from finding and documenting pollution.

Hang had built a career on high-profile environmental causes, beginning in 1976 when he was fresh out of college and a new hire with NYPIRG, a branch of a nationwide organization founded by Ralph Nader, the iconoclastic consumer advocate and champion of progressive causes. At the time, polychlorinated biphenyls (PCBs) in the Hudson River and solid-waste disposal were leading environmental issues in New York State. Hang's job was to organize citizens—with a focus on student bodies—to press NYPIRG's positions on environmental and commercial reform regarding these and other issues. In the 1980s, Hang and his peers at NYPIRG successfully campaigned to hold General Electric accountable for the pollution in the Hudson and to pass recycling laws intended to lessen the need for incinerators and dumps. He moved to Ithaca in 1988 and created his own company dedicated to supporting efforts to clean up the environment and prevent future damage. His twelve years as a professional environmental advocate prepared him well for the fight over shale gas policy taking shape in New York. He was a veteran of political action campaigns, with a mastery of the who's who list in matters of environmental policy. He was on a first-name basis with influential figures who would find themselves at the center of the shale gas policy debate in New York, including DEC Commissioner Peter Grannis, Grannis's successor Joseph Martens, Judith Enck, and Stuart Gruskin. Gruskin, Enck, and Martens were Hang's contemporaries and also former staffers at liberal environmental advocacy agencies; and Grannis had been a pro-environment ally in the New York State Assembly prior to his time as commissioner.

Explaining his lifelong passion for public interest causes, Hang told me the story of his father, a Chinese national, who, with only 47 cents in his pocket, jumped off a ship docked at a Brooklyn pier and dog-paddled across the East River, where he found refuge in Harlem. That was in 1929. Hang's father eventually earned U.S. citizenship. He had earned his freedom the hard way, and he impressed upon his son the importance of public involvement in government and civic affairs. Hang, a slight man with a brush cut, has the polish of a politician and projects a cheerful aspect even during confrontational demonstrations, which he usually attends smartly dressed in a suit and tie. He is a disciple of Nader's visions of open government, and he relishes the role of public watchdog and community organizer. An informed and active citizen base, Hang believes, is a primary tool against unchecked corporate exploitation to benefit a few at the expense of many. "People think the government will sort everything out and protect them, because that's what it's there for, right?" he said, describing his outlook to me. The debate over gas "is a very complex situation, and it comes down to this: Some people want the money and some people want the environmental protection."

When he investigated claims regarding the solid track record of the oil and gas industry, Hang found the lack of problems documented in New York wasn't due to a clean, well-regulated industry. Rather, it stemmed from a reporting system that was obscure, archaic, and incomplete. The confusion started with the lack of a requirement for companies to report contamination of private water wells. As with Pennsylvania, the regulatory agencies were directed by statutes, drafted decades earlier, that let the companies handle complaints directly. In so doing, they were free to require that landowners sign confidentiality clauses as part of a settlement. Such clauses would keep some cases of contamination forever off the record. In cases in which complaints did reach the New York DEC, the system worked against their public disclosure. The DEC's Spills Division, which policed cases of pollution and logged complaints on a comprehensive database, lacked jurisdiction over gas wells. Problems with wells were investigated by the Mineral Resources Division or referred to local health departments, which had no regulatory authority over gas drilling. This had been an internal protocol for decades in New York State, and Hang traced its genesis to memorandums between the DEC Spills Division and Mineral Resources Division.

One of these documents spelled out the rules for fielding complaints as drilling progressed in the late 1980s in western New York: "For major spills . . . Mineral Resources will contact the spills unit. A decision could then be made at the site or over the telephone about who handles the spill. Mineral resources will check the spill in most instances, and let the spill unit know of their findings if the lead DEC unit responsibility was in question."

Another memorandum, also written during the period of well development in western New York in the 1980s, instructed DEC Mineral staff to refer complaints of private water-well contamination to county health departments. "Any requests that come into this office must be referred to the local County Health Authorities 'LHU' rather than us log[ging] in the report," the directive stated. For cases that the health department, in turn, referred back to the Mineral Resources Division, staff was to then "proceed in our usual manner to try to determine the cause of pollution."

Lack of a centralized record-keeping system was one of several problems that compounded efforts to access public records. Another was the lack of a searchable database cataloging problems and complaints investigated by the Mineral Resources Division. Hard copies of records—where they existed—were available through written requests under the Freedom of Information Law, but only on a well-by-well, permit-by-permit basis. These limits made attempts to itemize and tally trends within the data set—such as methane migration or water contamination—impracticable, if not impossible. Still, some drilling problems involved issues broad enough to fall within the jurisdiction of the Spills Division. Those, Hang knew, might show up on the Spills Division electronic database, which included more than 300,000 records encompassing everything from possible gas station violations to sewage overflows.

Recognizing that as a start, Hang began analyzing the records and immediately found the data lacked a field that uniquely identified natural gas–related problems. Using various search terms, he began finding some records of natural gas and oil drilling complaints under a field labeled "Category," with this description: "Known release which created a fire/explosion hazard (inside or outdoors), drinking supply contamination, or significant releases to surface waters." Using this key term, he culled 270 records documenting mishaps with this and similar descriptions. They included fires, blowouts, methane migrations, and spills relating to wells or infrastructure. Some were from current drilling, some were from legacy sites or long-forgotten "orphan wells," and some were related to compressors and pipelines. Often the cases were unresolved. On January 1, 2005, for example, a resident of Allegany called the DEC to complain of petroleum in her water well, which was next to waste pits that serviced an oil well field. Tests had shown a history of compounds related to gas and oil drilling in her well, including benzene and naphthalene. A subsequent investigation found pollution from the pits on her property, but the water tested okay. The inspectors concluded that it was unnecessary to filter the water, and they closed the case on November 4, 2005, although the spills database record indicated the site did not meet cleanup standards. Under the heading "Class" some spills had, "Unable or unwilling RP (responsible party) . . . DEC Corrective Action Required." That status applied, for example, to spill no. 9610441, reported on November 20, 1996, which documented a problem with an uncontrolled gas release from a nearby drilling operation:

> Natural gas escaped through fault in shale. Affected properties approximately 1 & 1/2 miles SW on Weaver Rd. Town of Yorkshire. Gas bubbling in Ron Lewis's pond, bubbling in ditch west side of Weaver Rd. 12 families evacuated. Gas in Lewis's basement (built on shale). Farmers well in barn 11708 Weaver Road vented to outside. Gas in ground coming up through yard.

Inspectors found the disaster had been caused by equipment failure on a drill rig, although no fines or penalties were recommended, and the case was closed.

These files, Hang asserted, showed a fundamental flaw and lack of transparency in the DEC's system that enabled regulators to downplay serious problems. I asked officials at the agency for a response to Hang's assessment for a story I was writing about the issue of oversight, and they told me he was wrong. Dennis Farrar, chief of the DEC Emergency Response Spills Unit, said fewer than three hundred oil and gas issues out of the more than 300,000 complaints logged in the spills database over thirty years represent a disproportionately small fraction (less than 0.1 percent) of overall environmental concerns. "In the scheme of things, this is not really a problem," he said. Jack Dahl, director for the Bureau of Oil and Gas Regulation, also responded to my queries for publication. He pointed out that the agency tracks problems through its Oil and Gas Bureau in addition to the Spills Division. But he couldn't quantify the complaints that the bureau had responded to or their outcome.

The resulting story drew criticism from the industry and the agency. Chris Tucker of Energy in Depth issued a press release titled "Lies, Damn Lies, and Walter Hang's Statistics: Ithaca Activist Scores Lots of Coverage over Claim of '270 Oil and Gas Spills in New York'—but What Do the Data ACTUALLY Say?" Tucker checked Hang's work and, by his count, the spills database contained 161 spills over thirty years related to oil and natural gas exploration and production, representing 0.045 percent of the total database. "Catch all that?" he wrote in a press release. "The process of exploring for, and eventually producing, oil and natural gas in New York over the past three decades is responsible for one-forty-thousandth of one percent (!) of the total spillage recorded over that time."

The New York DEC also sent a formal rebuttal in a letter signed by Commissioner Peter Grannis to elected state officials in the drilling fairway alarmed by the article. Hang's analysis, Grannis wrote, was "overblown and taken out of context." More than half the cases were unrelated to natural gas drilling, and they had occurred while the DEC was overseeing 10,400 wells. Overall, Grannis concluded along the lines of Tucker, the number of problems related to drilling were disproportionately small compared to other causes. Moreover, he wrote, "Requirements in place since the 1980s have successfully rendered drilling associated methane migration so rare that there has not been a reported incident since 1996. . . . When problems do occur, they are promptly and effectively addressed by DEC's spill response and Oil & Gas regulatory programs and staff."

Hang dismissed the Grannis letter as an expected response from the leader of an agency under pressure. The continued denial of any concerns by rank-and-file regulators nagged him, however. He knew the spills database reflected an incomplete picture of the actual number of problems; so he pressed on with his investigation. A search of records of health departments in places with a history of gas drilling turned up, among other things, a memo from William T. Boria, a water resources specialist at the Chautauqua County Health Department. Boria reported that his agency received more than 140 complaints related to water pollution or gas migration associated with nearby drilling operations. "Those complaints that were recorded are probably just a fraction of the actual problems that occurred," Boria stated in a 2004 memo summarizing the issue. County health officials tabulated information on fifty-three of the cases from 1983 to 2008 on a spreadsheet, including methane migration, brine pollution, and at least one in which a home had to be evacuated after the water well exploded. "A representative I spoke with from the Division of Minerals insists that the potential for drinking water contamination by oil and gas drilling is almost non-existent," Boria wrote in his memo to a party whose name was redacted. "However, this department has investigated numerous complaints of potential contamination problems resulting from oil and gas drilling."

Without a centralized reporting system, assessing these complaints comprehensively was difficult; yet Hang found signs of them throughout western New York. In spring 2009, the Allegany County Health Department documented a case in a report to a county legislator. Workers had been fracking a non-Marcellus well near a home where Dave Eddy lived with his wife and two young children. The water well at this location, 360 feet deep, "should provide water of very high quality as long as the integrity of the well or aquifer from which it draws water has not been compromised," the report stated. Soon after the operation began, according to the memo, Eddy reported the water turned muddy and smelled like gas. (Eddy told me in a subsequent interview that the family found the problem one night when his wife drew a bath for the kids and the faucet produced a foamy, chocolate-brown stream.) Testing by the company found the well was polluted with petroleum. The company subsequently installed a filter on the home, put the family up in a hotel, and offered compensation for the pollution, the letter states. The public record of the case after that is obscured by litigation.

Long before that, newspaper records reflected public frustration over a number of methane problems in western New York throughout the 1980s. In spring 1984, the *Post Journal* in Jamestown began covering a story of gas leaking into the homes of more than fifty families living near natural gas wells in Chautauqua County. The supply of water that sustained Tim Short, his wife, and four children began to bubble like seltzer and later, in November of that year, an explosion blew the cover off his water well. Emergency responders cut the power to their home and evacuated the family. They returned when levels dropped, only to find the problem was recurring. Short drove pipes into his yard to release the gas from the ground and rented a backhoe to dig a trench around his basement. He bought a gas detector that sounded an alarm when levels get dangerous. Other families were also affected, and some had to evacuate. The newspaper printed a photograph of a flame fueled by gas streaming through cracks in the basement wall of one home.

Angry residents turned to the town board for help, but they were told that the power of the town government to regulate gas drilling was preempted by state environmental conservation laws. Back then, as now, a gas well was not considered a source of pollution unless residents could prove it, and local governments lacked jurisdiction in the matter. "The burden of proof is with the landowner," Town Attorney Paul Webb told the *Post Journal*. "It's an easy way out for the DEC." Town Supervisor Barry Leyman told the paper, "Our hands are tied. The board wants to help, but can't."

The problem dragged on unresolved for years with a series of investigations by the New York DEC and the attorney general's office. From the beginning, DEC investigators attributed the problem to natural conditions—a conclusion challenged by Attorney General Robert Abrams, whose office oversaw radiocarbon tests of gas samples. "The bottom-line conclusion is that the gas is of a petrochemical origin. Not swamp gas," said Nathan Riley, Abrams' press secretary. "The tests indicate to the attorney general's office that the industry is going to have to take greater responsibility for the effects of drilling." Five years after the problem began, the DEC issued its final report. Its position remained unchanged: the methane problem that suddenly and persistently affected the Short's home and dozens of other nearby residents after the drilling began was from naturally occurring seeps.

The Shorts eventually abandoned their home and moved into a trailer park. The source of gas, beyond being identified as a natural phenomenon, remained officially undetermined on the DEC record. "You've had these problems for twenty-five years," Hang told me. "Local communities have been struggling with the problem on their own. . . . The DEC couldn't even oversee drilling in these conventional formations. How are they going to handle unconventional shale gas development? We need assurances that when these concerns come up, they will be addressed in a more comprehensive fashion."

The SGEIS—the instrument for the environmental review (and de facto moratorium) ordered by Paterson's administration after the meeting in Greene in 2008—was intended to adjust the existing regulatory format to accommodate unconventional drilling. It was not intended to review the laws already in place. These rules had no provisions to shift the burden of proof away from local residents regarding water pollution or to require gas companies to report problems along these lines; and Hang saw that as a fatal flaw. For these reasons alone, according to Hang, any amendment built on the old regulatory format would be insufficient and possibly disastrous. Rather than a "supplemental" review, the entire regulatory format would have to be rebuilt based on a full and transparent understanding of previous problems. Even before a draft of the SGEIS was released, Hang would begin organizing efforts to defeat it.

"MISINFORMATION"

The drilling industry in New York was chiefly represented by Brad Gill, executive director of the IOGA New York. Gill was a distinctive figure at public forums and hearings, where he appeared frequently as the Marcellus controversy intensified in 2008 and 2009. A muscular man, sometimes sporting a finely trimmed mustache and goatee, he often wore jeans, cowboy boots, a sport coat, and a stud in his left earlobe. He invariably dismissed opposition to drilling that, according to his assessment, was based on overblown concerns, misinformation, and propaganda from self-serving liberal interest groups and biased media. Such was the case on October 15, 2009, when he testified in front of the New York State Assembly Committee on Environmental Conservation. "You—as lawmakers, along with the general public—are being misled," he told committee members. "We have to respond to those who have worked to instill fear in the public, in the media and in some elected officials by spreading reckless misinformation in an attempt to block the expansion of natural gas exploration in New York—or regulate us to the point that we abandon our hopes of drilling in this state's Marcellus Shale reservoir." If elected officials follow the false lead of activists, he added, "New York will lose a golden opportunity." Gill used the platform to deliver the industry rebuttal to the claim that fracking agents are secret ingredients. Anybody who wants to know can refer to any industry website, where they will learn that fracking agents are made of a "small amount of dilute, benign additives" found in common household products: a friction reducer, "similar to Canola oil," that allows fracking agents to penetrate deeper into a well, and an agent to prevent bacteria growth from clogging up a well, which works "in the same way chlorine is used in our drinking water supply." The final ingredient, according to Gill, is "0.1 percent portion of a micro-emulsion element, similar to those found in personal care products."

Gill delivered this same message at meetings he scheduled with the editorial boards of upstate newspapers. He urged them to devote more coverage to the opportunity the Marcellus presented rather than spreading "negative publicity" by dwelling on overblown reports of accidents.

Hydraulic fracturing was being depicted as something that was new and threatening, he stated, when in fact it had been around for decades, and it had never been proven to cause groundwater contamination.

It is true that hydraulic fracturing has been employed for decades to stimulate conventional wells. It is also true that the volumes for each well were relatively small and that conventional gas extraction was a secondary component to the overall energy outlook behind oil and coal. Stimulation for conventional formations typically required less than 100,000 gallons of fluid per well—one-thirtieth to one-fiftieth the amount required for a shale gas well; and some reservoirs, including certain shallow formations and sandstones in New York and Pennsylvania, were produced without hydraulic fracturing. But full-scale shale gas development changed these dynamics by increasing both the number of wells in any given region and the quantity of fluid required to stimulate each well.

In his appeals to editorial boards and state lawmakers, Gill was articulating an industry argument based on an EPA study of hydraulic fracturing risks under the George W. Bush administration. At that time, in 2004, early success in the Barnett Shale in Texas signaled the beginning of the shale gas boom and piqued regulatory interest in the high-volume methods that made it possible. As with any transformational manufacturing technology, there was a lot riding on how this new generation of well stimulation—and the set of environmental challenges that it presented—would be incorporated into standing policy and law. Few companies had a more vested interest in this policy review than Halliburton, one of the world's largest well services companies. A notable person with direct influence over the process was Dick Cheney, who served as Halliburton's CEO after he was defense secretary under George H. W. Bush and before he became vice president under George W. Bush.

Halliburton prospered greatly under both Bush administrations. In 1992 when Cheney was secretary of defense, Brown and Root, a Halliburton subsidiary, earned more than $2.2 billion for a U.S. Army contract for logistics and engineering services in the Balkans. A dozen years later, as high-volume hydraulic fracturing was poised to become integral to national energy production, Cheney was vice president of the administration that engineered the 2005 Energy Policy Act, an omnibus bill for energy policy that included, among other things, tax incentives for producers of alternative energy sources. Buried in the act, which included more than 1,700 sections, was a provision that exempted the natural gas industry from the Safe Drinking Water Act, which governs what can be injected into the ground. The two-paragraph clause didn't command much attention prior to Marcellus development, but later it became known among critics as the "Halliburton Loophole" and a central target for reform efforts.

Drilling proponents held that the exemption was justified by the EPA study completed the prior year. The study took into account the effect of hydraulic fracturing when used to stimulate gas production from wells drilled into coal beds (coal-bed methane, CBM, wells). Because these wells were shallow, they were perceived to present the worst-case scenario for problems related to hydraulic fracturing. The study found "no confirmed cases [of polluted water wells] that are linked to fracturing fluid injection into CBM wells or subsequent underground movement of fracturing fluids." Working from the premise that, if fracking doesn't cause problems in coal-bed wells, it shouldn't cause problems in other formations, the authors of the EPA study concluded that fracturing poses a "minimal threat" to water supplies.

Some EPA staff members later criticized the conclusions of the report as unsupportable and a flawed tool for shaping long-term policy decisions. Weston Wilson, an environmental engineer for the EPA and a Colorado resident, wrote to his congressional representatives about the matter on October 8, 2004. Invoking protection as a whistleblower, he publically criticized the use of the report. "The EPA has conducted limited research reaching the unsupported conclusion that this industry practice needs no further study at this time," he stated in the letter to Senators Wayne Allard (R-Colo.) and Ben Nighthorse Campbell (R-Colo.), and Congresswoman Diana DeGette (D-Colo.).

As the shale gas debate intensified, so did the criticism over the EPA study of CBM wells. EPA higher-ups stood by the integrity of the 2004 study, although some agreed with Wilson's assessment that it was ill suited for policy development. "It wasn't meant to be a bill of health saying 'well, this practice is fine. Exempt it in all respects from any regulation,'" Benjamin Grumbles said in an interview with Abrahm Lustgarten, *ProPublica* reporter. Grumbles had overseen the release of the report as the EPA assistant administrator for water under the Bush administration. "I'm sure that wasn't the intent of the panel of experts, and EPA never viewed it that way. . . . Our view was, we had concerns about the scope of the language, we provided technical assistance and information, and ultimately Congress decided not to include the language that we had suggested. I was disappointed by that, but there is always tomorrow, and there is always the opportunity for additional facts to get Congress to revisit the exemption."

The 2004 study noted one exception to its findings that hydraulic fracturing posed minimal risks to groundwater. Companies sometimes used diesel fuel to frack wells, "which may pose some environmental concerns," the report noted. To address this, the EPA "worked with the three service companies that perform 95 percent of the hydraulic fracturing projects in the U.S. to voluntarily remove diesel fuel" from fluids used to fracture coal bed methane wells. The three companies, including Halliburton, agreed, and signed a memorandum of understanding with the EPA to stop the practice. But the document meant little when it came to enforcement; and when the companies found that diesel suited their needs in the field, they went ahead and used it anyway. According to an ensuing congressional investigation, oil and gas service companies injected over 32 million gallons of diesel fuel or hydraulic fracturing fluids containing diesel fuel into wells in nineteen states between 2005 and 2009. For this, there were no permits issued, nor were fines levied.

A NATIONAL ISSUE

As shale gas development progressed in western, midwestern, southern, and eastern regions of the United States in the first decade of this century, cases implicating drilling operations in explosions, pollution, and illness began raising questions by advocates who demanded that federal policy be revisited.

On December 15, 2007, a house exploded in Bainbridge, Ohio. Nineteen other homes were subsequently evacuated due to high methane levels. The Ohio Division of Mineral Resources Management attributed the problem to the English 1 natural gas well, owned by Ohio Valley Energy Systems Corporation. According to the agency investigation, the problem arose from the operator's decision to stimulate a well without properly cementing the production casing. When operators fractured English 1, the casing failed.

Not long after the Ohio incident, drilling fluids became the focus of an EPA investigation of contamination of groundwater on the Wind River Indian Reservation in Pavillion, Wyoming. Agency officials noted that there was little development in the area outside of natural gas production and that operators used methods that "involve injecting water and other fluids into the well and have the potential to create cross contamination of aquifers." The investigation found methane in seven wells and "widespread incidence of low levels of organic compounds in drinking water wells." Overall, in seventeen of nineteen drinking wells levels of total petroleum hydrocarbons, including naphthalene, phenols and methane, were detected. The levels were high enough to pose health concerns, based on the assessment of the Agency for Toxic Substances and Disease Registry (ATSDR), which issued a bulletin advising Pavillion residents not to use the well water for drinking or cooking. Monitoring wells installed by the EPA in a shallow aquifer connected to drinking water supplies showed high levels of petroleum compounds, including benzene, xylene, methylcyclohexane, naphthalene, and phenol. The report recommended a follow-up investigation to trace the chemicals found in the aquifer to their source. "We're at the initial investigation, and we're just trying to get our hands around this," Greg Oberley, Aquifer Protection Team coordinator with the EPA in Denver, told me after the initial results were released in 2009. Representatives of Encana Oil and Gas, a company that operates about 250 gas wells around Pavillion, said their operations were not responsible for the pollution. Even so, the company undertook the task of cleaning polluted wastewater pits inherited by Encana from a previous operator, according to a company spokesman.

Just months before the situation in Pavillion came to light, an incident in an emergency room in Durango, Colorado, intensified the debate over disclosure and regulation. Cathy Behr was treating a roughneck who had been exposed to a chemical spill during a hydraulic fracturing operation in spring 2008. The hospital, Mercy Regional Medical Center, enacted its emergency protocol for dealing with victims of chemical spills—evacuating and locking down the emergency room and instructing staff to don protective masks and clothing. As Behr helped strip and wash the victim, vapors wafted from his clothes and boots. A day later, Behr's health began to fail. Her skin turned yellow, and within days, she became critically ill. Treating her was complicated by the fact that doctors didn't know what they were dealing with. After doctors ruled out the possibilities of hantavirus, Legionnaires' disease, and other infectious diseases that would explain her symptoms, blood tests indicated she was suffering from chemical poisoning. After more tests and examinations in Mercy's intensive care ward, Behr's doctors concluded she had been poisoned from exposure in the emergency room on the day of the fracking accident. With this diagnosis, they began treatment, but more information about the chemistry of the poison was needed. For this, doctors consulted a form called a Material Safety and Data Sheet (MSDS)—required by the Occupational Safety and Health Administration (OSHA) at all workplaces that handle hazardous chemicals. The five-page data sheet provided by the drill operators for the roughneck's treatment listed the product to which Behr had been exposed as ZetaFlow. It warned that people handling ZetaFlow should wear goggles and chemical-resistant clothing and boots, and that inhalation of the chemical can lead to headaches, dizziness, low blood pressure, and low oxygen levels in the blood. The data sheet didn't specify the chemical composition of the product.

Behr did recover, but her treatment was hindered by this lack of information. This story of exposure and risk was told in the regional and national press. In August, *Newsweek* reported, "Throughout the Rocky Mountain states, Behr's run-in with fracturing fluid is getting a lot of attention and exacerbating already frayed nerves. After nearly eight years of some of the most intense oil and gas development ever recorded in the American West, concerns over the environmental and health impacts are bubbling over." The persistence of these concerns led to calls by reformers for the EPA to revisit the 2004 study and eliminate the Halliburton Loophole. Without federal oversight, there was no comprehensive documentation or data sets on which to accurately assess and analyze the risks related to natural gas operations. Records were kept by the states, and in many of those states (as with New York) the paper trail was decentralized, unorganized, incomplete, and, in some instances, missing entirely.

All the while the industry stood by its position: the cases reported by the media were sensationalized, irrelevant, anecdotal, and an unfair representation of the industry safety record. A year after Brad Gill chided New York lawmakers for being misled, Thomas Pyle, president of the Institute of Energy Research, responded to a question put forth in the blog of Amy Harter, *National Journal* reporter: "Should the federal government seek to regulate a controversial extraction method for natural gas known as hydraulic fracturing?" Expressing a view that had by then been thoroughly articulated by industry lobbyists and advocates in hearings and demonstrations from Albany to Washington, Pyle responded:

> Let's get a few things straight right off the bat. Hydraulic fracturing is not new; it's been widely deployed as a safe extraction technique over 1.1 million times and has been around since 1949. Secondly, and a point not mentioned in the question by the *National Journal*, fracking fluids are more than 99.5 percent water and sand. The additives used amount to less than one half of one percent—many of which can be found under your kitchen

sink. . . . Every well site, in every state is required by law to have a list of additives contained in the fracking fluid on site in case of an emergency, they're called material safety data sheets, or MSDS. In Pennsylvania, the top environmental regulator has these additives listed on their website. Seems like disclosure to me.

The MSDSs are indeed an OSHA mandate for well service companies or any industry that handles dangerous chemicals. They are designed to provide HAZMAT teams and public safety responders with the information to manage chemical hazards they might encounter at a particular spill or fire, as was the case at Mercy Regional Medical Center. The sheets have fields for identifying the chemicals and their properties—what they emit when they burn, whether they dissolve in water, their breakdown products, what protective gear is needed, how volatile they are, and what to expect when they are ingested or emitted to the environment. When well services companies fill out these forms, it is accepted practice to list the compounds used in hydraulic fracturing by their trade names, without specifying their chemical ingredients. The MSDSs identifies ZetaFlow, for example, as methanol and a proprietary phosphate ester, but it doesn't itemize its chemical recipe. Together, the two ingredients make up 22–53 percent of ZetaFlow's ingredients, according to the MSDS. The other constituents aren't specified.

WHAT COMES OUT

The controversy over shale gas development had a lot to do with what goes into the well to stimulate gas production. Perhaps, even more so, it had a lot to do with what comes out and what remains indefinitely in the ground. A main byproduct of shale gas production—flowback—is a sum of unidentified compounds injected by operators, plus a host of variables that nature contributes to the mix. What goes into each well—biocides, acids, anticorrosives, lubricants, and friction reducers—comes out with millions of gallons of brine, metals, and naturally occurring radioactive material (NORM). In water-quality terms, these, for the most part, are collectively defined as total dissolved solids (TDS)—a unit that generically quantifies concentrations of all soluble material in a given water sample. These include chlorides, sulfates, and other ions that are a natural and vital part of ecosystems. When concentrated or redistributed through industrial processes, they can clog machinery; change the color, taste and odor of drinking water; and throw natural systems fatally out of balance. Table salt is not toxic, but if you pour half a cup of it into a 5-gallon freshwater aquarium, you will see dire results; and if you eat a half pound of salt, you will feel the impact yourself.

In summer and fall 2008 (when Ken Ely began complaining about leaking tanks stored on his land), TDS overloads began degrading Pennsylvania watersheds. TDS levels spiked in the Monongahela River, which bisects the heart of Marcellus development in southwestern Pennsylvania, as municipal sewage treatment plants along the river accepted tankers loaded with wastewater from drillers throughout the state. Workers at a steel mill and a power plant in Greene County were the first to notice something strange—river water began corroding equipment at U.S. Steel in Clairton and at Allegheny Energy Supply in Greene County. In homes, dishwashers began collecting residue and the tap water tasted bad. By October 2008, TDS levels exceeded water-quality standards at all of the seventeen potable water supply (PWS) intakes from the border with West Virginia to Pittsburgh, prompting an advisory to use bottled water that affected 325,000 people. In response, the DEP ordered sewage treatment plants to cut back on the amount of waste from drilling operations, and officials set to work on amendments to the state water quality regulations to address TDS pollution.

Those regulations were still being drafted a year later when a more dramatic problem arose. It began in rivulets flowing over the remote countryside of northern West Virginia and southwestern Pennsylvania. These were the headwaters of Dunkard Creek, which come together upstream of Brave, Pennsylvania, a hamlet of 412 people in Greene County. From there, the creek winds 43 miles across Pennsylvania's border with West Virginia and back to the Monongahela. Teeming with more than 161 aquatic species—ranging from freshwater mussels to 3-foot muskellunge—Dunkard Creek was at the heart of one of the most biologically diverse watersheds in the region. Its clear, green eddies and swimming holes, shaded by hemlock and sycamore trees, attracted generations of anglers, paddlers, picnickers, and nature lovers. On September 4, 2009, as families along the creek readied for their Labor Day gatherings and children played along the banks, dead fish began collecting in a pool below the Lower Brave Dam. Over the next week, fish floated to the surface, sank to the bottom, and washed up on the banks along the entire length of the creek. By the end of the month, almost everything in Dunkard Creek was dead. The only exception was an invasive microscopic alga—common in Texas estuaries—that had somehow migrated into the creek and now thrived in its suddenly brackish water. A stench hung over the watershed.

Subsequent investigations by the EPA, along with West Virginia and Pennsylvania regulators, found that Dunkard Creek had been ruined by this bloom of *Prymnesium parvum,* or golden algae, which is fatal to fish. The bloom was encouraged by TDS levels "many times higher than levels known to cause aquatic life use impairment," according to an analysis of the EPA. Low water levels, partly due to drought and partly due to the extensive withdrawals to support drilling operations, had exacerbated the problem. TDS overload in Dunkard Creek was tracked to coal mines, operated by Consol Energy, that had been granted variances for discharges with high TDS concentrations. Illegal dumping of drilling wastewater had also been identified as a suspect at several locations, including a brine disposal well, called Morris Run. This well had come under scrutiny for lax security and possible environmental problems. In 2009, the EPA fined Consol Energy, the owner of the well, $158,000 for failing to keep gates locked or to properly log the trucks coming and going from the site.

Because unhealthy TDS loads from sources ranging from sewage treatment effluent to run-off from farms, roads, and suburban developments were recognized as a growing problem in freshwater lakes and tributaries throughout the country, Dunkard Creek served as a dramatic warning of how quickly and unexpectedly TDS pollution—which, until 2009, had been considered a water-quality standard of secondary importance—can tip a thriving watershed into crisis. The events in the Monongahela River and Dunkard Creek in 2008 and 2009 raised public awareness about the TDS issue just as Norma Fiorentino's well explosion drew the public eye to methane migration. Drilling was not the only cause of methane hazards or TDS pollution, but it was a significant factor; and fairly gauging impact of drilling in the Marcellus era became central to the discussion and policy of environmental protection.

In December 2010, because of persistently high levels of TDS, the DEP listed the Monongahela as "impaired" in its annual report to the EPA. In his last month in office, DEP Secretary John Hanger summed up the problem: "Our extensive research clearly shows TDS levels in the Mon are close to the upper limits of the safe drinking water standard. This river is stressed, and TDS must be addressed. Any further increases in TDS loads will ensure that the river becomes impaired, adversely affecting all dischargers in the watershed and those businesses and industries that rely on clean Monongahela River water."

The Monongahela River also reflected a broader problem. Pennsylvania regulators, working to update water-quality regulations to address growing

TDS loads, concluded that "The Monongahela is not an anomalous situation. Recent reports on the water quality of the Beaver and Conemaugh Rivers in southwestern Pennsylvania and the West Branch of the Susquehanna River watershed have documented that these rivers are also severely limited in the capacity to assimilate new loads of TDS and sulfates."

Industry proponents didn't see it that way. Stephen Rhoads, president of the Pennsylvania Oil and Gas Association, told Joaquin Sapien, *ProPublica* reporter, that the Pennsylvania waterways "are not anywhere near" their capacity to handle TDS and that the DEP estimate of how much wastewater the industry produces is "completely exaggerated." Rhoads cited a study commissioned by the Marcellus Shale Committee characterizing natural gas development as "only a minor contributor" to elevated TDS levels in the Monongahela River in fall 2008. According to this assessment, most of the problem was due to discharges from abandoned mines. Drilling activity accounted for approximately 7 percent of the total TDS concentrations detected in the Monongahela River in October 2008 and decreased to less than 1 percent by December 2008.

This particular dispute over TDS discharges was eventually won by advocates who lobbied for regulatory changes to outlaw the disposal of high-TDS drilling waste in Pennsylvania waterways. This came in summer 2010 in the form of updates to Chapter 95 of Pennsylvania's Clean Streams Law, which outline wastewater treatment provisions. The new rules limited the discharge of drilling wastewater effluent to TDS levels of 500 milligrams per liter—the federal and state standard for drinking water. With this, the drilling industry was held to a higher standard than other industries, which were allowed a TDS ceiling of 2,000 milligrams per liter. The DEP cited several reasons: "The most significant rationale . . . is the fact that wastewaters resulting from the extraction of natural gas are of much higher concentration and represent higher overall loadings when compared to those from other industries." In addition, "The potential for growth within this sector is enormous and should that growth be realized, the potential impacts are just as massive." While the new regulations specified that drilling wastewater could not legally be discharged into rivers without treatment, they did not address what should be done with it.

Even before Pennsylvania's tougher Chapter 95 standards eliminated a disposal pathway for drilling companies, there were signs that waste production from the nascent shale gas industry in West Virginia and Pennsylvania was outpacing the development of treatment options. Some of the waste was being trucked out of state for treatment or injected into abandoned wells, some was entering Pennsylvania waterways through legal discharges (prior to the Chapter 95 amendments), and some was being dumped or was leaking into remote tributaries. In August 2009, Tapo Energy discharged an unknown quantity of a "petroleum-based material," polluting a 3-mile section of a tributary of Buckeye Creek in Doddridge County, West Virginia. One of the residents reported the problem after finding the creek, a favorite fishing spot, covered with a red gel. In October of that year, approximately 10,500 gallons of fracking fluid spilled from a broken joint in a transmission line from a Range Resources well pad into an unnamed tributary of Brush Run, a fishery receiving special protections for its rich biodiversity in Hopewell Township, Pennsylvania. The spill killed fish and other aquatic life and covered the water with a sheen. That event was followed by another spill in December, when an unknown quantity of waste overflowed at a pit, operated by Atlas Resources, and polluted Dunkle Run, a watershed in an unspoiled area of Washington County, Pennsylvania. Atlas didn't immediately report the problem, which eventually led to a $97,350 fine from the DEP. Investigators were faced with another mystery in late November 2010, after an inspector found liquid pouring from an open valve on the bottom of a 21,000-gallon tank and draining into Sugar Run. About 13,000 gallons drained from the tank, operated by XTO energy, but few details—including why the valve was open—were known about the incident in Penn Township, an area of fewer than 900 residents in a remote part of Lycoming County.

Inspection reports suggested a pattern of operators routinely disregarding regulations. In the first six months of 2010, DEP staff made 1,700 inspections of Marcellus Shale sites and found more than 530 violations, ranging from poor erosion and sediment controls to administrative violations to spills and leaks from improperly managed or constructed containment pits. Record-keeping and oversight remained a problem. An investigation by the Associated Press, based on a review of Pennsylvania DEP records on the disposal of 6 million barrels of drilling waste in 2009 and 2010, found the state couldn't account for 1.28 million barrels, about one-fifth of the total reviewed, because of a weakness in its reporting system and incomplete filings by some energy companies.

In response to the concerns about wastewater disposal in the later part of the decade, the industry began refining technology to recycle drilling waste fluids through a process called closed loop drilling. Rather than collecting and storing drilling waste in open lagoons and carting it off-site to be disposed of, closed loop drilling involves reclaiming the flowback in steel tanks and then recycling it on-site. Although the practice required costly technology, logistics, and engineering, the industry promoted it as saving the environment and reducing the costs and red tape associated with withdrawing and trucking millions of gallons of freshwater to remote sites. Well drillers, such as Chesapeake and Cabot, liked to showcase closed loop drilling systems when they took the media and politicians on well tours. As good an idea as the practice appeared, the industry objected to making it a mandate; and due to this objection, it was never incorporated in the regulatory update addressing the TDS issue in Pennsylvania. According to policymakers, Pennsylvania's new TDS ceiling on the drilling industry, enacted in 2010 through the Chapter 95 upgrades, effectively served as an incentive that "promotes the reuse of fracturing fluids." Like the diesel fuel agreement, closed loop recycling was voluntary; it was taken on faith that it would be employed to serve both public and corporate interests.

5. ACCIDENTAL ACTIVISTS

The residents of Dimock lacked the benefit of economic studies, investigations, exposés, and public hearings when they signed leases in 2006 and 2007. They learned about shale gas development on their own.

By fall 2008, soon after the Carters' and Sautners' water went bad, stories began circulating about other suspect wells next to drilling operations. Pat Farnelli said her water had a foul taste and that it was making her and her kids sick. Michael Ely, Scott Ely's cousin and next door neighbor, could light his water on fire when he placed it in a jug. So, too, could Bill Ely, Scott's uncle and Ken's brother, who lived in the ancestral farmhouse on the opposite side of Burdick Creek. Ken Ely was worried that chemicals from the waste pits and storage tanks were polluting the spring that supplied his water and fed his pond; and he was growing impatient with Pennsylvania DEP inspectors, who were unable or unwilling to validate his complaints. The inspectors always seemed to give Cabot the benefit of the doubt.

Here, Ken saw the Switzers as potential allies. Victoria was a woman of poise, education, organizational wherewithal, and, perhaps, political connections. The Switzer house, which was progressing with painstaking attention to detail, was a testimony to Victoria and Jimmy's capabilities and gumption. They knew how to get things done—sometimes enlisting the help of friends and neighbors.

Victoria and Jimmy were concerned about the stories they were hearing of water problems, but they remained hopeful (naiveté was what Victoria later called it) that the impacts from drilling would be temporary. They were used to neighbors stopping by to admire the house, lend a hand, or offer some advice; so when Ken pulled up in his Ford Bronco one day, Victoria was only mildly surprised, given his interest in stonework and timber. She gave him a tour of the soaring, unfinished structure, and he considered the hewn lacquered beams and listened to their plans for the bluestone hearth and patio. He knew of a good source for the bluestone, he told her. He could help her and Jimmy with that. And maybe they could help him with something. He explained the problems he was having with Cabot and his frustration with the DEP, and he asked her to come and see what was happening on his land.

It was late November—a time of year Ken would normally be surveying the land from a tree stand. With the exception of some scattered stands of oak and poplar, the hardwoods had shed their leaves, opening views obscured by foliage a few weeks earlier. Ken parked his Bronco, with Victoria and her camera in the passenger seat, in a clearing on the ridge that provided a commanding view to the operations at Ely 6H, which sat below their vantage point and above Ken's cabin. A fog of dust and exhaust rose from a dense cluster of tanks, hoses, platforms, generators, turbines, and compressors on the site producing an even, high-pitched drone. The machinery was mixing silica sand with several million gallons of a chemical solution and then forcing it down the hole at pressures exceeding 10,000 pounds per square inch. It was one of six wells on three pads in various stages of completion on Ken's property. Ely 2 was being drilled on a pad a third of a mile north, near Ken's quarry. Less than a half mile beyond that, a pad was being prepared at the north end of the property, which bordered a nature conservancy. There, Cabot was preparing to drill Ely1, Ely 7V, and Ely 5H.

Once a well was started, pits needed to be dug to support the operation. These served as both staging areas and repositories for various materials and waste that went into or came from the hole, including something known in the industry as "drilling mud." Although it looks like its name suggests, the substance is engineered to precise chemical specifications to lubricate the drill bit, float cuttings to the surface, and maintain hydrostatic pressure to prevent blowouts. It typically contains agents that are inert, such as clay, and some that are toxic, such as barium. The pits also captured the afterbirth of production pushed from the hole with the first rush of gas. This included fracking solutions, with chemical concentrations and profiles unknown

The Kenneth R. Ely property and adjacent properties in Dimock, Pennsylvania

Source: Cabot Oil and Gas Corporation public submission in Case 3:09-cv-00547-ARC to Ken, and brine, which Ken knew was salty enough to foul his land and water if not contained. Cabot had dug seven of these pits on Ely's mountain, and one of them sat directly upstream from Ken's spring. At each pad, Ken pointed out details to Victoria that could be easily missed amid the noise and drama of operators hitting the pay zone. A fountain of pressurized gas erupting from a well bore into a 60-foot tower of flame—flaring—is indeed a riveting spectacle, but a barely visible tear in a pit liner was a more troubling sign for Ken. That was the sort of thing he had an eye for, along with other particulars, such as fluid trickling from a tanker, a coffee cup floating in a puddle with an unnatural sheen, and pools of water on the ground that didn't freeze in the winter. Ken read these details—more than the truck convoys and equipment—as signs of a breach between humans and nature. He told Victoria he was sure the liquid in the pits was leaking into the water table, and he wondered how he, with her help, might go about testing that liquid.

Ken was seriously ill at the time, but he didn't volunteer this information to Victoria, or anybody else for that matter. "My fish think they're sick," is what he said. "They aren't dead yet, but I can tell by looking at them they don't feel too good. I guess the DEP likes them to be dead and floating before they see a problem and, you know, I just hate to wait that long." Ken's words were prophetic—fish would be floating down polluted streams while the Pennsylvania DEP considered enforcement action; pits would burn and leak. Ken was running out of time, although he probably didn't realize just how little time he had left.

"He wasn't much for being in the limelight," his son Scott told me much later. "And he saw Victoria as capable and willing to stand up for the environment."

LOSING HOPE

As 2009 approached, Jimmy and Victoria spent more of their days consumed with questions posed by the expanding drilling operations: What wells were coming on line? Where would the next pipelines and compressor stations be built? Who was getting royalties? Who was having water trouble? What was Cabot's response? What was the DEP's response? What rights did they have? And what was in those pits?

The stakes rose when Norma's well exploded on New Year's Day 2009. Within days, DEP technicians and Cabot contractors began collecting water and air samples from properties along Carter Road and Route 3023, including the Switzers'. They found explosive levels of methane in more than a dozen water wells. This, Victoria thought, was it for Cabot: no more drilling. At the very least, she believed, the discovery would prompt mandates for immediate and vast restrictions and improvements. She asked Michael O'Donnell about this one morning when he came to collect samples at her home. O'Donnell was a DEP water-quality specialist and, as a front-line agency person, in a good position to put things in perspective; and so he did. He told her "enforcement actions" had to be weighed against the industry's reaction to them and much of the time the enforcement was determined through negotiation. Gas companies had plenty of wherewithal to mount legal challenges against penalties, and industry officials might stop reporting spills altogether, he added, if they felt regulators were coming down too hard on them. O'Donnell was one of two inspectors overseeing the industry proliferating in five Northern Tier counties, and he counted on the industry's cooperation in reporting problems. The conversation, which Victoria recounted, squared with accounts from both regulators and activists that inspections rely heavily, sometimes exclusively, on industry reports. Many of these reports were filed away or incomplete; and in the case of Cabot, some reports were not filed until after problems had appeared.

In addressing the methane problem at the homes along Carter Road and Route 3023, the approach of the DEP was to work with the industry rather than against it. Cabot operators agreed to vent gas wells to relieve excessive pressure in well bores; and the DEP advised homeowners to vent their water wells to prevent explosions. At least four private water supplies were found to pose especially high combustion risks, and Cabot disconnected these wells. Cabot also hired a local contractor to deliver water to the households with decommissioned wells. This gesture, according to Ken Komoroski, wasn't an admission of guilt. Rather, it was an example of the company being "a good corporate neighbor." To address risks at twenty other homes connected to the aquifer, the DEP sent out fact sheets that explained that methane is not toxic to consume but that it does pose incendiary hazards. The bulletin recommended venting as "an inexpensive and effective way" to prevent explosions. At press conferences, Mark Carmon, DEP spokesman, stressed that tests found no gas in the basements. All in all, DEP officials and Cabot publically conveyed a sense that the problem was being taken care of.

The methane migration proved to be far more problematic for the residents than for Cabot. As the company progressed toward completing one hundred Marcellus wells in Susquehanna County by the end of 2009 (the schedule Dinges had outlined to investors), I learned in interviews at people's homes that some wells tested positive for methane only and some tested positive for methane, iron, pathogens, metals, and other elements that suggested a problem beyond methane migration. The filtering system that Cabot provided for the Sautners filled their basement with tanks and filters. It required routine maintenance and ended up being unreliable and ineffective. A similar system of tanks and filters that packed a closet in the Carter home (and which they purchased at their own expense) didn't work for methane. A system that was effective for methane would require more expense and space. No matter the type of system, all were susceptible to power outages and malfunction. Not everyone with a system felt confident that his or her water was safe. Other homes lacked any purification system because one typically cost $5,000 or more.

To restore water to the four homes disconnected from the aquifer, Cabot delivered large plastic tanks—water buffalos—on flatbed trucks and hooked them up outside the homes. These tanks tended to freeze in cold weather, and accordingly they required heaters that drew a lot of energy and added to utility bills. When it was hot, the water became warm and musty. The water could be used for some things, such as gardening or irrigating, depending on your confidence level in it, but, particularly in the summer, the system was prone to bacteria and not suitable for drinking. The inconvenience and expense of buying and storing bottled water for drinking was compounded ten- or a hundredfold when residents tried to arrange clean water for laundry, cooking, and bathing. Norma's plumbing was off-line for six weeks after the explosion destroyed her well and her confidence in the water supply. Cabot eventually repaired Norma's plumbing (which had been prone to freezing before the explosion), installed a vent to address the methane problem, and told her the water was good enough to drink, but she remained fearful of drawing water from the tap.

Switzer found Cabot's response to her water problems and those of her neighbors to be unfair and humiliating. Residents were now beholden to the company for their water, and the water service from the tanks was substandard for those lucky enough to receive it. "These are used for cattle," she said about the tanks. "We are not cattle." It was the water issue that triggered her role as what she later described as an "accidental activist." In Victoria's mind, water delivery was an unsatisfactory long-term solution to the problem; but it was Cabot's answer for now, and appealing for water seemed to be the first step in the grievance process.

Victoria, a methodical and disciplined learner, was disinclined to offer public opinions that could complicate and inflame an issue she had yet to master. So when Dimock became a popular spot for reporters immediately after the explosion of Norma's well, she politely declined interviews while holding out the possibility that she might change her mind at some future time. She had spent the months following her tour of Ken Ely's land organizing information via Internet searches. One item that piqued her interest was an open letter to DEP Secretary John Hanger posted by a group called Clean Water Action. It itemized concerns the group had with drilling: de-watering of streams, creeks, and rivers; handling of chemicals; treatment and discharge of flowback; habitat destruction; erosion and run off; pipeline construction; air emissions; and traffic. Victoria found the letter interesting because it presented such a broad set of environmental issues but somehow lacked mention of the waste pits that were so concerning for Ken. After alerting the group to this omission in an email, she received a prompt reply from Brady Russell, a community organizer who stated that he was interested in learning more about her story. Switzer invited him to come to meetings with local residents that she had begun organizing. That would give him a better account of what it's like living in a gas field.

Russell particularly remembered a February 25, 2009, meeting at the home of the Seymours, the organic garlic farmers who lived next to Switzers. About twenty people, including Ken Ely and Norma Fiorentino, sat in chairs in a semicircle in the large walk-in basement finished to serve as a kitchen for roasting garlic. The group also included state Representative Tina Pickett. Victoria had invited the politician to hear firsthand accounts of the impacts from drilling, despite Russell's advice not to. Pickett was among a majority of local and state elected officials in northern Pennsylvania who supported shale gas development for economic reasons. (At the time, she was preparing a plan that would be unveiled in a matter of weeks to expand natural gas drilling on 390,000 acres of state forest in the interest of fostering economic development.) The concerns of the group mostly involved problems with their water, and people there were at a loss regarding how to deal with the kind of day-to-day, let alone long-term, problems they were now facing. Debbie and Tim Maye, a couple on Carter Road who owned horses and cats, said their animals were vomiting after drinking the water. Getting an alternative water source for household consumption was one thing. Getting it to supply a farm was something altogether different.

At one point during the meeting, a few of the residents recounted the community effort a decade earlier to fight plans for a low-level community landfill in Susquehanna County. Organized through an organization called Return Susquehanna County Under Ecology (RESCUE), activists had orchestrated a political action campaign that defeated the proposal. At the mention of this, Ely stood up and offered his 2 cents: "We should do it like they did with the landfills," he exclaimed. "There should be marches, and rallies, and signs to really get this out in the public." This was just the development Russell was hoping for, but it didn't last. Pickett, who had been quiet for the first part of the meeting as people aired their problems, soon stepped forward and launched into a fifteen-minute elaboration on the economic gains that drilling would bring, with reassurances that the DEP was on top of the problem. For every person she heard with a complaint about the water, she reported, she heard from many more people asking how they could lease their land. The problems people were experiencing here weren't to be diminished, she added, but they were anomalies; and DEP officials, working with Cabot, were well equipped to handle them.

As the meeting broke up, Russell complained to Victoria about Pickett. "It's any elected official," he told her. "They have a way of sucking all the air out of a room full of people who want to get things done." Victoria, unperturbed by Russell's frustration, was willing to consider Pickett's reassurances. With a temperament leaning more toward facilitator than activist, Victoria trusted the system and adhered to the lesson she used to teach her social studies class: desirable change was driven by informed civic involvement, and that meant engaging elected representatives in grassroots and community affairs. She was willing to give Cabot and the DEP a chance to make things right and was reluctant to undermine these early efforts with actions that could polarize and alienate. After the meeting, she pressed ahead with writing letters and holding community forums. Her plea in a follow-up email to Komoroski was simple: "We want water—safe clean water—flowing into our homes as it did before your company came into our lives. . . . We await your response and your offer to improve the situation here in Dimock." She copied Pickett, O'Donnell and his superiors at the DEP, and state Senator Gene Yaw. In a "Reply All" response, Komoroski told her that the company was taking her concerns seriously and that he would be in touch.

In the meantime, Victoria and Jimmy continued to pour their energies and resources into their home. They vented their water well as DEP officials had advised, and cooked with bottled water. After a full day of work at school, Jimmy spent evenings building maple cabinets, laying tile, or framing, wiring, and plumbing. Victoria continued to help him, although she found herself spending more and more time writing letters or surfing the Internet for information about drilling, talking on the phone, and arranging and attending meetings with neighbors, DEP staff, and Cabot officials. As weeks turned into months with nothing but reassurances about the water, she and Jimmy sometimes talked about abandoning their dream and starting over someplace else. But the talk didn't last long or go far. There was little market for a home—even a magnificent one—in the middle of a working gas field. "I can see the ad now," Victoria told Jimmy. "Beautiful, contemporary resort home in the Endless Mountains. Bring your own water."

KEN ELY'S STAND

Ken Ely was making enough in royalties to move just about anywhere he liked, but he wasn't going to abandon his mountain. And even though he didn't subscribe to the politics of Nader-inspired reform groups or environmental organizations such as Greenpeace or Earth First!, in late winter 2009 he was stirred by the sort of passion that motivated the members of those groups.

The Pennsylvania DEP wouldn't shut down Cabot, and there were no environmentalists stepping in to rally or chain themselves to trees. So Ken decided to take matters into his own hands. By March, thoughts along these lines solidified into a plan. He chose the week that Scott took his family to Walt Disney World in Florida to save his son from being pulled into the conflict that he was sure would ensue. On the morning of March 24, a mild late winter day, Ken pulled a piece of plywood from a log bunker built into the hill that he used for storage. He painted a message to the company—officials, crew managers, and hired workers alike—in large red letters. He then loaded the sign into his Bronco and drove up the access road along the ridge of his mountain. At the midpoint of the ridge, halfway to the well pad on the northern boundary of his property, he pulled off into his quarry and parked next to a skid steer and several pallets of bluestone. He transferred the sign—along with some bottles, rags, and a gasoline tank he had with him—into the cockpit of the skid steer; and he then climbed on board and settled in behind the controls. He turned the key, waited a minute for the glow plug to heat, and then cranked over the engine. He maneuvered the machine's forks under a pallet of bluestone and, once loaded, he steered toward the road along the ridge, where tanker trucks were hauling fluid to the north end of his property. He waited for a break in traffic, then pulled the machine—loaded to the capacity and teetering—on to the road and squared it to oncoming traffic. He lowered the pallet and cut the engine. He climbed out with the sign and propped it on the pallet: Road Closed until Further Notice.

An approaching tanker downshifted and rolled to a stop in front of the skid steer, which looked toylike compared to the truck. The driver set the brake with a blast of compressed air and climbed out of the cab to talk with Ken. After a brief exchange, the driver looked at the gasoline tanks and rags in the cockpit of the skid steer, climbed back into the truck, and radioed the tool-pusher (the on-site crew boss). Another truck rolled to a stop behind him, and then another. A few minutes later, men arrived in pickups and talked to Ken. They told him he would have to move, one way or another; and it would have to be sooner rather than later.

Just above the quarry-stone blockade, a road from the main route along the north end of the ridge branched to the well site in the middle of Ely's property. This allowed an indirect detour for vehicles to exit the well pad on the north end of the ridge, but impractical access for incoming traffic. Ken had effectively blocked access to the wells on his property and shut down Cabot's operation. The next day, March 25, Cabot filed a complaint in the U.S. District Court seeking an injunction forcing Ken to move. The complaint claimed that Ken had threatened to firebomb the equipment if anybody tried to move him out of the way. Ken denied that intention but made no effort to dispel the belief. Let them think what they would, his inaction seemed to imply.

The complaint filed by the company cited Ely's lease with Cabot, which granted the company "the right of ingress and egress along with the right to conduct such operations on the leased premises as may be reasonably necessary for such purposes, including but not limited to geophysical operations, the drilling of wells, and the construction and use of roads, canals, pipelines, tanks, water wells, disposal wells, injections wells, pits, electric and telephone lines, power stations and other facilities to discover, produce, store, treat and/or transport product." In other words, the gas company had broad legal standing to do what it found necessary to find and remove gas or other product from his land. The court granted the injunction, which was promptly issued. Ely complied. The event, which lasted less than two days, cost the company about $3,000 per hour in downtime.

Ken didn't share this story, but word got around. Roughnecks talked about it with locals, and soon the tale was being passed through the community. I heard various versions from residents in the months that followed. The particulars of some of these stories differed from the account spelled out in Cabot's complaint filed in the court as well as from recollections conveyed by his son, Scott. In some versions, the gas cans and rags were dynamite, and Ely was armed with a shotgun. But the general image was the same: an armed Ken Ely standing up to a fracking convoy. To some who knew him as independent and contrary, the act of defiance was an example of Ken being Ken. In the minds of others, it brought him folk hero status.

A radical environmentalist may have applauded Ken's actions, but there was a difference between this standoff and those by activists putting themselves between a whaling boat and a humpback whale or chaining themselves to trees to impede a clear cut. Actions by Greenpeace and Earth First! are coordinated by a group bent on drawing attention to an issue. For Ken, the conflict was deeply personal. A sense of frustration and finally, desperation, had compelled him to invite a reporter and then Victoria to view operations on his land months before the standoff. Ken didn't share his intentions with the press, and it's impossible to know whether he ever intended to. On May 20, 2009, fifty-seven days after the incident, he died suddenly of a heart attack, alone in his two-room cabin.

THE WHISTLEBLOWER

Two years after I met Ken at his pond and recorded his positive expectations for gas development, including a promising career for his son, I drove out to meet Scott. It had been a year and a half since his father died. He lived with his wife, a dentist, in a small single-story house alive with the commotion from three young children. The place was set well back and out of sight from Carter road, with a sprawling well-tended front yard that encompassed a large pond, an above-ground pool, and outbuildings, including a coop for ducklings, a gazebo, and a big deck flanked by decorative stone work and several statuettes. At the time, his friends and numerous family members were helping him build a much larger house across the driveway.

Scott is a difficult guy to reach by phone, so I arrived at his house unannounced. He listened to my questions about his personal and working life, absently pulling a cigarette from a pack and holding it for a long time before lighting it. He fully considered each question before answering. He has a deep, slow voice and, like his father, he uses words economically.

When Scott was a boy, his father taught him chess, and they had many epic battles over the years. "We gave each other headaches," he said. "Chess is not a game of chance. You have to think through problems and anticipate what the other guy's going to do." Ken savored the tactical complexity of the game and offered it to his son as a metaphor for life. Scott nodded when I asked if the situation they found themselves in with Cabot applied.

Scott, who looked to be in his early thirties, was operating heavy equipment—something he knew well—while learning other aspects of the drilling industry, with plans for advancement. Ken had wanted to protect his son from fallout of the conflict that Ken was having with the company. Even with all the problems in Dimock, Ken had held to the belief that the industry was a good thing for the country and its citizens; he believed the difficulties that he and the community were having were the unnecessary consequences of shoddy work. "Dad knew there was no stopping this, as far as the industry itself, and he didn't want that," Scott said. "He told them, 'You can do this without destroying everything. You're a multibillion industry. There's got to be a better way.' . . . He didn't sign on to bury shit in the ground and contaminate land for years to come." I asked Scott to elaborate on this last point. There were things he would like to tell me but couldn't, he said, because they involved lawyers and his employer.

I found out more about the things that Scott had alluded to, at least some of them, in a review of DEP records in Williamsport. After reading through the paperwork, I also understood why Scott was "on leave" from his work with Cabot when I interviewed him in summer 2010.

On November 2009, six months after Ken died, Scott contacted Cabot corporate management through his attorney, Paul Schmidt, and told the company he had firsthand knowledge of spills and other problems unreported by contractors. He wanted to see those things addressed. Cabot officials in Houston relayed Ely's complaints to the DEP and arranged to investigate them with the agency and an independent contractor. That investigation began in December, with a field tour of the problem sites with Ely, Schmidt, Komoroski, Cabot consultants, and a team of DEP inspectors. They started at the Ely 2 site, above the spring that fed his father's cabin. It had been reclaimed, and little was visible other than valves sticking up from a field. Chemicals in the pit underneath were flowing from the site with groundwater, Scott said. Before the pit was filled in, a supervisor had ordered an employee to throw stones through the liner, according to Scott's account. If anybody asked, they would blame it on Ken, who was so eager to derail the drilling operation on his land that, this supervisor asserted, he could be easily framed as a saboteur. When the pit was eventually buried, Scott continued, it was not encapsulated, and chemicals were seeping through and threatening the spring water used for drinking and bathing. Scott then took the party of officials to the north end of the ridge where, he said, diesel fuel had been spilled and a leaky pit had been backfilled. Mike O'Donnell noted in a DEP file that he had received a copy of the analytical results of samples sent to the DEP by one of Ely's neighbors, and they showed elevated levels of barium (a common constituent of drilling mud and a toxicant that can cause arrhythmia, respiratory distress, and muscle weakness).

The allegations, itemized in correspondence between Cabot and the DEP in a folder labeled "Ely Investigation," outlined problems on the properties of Teel, Lewis, Gesford, Costello, Brooks, and Black. Many involved Scott's assertion that drilling pits had not been properly closed and that spills that had gone unreported or underreported were threatening aquifers and streams. A diesel spill at Teel 5 reported to the DEP as being 800 gallons in June 2008 was, in fact, closer to 3,000 gallons, Scott said. The spill, from a faulty tank, had happened in the middle of the night and wasn't reported until 6 a.m., and a supervisor "intentionally moved a reference point hay bale so the DEP would incorrectly obtain a clean post-remediation sample." Scott suspected that diesel 2 feet below the surface was still leaching into Meshoppen Creek.

After the tour, Phillip Stalnaker, a vice president and regional manager for Cabot, wrote a letter to Jennifer Means, environmental program manager at the DEP, assuring her that the company was taking the allegations seriously. It had hired URS, an independent contractor, to follow up with soil and water tests at the sites. Cabot was also following the DEP recommendation that the company "take appropriate measures related to Mr. Ely's allegations concerning general work practices, training and personnel matters; to make sure operations are being well managed." The letter from Stalnaker concluded that Cabot "has carefully reviewed all of Mr. Ely's allegations, has taken steps to investigate the allegations and, where appropriate, to address those allegations."

Publicly, Cabot denied claims that operations on Ely's land or elsewhere had caused environmental damage or that the methane migration in Dimock was related to drilling. "We don't have an answer," Komoroski said in January 2009, addressing a question about how methane had gotten into water wells. "We've checked our pipelines and equipment, and they are not leaking." As events unfolded, correspondence between the DEP and Cabot officials would point to the stuck drill bit in Gesford 3 as a likely cause. DEP inspections subsequently identified a dozen other defective wells contributing to the problem, and the agency ordered the company to fix them. As these facts became public, the company's explanation remained ambiguous, at first holding out the possibility that problems could be related to drilling and, later, denying it. In March 2010, Komoroski described the collapsing hole that had led to the drill bit being stuck in Gesford 3 as "at least a theoretical possibility that allowed the migration of gas . . . and allowed a pathway for it to get to the water supply." He continued, "There are measures in place to ensure that cannot ever happen again, whether or not that was the cause. It's possible that wasn't the cause. But if it was caused by Cabot, that is our most likely culprit. We're continuing to focus our efforts there. We're continuing to work with the Department and to do the right thing."

In addition to the complaints about pollution, there were allegations of other problems—reported by an unnamed whistleblower—plaguing operations that fell outside DEP's jurisdiction, and Cabot handled those on its own. After receiving a tip from "an employee who was concerned about the use of drugs and alcohol at well sites in Dimock Township," the company used a private security firm with drug-sniffing dogs to conduct surprise sweeps of workers on the rigs in Susquehanna County in early December. Following the raid, Cabot fired an undisclosed number of workers found to be using illegal drugs or alcohol.

PASS GAS

Each new Marcellus well in Pennsylvania increased production by several million cubic feet per day, encouraging further what Dinges had called "the greatest land grab in decades." That land grab extended into New York, but politics and policies on the northern side of the Twin Tier border were shaping a different story there. By summer 2009, Chesapeake, Fortuna, and Hess Corporation began negotiating for rights to several hundred thousand acres controlled by coalitions in the prime part of the northern fairway. The unleased area that was the focus of attention began with acreage just northwest of the Cabot gas fields in northern Susquehanna County and extended well into the heart of Broome County; and it included acreage controlled by coalitions lead by Dan Fitzsimmons and Jim Worden. The coalition's approach to dealing multiple tracts of unleased land through a single negotiation had proven effective in bringing some of the largest energy companies in the country to the table. Now drilling advocates in New York faced another problem—the release of the New York DEC's review of the impacts from high-volume hydrofracturing, on which Marcellus permitting hinged, was behind schedule. Until the review was complete, there would be no

Marcellus permits; and some coalition leaders felt that uncertainty about the time frame was eroding their bargaining power.

What was a problem for the coalitions was also a problem for Barbara Fiala, Broome County executive. Fiala was a drilling proponent determined to make the responsible development of the Marcellus Shale her legacy by establishing Broome County as "a model to the rest of the world for how this is done." Fiala had employed a keen talent as an administrator and a popularity borne of working-class appeal to rise through the regional political scene. She was the first woman in the history of Broome County to be elected to county-wide office when she became county clerk in 1997. In 2005, she ran for county executive and unseated the Republican incumbent, and in 2008 she was reelected on a platform of economic development. Her campaigns were endorsed by prominent party figures, including Senators Hillary Clinton and Chuck Schumer, Governor David Paterson, U.S. Representative Maurice Hinchey, and Attorney General Andrew Cuomo. Cuomo would appoint her to his transition team when he was elected governor and later as head of the New York Department of Motor Vehicles.

The Marcellus embodied paramount issues facing government: the need to balance environmental and economic interests amid complicating jurisdictional issues involving municipal, state, and federal governments and agencies. Fiala felt her administration had something special at stake among these issues. Marcellus development, she told me, represented economic opportunity that could "match or surpass" anything the area had seen. To her thinking, that included the impact of IBM Corporation during its heyday in the 1970 and 1980s, when it employed more than 11,000 workers at its Endicott headquarters and thousands more in its contracting facilities and satellite offices. All those jobs had been secure for a two-decade run and came along with top-notch benefits. Since IBM had departed, however, no business had come to take its place. Shale gas development promised to fill that void, according to Fiala.

With Marcellus development stalled at the doorstep of the Southern Tier in 2009, Fiala's administration commissioned two University of North Texas economics professors to quantify the worth of the Marcellus to Broome County public and private sectors. The resulting white paper, by Bernard Weinstein and Terry Clower, was issued the same month the analysis by Considine and Watson was published by Penn State. The findings of the Texas team were also premised on Engelder's predictions of the Marcellus potential, and their conclusions were even more spectacular. With Broome County "fortunately located at the epicenter" of the play, the economic impact of drilling alone could exceed $15 billion over ten years, supporting more than 16,000 jobs, generating $792 million in salaries and wages and $85 million in state and local taxes. Added to that was wealth that would be generated by "ongoing production" from completed wells. That, the authors of the report concluded, "would," "could," "should," and "will" provide something for everybody:

> Our model predicts as much as $4.1 billion in new economic activity per year over a 10-year period supporting over 4,000 jobs and $314 million in salaries and wages. State and local tax receipts could be boosted by $52 million per year, with slightly less than half accruing to Broome County taxing jurisdictions. Local revenues will also be enhanced by bonus payments and royalties from wells located on county-owned property as well as new ad valorem taxes on wells located on private property.

Finally, should Broome County and the City of Binghamton evolve into an administrative center for the natural gas industry in New York State, ancillary development could be anticipated, with attendant income, employment, and fiscal benefits to the region. The report concluded with this familiar warning:

> Of course, all of these economic and fiscal benefit calculations are predicated on the assumption that the State of New York creates and maintains a supportive regulatory and tax climate toward the natural gas drilling and production industry. . . . Excessive regulatory or fiscal burdens could significantly limit New York's prospects. In particular, the state should avoid the temptation to levy a severance tax on natural gas production as this would only serve to drive the industry to Pennsylvania or another state.

Like the Penn State Study, the Broome County study was immediately incorporated into the rhetoric of drilling advocates. "At the center of all this excitement stands a critical energy technology known as hydraulic fracturing," Lee Fuller, policy director of Energy in Depth, wrote of the multibillion dollar benefits to "a single rural county in a single U.S. state. . . . With it, opportunities for new jobs, new revenues, and greater security abound. Without it, those opportunities will be lost."

Fiala was so confident about the Marcellus-based opportunities that, prior to a single offer being made, her administration budgeted $5 million in revenue in its 2009 budget for lease payments on 2,100 acres owned by the county around the periphery of its airport and landfill. The study was reviewed and accepted by the county legislature, which included three members who, as owners of large tracts, belonged to coalitions.

The Marcellus permitting moratorium, and a drop in gas prices, had cooled the market for natural gas leases some by mid-summer 2009, but it was still hot in the center of the fairway. In planning lease acquisitions, oil and gas companies have to direct their vision beyond year-to-year economic gyrations, keeping in mind investments that take decades to develop. In this context, they embraced the conventional wisdom that, sooner or later, New York would be open for Marcellus development. Even amid regulatory uncertainty, landmen were still making the rounds with individuals and coalitions in the Southern Tier.

In July, several coalition leaders announced they were finalizing a seven-year deal with Hess that would pay landowners $3,500 per acre for leasing, with a potentially much larger payout tied to production. That included 20 percent royalties, $15,000 for each well pad, and a separate deal for pipelines and infrastructure. The deal was collectively worth more than $100 million in lease payments alone, and possibly more than double that from royalties on production as well as income from easements and well pads. Scott Kurkoski was finalizing the deal for 800 property owners who controlled 19,000 acres in the towns of Conklin and Binghamton. Chris Denton was also at the table representing 450 landowners with 11,400 acres in the adjoining town of Kirkwood, and the Windsor coalition, which controlled 60,000 acres owned by 1,400 people between Kirkwood and Deposit.

Dan Fitzsimmons, leader of the Conklin Coalition, saw a problem, however, and he shared his concerns with me on a regular basis. Over the course of the year, the attention of the local news media had shifted from the monumental deal in Deposit and the promise of jobs and money to things that had gone wrong in Dimock and elsewhere. Dan saw this as "bad press" fostered by environmental campaigns and people such as Walter Hang—and intended to smear the industry record. Dan was among drilling advocates who saw criticism as malice and opposition as a distorted liberal attack on an economy that could be saved with gas development. The opposition included just about anybody who highlighted problems that could be presented as part of an argument against drilling. Big environmental agencies egging on this opposition were, in fact, using the cause to line their own pockets, Dan believed. Getting information from these sources, he told me, was like "going to an informational meeting on air travel only to be shown pictures of every plane crash from the last decade."

Although the coalitions had not been intended as political organizations, under the leadership of Dan and others they developed into an extensive and cohesive grassroots infrastructure that could wield significant political force. The members of the coalitions now intended to take a more active role in controlling the message. They chose a Sunday in mid-August 2009 for their breakout rally on the sprawling grounds of Bainbridge Park on the Susquehanna River. In addition to landowners, the event included bankers, lawyers, and gas industry representatives who sponsored their cause. It was a scorching day, and thunderheads were mounting as Ellen Harrington and the Smoking Guns, a country band, jammed on a large stage that sponsors had set up in the center of the park. Rally-goers, perhaps 1,500–2,000 of them, set up folding chairs and umbrellas on the lawn and circulated among makeshift booths in shelters and pavilions, where lawyers and financial firms offered their services and advice to landowners, some on the verge of suddenly becoming wealthy. Vendors sold bright yellow t-shirts and bumper stickers stating, "Pass Gas. It's a movement." Aaron Price, a filmmaker from Broome County, worked his way around the park with a camera crew shooting footage for *A Gas Odyssey*, a documentary featuring the stories of Don Lockhart, Dan Fitzsimmons, Jim Worden, and others in the Twin Tiers looking to Marcellus development to save them from financial straits.

I found Dan in a bivouac set up with tables, chairs, and coolers on the park lawn. He was wearing one of the Pass Gas t-shirts, and he wondered why I wasn't, suggesting perhaps I was "on the other side." We discussed the events leading up to the rally. "We have to go political," he said. "It's wasn't our intention, but we have to show the support for this to happen." Dan was sitting with Tom Libous, known to most of his constituents as "Senator Tom." Libous is a man with an easy smile and a self-deprecating humor that might include a joke about his baldness or misguided boyhood exploits. As a ranking member of the state Republican Party, which held power in the New York Senate for more than twenty years, he was one of the most successful and influential politicians in the Southern Tier, and a prolific provider of member-item funding from Albany. Seldom was Libous seriously challenged in his hometown, and in 2008 he ran unopposed. His platform invariably was based on upstate economic development interests as they competed with New York City and the counties of the lower Hudson River and Long Island in matters ranging from funding for roads and bridges to fights over tax money and public policy. It was an interminable fight and one that Libous (and his supporters) relished. Now this upstate-downstate divide was easily recast in matters concerning Marcellus development. The New York City Department of Environmental Protection (NYC-DEP) was building a case against development in its watershed while environmental groups were organizing to protect the Catskills. Owners of rural property upstate, who for decades had suffered a population decline accompanied by economic shifts away from farming, suddenly held the key to the production of one of the richest portions of the Marcellus; and they felt it was being taken away by inflated liberal concerns held by an urban and suburban population. Libous was scheduled as a featured speaker at the rally, and he was ready to capitalize on the opportunity presented by the Marcellus. That afternoon he would do some old-fashion stumping against a favorite target—the administration of Governor David Paterson.

Libous had spent much of the morning mingling with his constituents. A few minutes prior to the beginning of the program, he sat with Dan, resting in the heat and sipping chilled bottle water, discussing the rapid developments over the last year. In summer 2008, when Paterson signed his spacing bill, Libous had publicly accepted the accompanying environmental review by New York DEC as a necessary step to updating regulations. But now he feared it had become a tool to stall, perhaps indefinitely, the drilling on which his constituents were counting. "Our big push here is to make sure the DEC brings this to the top of the page," Libous told me. "I'm not sure it is."

The band concluded its set and speeches began, marked by patriotic themes centered on landowners' rights to do what they wanted with their property. State government in general and environmental regulators in particular were rhetorical targets. Carol Robinson, an unemployed pipeline worker from Steuben County, drew enthusiastic applause when she said, "We need to put the money in the hands of the people who own the land and the mineral rights. Not the politicians. It's our land, our royalties, and our lives. I am an American. We are the people. If we the people don't protect our rights, we have no one to blame but ourselves." After several such speeches from citizen stakeholders, Libous took the stage, joined by Assemblyman Clifford Crouch, a Republican ally from Guilford. "We are not on the Paterson team," Libous began. "We are on your team. That's why we are here. . ."

The negotiations with Hess that the New York coalitions found so promising in summer 2009 ended up stalling. The cause was not political issues or the amount of money offered. It was differences over how much control property owners would relinquish to the energy companies developing their land that brought negotiations to a halt. In addition to a minerals lease, the coalitions, following the counsel of Denton and Kurkoski, insisted on a detailed contract called a Surface Use Agreement, which limited the lessee's control of the land only to operations spelled out in the contract. "The issue becomes, what is left on the land for us when the oil and gas company starts to assert its right of priority over every inch of the surface?" Denton explained. "If you want to continue to control the destiny of your own land you need to control the nature, magnitude, and scope of the uses of your surface by the oil and gas company. Everything in the Surface Use Agreement is designed to protect the rights of the landowner to continue to have a house, a factory, farm, forest, quarry, etc."

As the New York coalitions waited for a more cooperative partner, operators continued to lease large sections of the fairway in northern Pennsylvania. In July, Hess reached a deal with several coalitions for the rights to 87,000 acres in Wayne and Susquehanna counties for $3,000 per acre and 20 percent royalties. Then, a few weeks after the rally, a coalition of 600 landowners based in the hamlet of Friendsville, Pennsylvania, just across state border from Broome County, announced a deal with Fortuna for rights to 35,000 acres for $5,500 per acre and 20 percent royalties. The deal, with upfront payments to the landowners totaling $165 million, would also be extended to members of the coalition who resided in Vestal, a town on the New York side of the border, when the moratorium was lifted.

With Fortuna paying $5,500 per acre, landowners in Broome County who were on the verge of accepting $3,500 per acre with Hess were suddenly looking for an upgrade. Fitzsimmons saw the Fortuna deal as a catalyst for competition and "exactly what we needed at just the right time." Even though New York's regulatory uncertainty remained a problem, coalitions in the towns of Kirkwood, Conklin, Binghamton, and Windsor were ideally positioned. They controlled land not only situated along the Millennium Pipeline corridor but between tracts now leased by XTO in Deposit to the east and Fortuna in Vestal to the southwest, Chesapeake and Cabot to the south, and Norse to the north. Other multinationals had also begun casting their lots in the play. StatoilHydro (now Statoil ASA), the largest oil and gas company in Norway, paid $3.37 billion for a 32.5 percent interest in Chesapeake's Marcellus leasehold of 1.8 million acres in late 2008 (including 12,000 acres in Broome County), with plans to develop 17,000 wells throughout Appalachia over twenty years.

Chris Denton had a message for any company that might be eyeing the unleased land in the center of the fairway that was controlled by coalitions, and he conveyed it to me during an interview: "You are not dealing with desperate widows or dumb farmers who need to sell the farm because the price of milk is down. You are dealing with an educated, business-savvy group of landowners." Negotiators for Hess, he added, had blown their chance to acquire the rights to this prime spot by failing to accept landowners' conditions to protect their land. He compared leasing with Southern Tier coalitions to dating a supermodel. "They could have had Christie Brinkley, but they didn't get her because she was too smart and educated," Chris told me, guessing correctly that his words would find their way into the next day's coverage. "They wanted a dumb broad."

POLITICS AND POLICY

The more money the deals commanded, the more news coverage they drew. With coverage came awareness, and with this awareness came varying degrees of opposition. Most of the population didn't own significant tracts of land or have large sums of money on the line. There were those who remained open to the idea of gas development but who were not eager to allow industry to move in without stronger assurances and safeguards. Others believed the cumulative economic returns would be less than the cumulative disruption, regardless of how the industry was regulated. Some people just did not care as long as there was not a gas well in their backyards or a pipeline bisecting their property. The most ardent opposition, however, came from those who saw shale gas—touted as a clean fuel alternative that would secure energy independence for the United States—as a dangerous lie encouraging an unsustainable future. "The biggest question of all is not: Can the industry do this right? Or would the industry do it right? But should the industry do this at all?" said Tony Ingraffea, the Cornell University engineering professor. "We don't need more energy, we need different energy."

Ingraffea's opinions, like those of his former research colleague at Penn State, Terry Engelder, were backed by strong academic credentials and a career of practical experience working with industry. In late 2009, Ingraffea stepped forward to challenge the claim that shale gas development is a bridge from dirty coal to a cleaner, better energy future. He decided to get publicly involved, he told me, when the industry responded to criticism with public relations campaigns—letters to the editor, press releases, and advertisements—that struck Ingraffea as "inaccurate, disingenuous, and in some cases outright lies." He said that he "kept hearing that fracking had been done since 1947 over a million times without problems. The type of fracking we are talking about is an unconventional application on an entirely different scale."

As events unfolded throughout 2009, politicians also began taking sides. Local and state representatives of upstate counties bordering Pennsylvania tended to support speedy revisions to state policies so permits for high-volume horizontal wells could be issued. Opposition flourished in the Finger Lakes region and central New York, and in urban areas in both Pennsylvania and New York. Although lacking jurisdiction to oversee drilling, some city governments brought to the debate what administrative or legislative leverage they could. Buffalo and Pittsburgh passed legislation to ban fracking; Philadelphia's city council hosted hearings sponsored by opposition groups; New York City issued a study critical of fracking upstate; and the mayors of Ithaca, Syracuse, Binghamton, and Elmira, among other city leaders, were outspoken in their stand against fracking. The federal government also lacked jurisdiction over shale gas development, and U.S. Representative Maurice Hinchey was intent on changing that.

Hinchey, a Democrat with a history of supporting progressive causes and challenging powerful institutional interests, represents the 22nd District, extending squarely through the New York drilling fairway, from Ithaca in the Finger Lakes region to Newburgh in the Catskills. Before he was elected to Congress in 1993, he chaired the Environmental Conservation Committee of the state legislature, and he took the lead on a report detailing the infiltration by organized crime into the waste disposal industry in New York. As a congressman, he was part of an effort to impeach George W. Bush in 2008; as a member of the House Appropriations Committee, he helped secure $3.2 million in funding for the U.S. Centers for Disease Control and Prevention to retrospectively study the cancer rates of workers exposed to chemicals at IBM's circuit-board manufacturing plant in Endicott. So it was not a surprise that in August 2009 Hinchey teamed with Congresswoman Dianna DeGette of Colorado to introduce the Fracturing Responsibility and Awareness of Chemicals Act (better known as the FRAC Act).

The legislation would eliminate the industry exemption—the Halliburton Loophole—provided in the 2005 Energy Policy Act, thereby applying the rules of the federal Safe Drinking Water Act to the drilling industry. Companies would be required to disclose what they were injecting into the ground, and this disclosure would open the door to controls on hydraulic fracturing formulas. The FRAC Act was a natural for Hinchey, who saw it as an opportunity to undo the environmental misdeeds of the Bush era, to serve green interests for the future, and to take on Big Energy interests. "If they're so sure that their process is safe, what do they have to fear from being held accountable to the same rules that every other industry has to play by?" Hinchey asked me. "We need safeguards against reckless and careless people who are looking out for their own interests and don't give a damn about anybody else."

The FRAC Act served as a rallying point for environmental groups across New York State that were organizing their own campaigns against unchecked shale gas development: Shaleshock, Damascus Citizens for Sustainability, Catskill Mountain Keeper, Gas Drilling Awareness Coalition, Frack Action Buffalo, Riverkeeper, and regional manifestations of Earth First! and the Sierra Club. The Pine Creek Headwaters Protection Group organized clinics to train people to become citizen watchdogs—or "waterdogs," as activists called them—to report drilling problems in waterways and well sites. National groups with full-time staff and significant budgets also took up the fight: Natural Resources Defense Council (NRDC), Earthjustice, Clean Water Action, and Earthworks. Like Walter Hang, these groups were eager to disprove the industry claims that hydraulic fracturing was well regulated and, when done right, problem-free.

With this goal, the Oil and Gas Accountability Project, organized through Earthworks, analyzed data kept by regulatory agencies and found three hundred records indicating water contamination related to drilling industry operations in Colorado and more than seven hundred instances in New Mexico between 2003 and 2008. Independent actors such as Josh Fox, a filmmaker and antidrilling advocate from Wayne County, Pennsylvania, also began documenting stories of water pollution in these and other places. *Gasland,* a film he began shooting in 2009, asserted that unchecked corporate exploitation and deceit had left residents with bad water and an industrialized landscape. The film, which premiered at the Sundance Film Festival in January 2010, provided a visual statement for those who saw shale gas development as destructive and reckless.

CAUSE CELEBRE

Stories of sudden wealth and bitter disillusionment began appearing more routinely in local and regional news outlets. As news of the Marcellus boom in Appalachia spread, Victoria, the "accidental activist," found herself not just engaged in a battle on behalf of herself and some of her neighbors but thrust onto the front line of a war she had never intended to wage. She was joined by a battery of career activists from near and far. Norma's well had become a cause celebre for activists. It provided a simple and compelling symbol of something much larger and more complicated, both socially and technically, underlying events related to a debate growing nationally over onshore drilling.

"It was a strange time looking back on it," Brady Russell told me later, reflecting on his trips to Dimock from the Philadelphia office of Clean Water Action in early 2009 within the larger context of the shale gas conflict. "We were all the proverbial blind men feeling the elephant with our hands, and each of us only getting a bit of it. I got in a car and went, but of course even then there were mixed messages because Dimock hadn't decided what it thought yet. Most people wanted to believe everything was okay." Michael Lebron, a New York City resident and supporter of several Catskill conservation groups, also received mixed messages when he attempted to facilitate a lawsuit against Cabot on behalf of residents. The Carters, Norma, Pat Farnelli, the Sautners, the Elys, and others listened to him and to the lawyers he put them in touch with, but not all were immediately sold on the idea. Like Victoria, they were drawn to country life in part because of its seeming simplicity and privacy. Some were unenthusiastic about inviting more complications into their lives. They questioned the chances of legally pressing a corporation the size of Cabot, which was well equipped to handle just these types of challenges. Victoria, especially, remained guarded about outside interests, especially lawyers. "That seems to be everybody's answer," she told me later. "Settlements are made behind closed doors and everybody walks away with money, but there is no shakedown. Nothing changes, and the industry does not get the spanking it deserves."

In its initial stages, at least, the fight against Cabot in Dimock remained a fundamentally grassroots endeavor undertaken by a loose network of locals with broadly divergent backgrounds. People like Ken, Norma, Victoria, and the Carters didn't seek out the fight, but with their homes and health on the line, they weren't going to back down, either. There were other residents of

Susquehanna County, including Vera Otasevic Scroggins, who felt compelled by a sense of righteousness to back them up. Vera—in her early sixties—is a grandmother, amateur videographer, and advocate for many causes, including home births, home schooling, and no mandated childhood vaccinations. She has voluminous gray hair that she wears down and retains traces of a Long Island accent, a place where she once lived and from which she "escaped" some time ago to pursue a new life in the Endless Mountains. In her adopted home in Brooklyn Township—not far from where Norma grew up—she could keep roosters and chickens without the restrictions of the zoning ordinances that had inhibited her farming ambitions in the suburbs of Nassau County.

Vera had not leased her mineral rights, and in early 2009, drilling operations were yet to significantly alter the landscape around her home. She recognized the conflict developing in the town to the west as one that would soon be playing out in her backyard and in the backyards of residents throughout the northeastern United States. She had come of age during the civil rights movement and had few inhibitions about committing acts of civil disobedience or trespass in the name of justice. Venturing onto private land to videotape drilling-related spills and accidents that might otherwise go unnoticed fell well within her ethical comfort zone. Vera explained this approach to Russell, who was becoming familiar with the Dimock landscape in early 2009. "We're extra eyes and ears for the DEP," she told him. "They don't have enough workers and we have to pick up the slack." Footage from some of Vera's vigilante patrols in 2009 shows encounters with roughnecks and pipeline workers, some reacting with amusement or annoyance to the woman with a home video camera showing up at these remote and often inaccessible work sites and peppering them with questions. Some called her "ma'am" and briefly addressed her questions; some directed her to the foreman, who almost always asked her to leave; and some simply ignored her or walked away. These brief encounters typically punctuate long unedited footage of vacuum trucks, excavation equipment, and hay bales. Vera also taped public forums and interviews with residents, such as Norma, recounting their experiences with gas development. These videos she posted online, where they joined a broad and growing collection of depictions of Susquehanna County gas development by other independent media, advocates (including Russell), and professional news outlets. They generally included interviews with residents and presented aspects of drilling that lent themselves to visuals: truck traffic, derricks, flaring, fracking, and heavy machinery cutting swaths through the countryside.

As Victoria Switzer allied herself with the Carter Road residents, she also formed new bonds with people who lived further away. Vera and Victoria became a frequent presence on an Internet forum dedicated to gas development issues in northeast Pennsylvania, called the Susquehanna County Gas Forum and organized by Lynn Senick, a Dimock resident and food bank coordinator. Although the forum had not been set up to be exclusively for women, it became the primary vehicle for communication by many women with dissimilar personalities, backgrounds, and political leanings, all of whom bonded through their convictions about how the drilling was progressing in their communities. They tended to be mothers and grandmothers in their fifties and sixties, with demographic similarities ending there.

There was Joyce Stone, a naturalist and wetlands specialist at the Woodbourne Preserve, a 648-acre parcel of old-growth forest and wetlands adjacent to Ken's mountain. Joyce and her husband had been instrumental in steering grassroots environmental causes since the 1970s, including opposition to the landfill and nuclear waste projects proposed for Susquehanna County. Joyce continued the work on her own after her husband died in the late 1990s. She found that organizing resistance against gas development, unlike taking on a proposal for a nuclear waste dump, was complicated by the fact that landowners in Dimock often had a personal financial stake in gas drilling and many were determined to see it through. This conflict was creating divisiveness within the community, which she found to be one of the most disturbing aspects of the gas rush. "The people who've known me and my children growing up and who loved my husband are just treating me like I'm the enemy or something," she lamented. Virginia Cody, a retired Air Force officer and political conservative who lives in Wyoming County, was also active in the forum, and so was Julie Sautner. When Victoria was feeling overwhelmed and hopeless, she found the esprit de corps among the women an energizing force. They were fighters, and they were fighting together. Relishing a comparison with Neytiri, the animated heroine of the movie *Avatar,* Victoria referred to the women bonded in their fight against drilling as "warrior women."

Through posts on the Susquehanna Gas Forum and personal email exchanges and phone calls, these women planned activities such as distributing leaflets on car windshields at community events, letter-writing campaigns, and community meetings. As events unfolded in the Twin Tiers, they kept each other apprised of news and unreported developments. They shared personal experiences, which sometimes Victoria posted on the forum in the form of poetry, vignettes, and short stories. They would not find out until events unfolded in 2010 that their activities and correspondence were being monitored by federal agents and gas company executives.

HEITSMAN SPILL

Victoria's efforts eventually produced the long-awaited meeting with Cabot. On September 16, 2009, Ken Komoroski and several other Cabot representatives sat on metal chairs at a folding cafeteria table in the main hall of Veterans of Foreign Wars (VFW) Post 5642 in Montrose. They faced three other tables, arranged on the tiled linoleum floor in a U-shape and occupied by a dozen or so Dimock residents. The topic was water. Victoria politely thanked Ken and his colleagues from Cabot for attending and then quickly got down to business. She asked the residents to stand up and introduce themselves one at a time. These people were all victims of drilling gone wrong, and they needed water, she said. "It's not the moon and stars we're asking for here," Victoria said. "It's clean water." Methane was not the only concern. Some wells showed spikes in concentrations of metals and bacteria that sometimes exceeded safety guidelines. They tended to vary in location, intensity, and time.

Ken told the assembly that he was sorry people were having water problems but the company had nothing to do with it. Cabot was not in the water business.

Norma Fiorentino, with the help of neighbors and her family, had been hauling jugs of water from other homes and the store on a daily basis for weeks after the explosion. After Cabot fixed her plumbing, she was afraid to use it. Through Lebron's efforts, she, the Sautners, the Carters, and some of the others had been talking with lawyers, who were assessing their case. She had been patient, but now she found Komoroski's indifference maddening.

"What do we have to do, sue you?" she asked.

"You can, but you'll lose," Ken replied.

Sitting near Norma was a man in jeans, a long-sleeved t-shirt, and sneakers. In keeping with the decorum that Victoria had established, he had introduced himself at the beginning, even though people knew him. His name was Doug Heitsman. He was in his mid-forties, but he had the appearance of being younger, with wide shoulders, a soft roundish face, and straight brown hair that swept across the top of his forehead in boyish bangs. He was a third-generation farmer who owned 210 acres on Route 29 in Springville, just south of the Dimock town line. He had taken over the farm as a teenager, shortly after his father died of cancer. He still limped from injuries received when he had been pulled into a haying tedder operated by his grandfather when he was eighteen. Farming was in his blood, he later told me, and he wouldn't give it up for anything. He had three gas wells on his property, divided into units directly next to the Carter Road gas wells. (Like Ken Ely, he had unknowingly signed away his rights for the minimum, but he was reaping significant royalty payments owing to the sheer volume of gas being pumped from his land.) Heitsman did not have a water problem, as far as he knew, but he had accepted Victoria's invitation to the meeting to voice another complaint—Cabot didn't keep him informed. He found out about accidents, including a drilling mud spill on his land that spring, too long after they had happened—sometimes while chatting with DEP officials who showed up to take samples or with crew members he found on parts of his property where he didn't expect them. He was unhappy about trash and litter accumulating along access roads and other spots; and he was getting fed up with what he saw as carelessness and indifference to his land.

"The landowners aren't told about what's going on," he said. "They [Cabot] keep us in the dark." Ken apologized for the information gap. The company would improve communication with landowners about all activities on site, he said.

Then a woman, who had introduced herself as Jennifer Carney, a Dimock resident, spoke up. Doug didn't immediately make the connection that she was the wife of U.S. Representative Christopher Carney and that she was there at Victoria's request. Her line of questioning focused on who knew what about spills and whether Komoroski himself was kept apprised of everything that went on at a well site. Ken said that he was.

Jennifer then asked him directly: "Was there a spill today?" Ken and his colleagues exchanged glances. There was a spill, Ken said, and it was being taken care of.

"Heitsman 4H?" Doug asked.

"That's correct," Ken said.

Doug was not a man given to raise his voice, but he made little effort to hide his exasperation: "That's just what I'm talking about."

The Heitsman spill was just one of many mishaps in Dimock, but it became a landmark in the fight over Marcellus development because it coincided with efforts of environmental groups to disprove the drilling industry's claim that hydraulic fracturing posed no threat to water. I first learned the fundamentals about the Heitsman spill through a press release issued by the DEP the next day, followed by conversations with representatives in the agency's press office and with Komoroski, who stressed that the spilled substance was innocuous. Industry representatives in New York, sensitive to criticism mounting from advocacy groups during the state's environmental review of shale gas development, also downplayed the accident. Brad Gill, executive director of IOGA New York, responded with a statement that explained the "incident" in Dimock involved a decoupling of a water line. The line contained "slick water," which is "99.5 percent water and sand and is not considered a toxic substance." He added, "The Pennsylvania DEP has been on hand and is satisfied with the remediation being completed."

In reconstructing the spill and its impact, I later reviewed internal reports and emails in Pennsylvania DEP files and drove to the site, where Doug gave me a tour. Although well pads are not legally allowed to be developed on wetlands, Heitsman 4H is on a pad built over a marsh that drains into the headwaters of Steven's Creek. Technically, an official explained to me, "it is surrounded by wetlands, but not in them." Extending from under the pad is a swale that collects water from springs running from the hill above to the creek below. Steven's Creek empties into Meshoppen Creek, just downstream from the mouth of Burdick Creek. About a half mile north of Heitsman 4H, the Heitsman homestead sits on a rise. It includes two small farmhouses—occupied by Doug, his mother, his wife, and his young son—two barns, a large pond, and a well-cared-for rose garden.

Doug generally kept off the drilling pad—his wife didn't like him down in that vicinity for fear that he would track home stuff that would be unhealthy for their preschool-age son. After he heard at the VFW post about the spill, he drove his four-wheeler to an open hillside above it. He saw no emergency vehicles and no flashing lights or barriers—just the work trailer, trucks, tanks, generators, hoses, and dozens of men working on the fracking job. The air was filled with the hum of electric motors, and occasionally Doug got a whiff of a smell like turpentine. A sheen covered wet spots and fluid-filled depressions. Steven's Creek was hidden in what was left of a stand of hemlocks that hadn't been cleared for the well pad.

According to information I pieced together from interviews and DEP files in the Williamsport office, an agency inspector by the name of Steve Watson had come and gone by the time Doug got to the hill. Watson was called to the site shortly after 2 p.m. He was met by Steven Jones of Consulting Solutions

Services, a spills contractor hired by Cabot. Jones provided these details: Halliburton, also under contract for Cabot, was fracking Heitsman 4H with a product called LGC-35 CBM, a lubricant that reduces resistance in the line. The substance was mixed at a location a mile or so up the hill—in the direction of the Heitsmans' house—and fed into a pipe that runs down the slope to the well pad. (Doug told me he had seen the mixing tank there and later pointed out a leaking coupling to an inspector.) The elevation of the fluid in the mixing station and powerful compressors raised pressure in the system necessary to push the fluids to the end of the lateral bore hole deep in the ground. In this case, it was too much for the equipment. A coupling at the base of the operation blew out, and 5,000 gallons of LGC-35 CBM solution gushed over the pad and into the wetland.

Jones told Watson the gel was water-soluble and that it posed "minimal safety issues." Workers would use hay bales and dirt to build a berm downstream, he proposed. From there, the run-off could be pumped back to tanks on the pad. Watson agreed with the plan and told the Cabot contractors they could continue the frack. He left the site that afternoon.

At 10 p.m., the DEP received a second emergency call. There was another spill at Heitsman 4H. This time a DEP inspector, who signed the emergency response report as C. Rogers, arrived at the site thirty minutes later and listened as Jones told him the same thing he had told Watson. The coupling had failed again, spilling several thousand more gallons of fluid into the wetland. Rogers noted in his report that the clean-up plan Jones had proposed before "appeared to be feasible." The agency oil and gas field staff would return to make further decisions and approvals, he added. He left around midnight.

Two days later, on Friday, Eric Rooney, Jonathon Ulanoski, and Michael O'Donnell, all of the DEP, inspected the site with Bryan Bendock, a conservation officer with the Pennsylvania Fish and Boat Commission. Dead tadpoles and minnows floated in the wetland, Rooney noted, and the water was covered with a sheen that produced "a strong solvent-type odor." The polluted water was backing up from a dam Cabot had built on Steven's Creek, where workers were trying to pump it back into tanks on the pad. Rooney told Kevin Bertrend of Cabot that they would have to collect samples to determine how far and how deep the pollution went. The dam violated environmental laws and also increased the footprint of the spill on the Heitsman property, so it would have to go. Rooney left the site after a few hours.

Nobody told Doug Heitsman any of this. What he learned (after hearing of the spill at the VFW post) was from news accounts, which reflected only information from the press offices of the DEP and Cabot. There had been two spills totaling between 7,000 and 8,000 gallons of fracking fluid, each caused by the failure of a pipe coupling. Komoroski was quoted as saying he didn't know specifically what was in the product but he did know that "this material is not hazardous or dangerous nor does it present any environmental risk."

The following Monday, five days after the spill, Rooney returned and spoke with Doug Heitsman, who complained that nobody was telling him anything. Another day passed. On Tuesday, Rooney asked Phillip Hill, of Cabot, "to meet Mr. Heitsman and update him on the release and the cleanup progress." By then, Hill had his hands full with other problems. The coupling feeding the fracking solution into the hole had blown apart a third time, flooding the pad with several hundred more gallons of LGC-35 CBM.

SECRET INGREDIENT

The MSDSs at the Heitsman site, following industry practice, were rolled up and stored in a plastic weatherproof container attached to the warning signs at the well gate. The white canisters, along with the crisp black and red lettering spelling out notices of warning, restrictions, and authorization at each gate, lent an air of authority and precision to each site. When DEP officials retrieved Halliburton's sheet for LGC-35 CBM at the Heitsman 4H, however, they found its catalogue too vague to be meaningful. The document listed the chemical family simply as "a blend." Its toxicity; its effects on reproduction and development; its mobility in water, soil, and air; the extent to which it decomposed or accumulated in the biomass; and its effects on plants and wildlife were answered with these words: "Not Determined." In this regard, the MSDSs at the Heitsman site was like many others, left incomplete or intentionally ambiguous. Lacking specifics, investigators would have to collect and analyze samples not only from the wetland and soil but of the fracking gel itself before they could begin documenting the impact of the spill on the ecology and public health.

The first lab tests came back a week after the spill. DEP officials learned that LGC-35 CBM contained, among other things, a collection of volatile organic compounds, including *n*-propylbenzene, *n*-butylbenzene, 4-isopropyltoluene, *sec*-butylbenzene, and isopropylbenzene. Exposure to these compounds and their byproducts—toluene, ethylbenzene, and xylene—can cause health problems ranging from brain damage to cancer.

Now came the next set of questions: How much of these chemicals remained in the stream and the ground, and how far had they spread? DEP officials expected clues in the analysis of the samples Cabot had been ordered to take from the wetland. When the results arrived from the lab a week after the spill, Rooney looked them over and found that Cabot's contractors had not collected samples from the soil in the main body of the wetland or at the site of the first spill. Nor had the water and soil samples they had collected in other areas been analyzed for the chemicals the DEP had found in the fracking gel. These omissions made correlations and comparisons impossible. Rooney called these problems to the attention of Kevin Bertrand, Cabot's manager of environmental and regulatory affairs, in an October 28 letter: "The laboratory analysis for soil samples collected from the area of concern were (*sic*) not representative of the constituents contained in the frac gel. Additional soil sampling is needed in the areas previously sampled." Three weeks later, Bertrand discussed the matter with Rooney in a conference call. After some negotiation, Cabot agreed to collect more samples and analyze them for the compounds found in the fracking gel. The conversation then turned from determining the public impact of the spill to a provision under Pennsylvania law that would protect Cabot from liability that could arise related to cleaning up the pollution.

I couldn't find the results of the follow-up tests in the stacks of files laid out on a table for my review on December 9, 2010, in the DEP's newly opened Northcentral Regional Office in Williamsport. Finding particular files was proving difficult for other reporters, as well. An investigation by the *Times Tribune* found the files detailing spills and enforcement for Marcellus development to be incomplete, disorganized, and sometimes randomly filed in the four different regional offices of the DEP. The paper submitted a Right-to-Know Request for the agency's well inspection reports and violation notices that detail spills, leaks, and seeps. But "inconsistent responses and record keeping from the four regional offices that regulate drilling made finding an exact count of spills impossible." Some offices gathered only "industrial waste" violations; some included erosion and runoff violations. Few of the offices included waste pit violations, even when plastic liners meant to protect the soil fell in or were breached. At the Southwest Regional Office, the *Times Tribune* investigation found more than a dozen files were missing for spills of diesel, wastewater, and other fluids at well sites. Sometimes the paper trail would begin in the middle of an investigation, with no record of the original spill, and sometimes it would include a record of a spill with incomplete follow-up. The newspaper quoted Ed Stokan, an assistant counsel at the Southwest Regional Office, who blamed problems partly on budget cuts, "The department staff has been so stripped that we don't have the personnel to go through the files."

Internal DEP emails sent days after the Heitsman spill, uncovered in the *Times Tribune* investigation, suggested a regulatory staff that was overwhelmed and mismatched. In an email to her colleagues, Jennifer Means, the oil and gas program manager in the agency's Williamsport office, wrote that she "wholeheartedly endorsed" revoking the drilling permits already issued to Cabot or halting pending permits "to slow down their future activity." That action "would go a long way with the public" whose "biggest frustration . . . is the rate at which they are allowed to continue given all these incidents. . . . Also—we could certainly stand to catch our breath." Means's idea was abandoned, however, after the top lawyer for the agency warned of formidable legal hurdles to suspending permits under state law. Instead, the agency ended up ordering Cabot to suspend hydrofracturing until the company evaluated its equipment, operations, and protocols for handling an emergency and made the necessary adjustments in a written plan to be approved by the agency—a task that took three weeks. The suspension was unnecessary, Komoroski said, "because we felt we were responding appropriately." He added, "The DEP is doing what it thinks needs to be done."

RIGS AND REGS

Although DEP officials publicly characterized the problems in Dimock, both the spills on the surface and disturbances under the surface, as anomalies, privately they began taking stock of the broader issue of methane migration related to drilling operations statewide. In September 2009, the DEP presented its compilation of known cases related to gas wells to the Pennsylvania Oil and Gas Technical Advisory Board, a group of state government–appointed industry representatives—and chaired by Bob Watson, author of the Penn State report—that advises the agency on technical and policy matters. Methane migration from gas drilling, according to the briefing, had "caused or contributed to" at least six explosions that killed four people and injured three others over the course of the decade preceding full-scale Marcellus development. The threat of explosions had forced twenty families from their homes, sometimes for months. At least twenty-five other families had to deal with the shut-off of utility service or the installation of venting systems in their homes. At least sixty water wells (including three municipal supplies) had been contaminated. This summary of damages was accompanied by proposed regulatory updates intended to address the problem. The suggestions included issuing mandates to test the integrity of wells (including casing seals), incorporating higher standards for cementing well bores, and "establishing a protocol for the regulated community if faced with alleged gas migrations which focuses first on stabilization of the situation to protect people and property." This last suggestion, in other words, was for companies to accept responsibility and fix the water supplies ruined by methane migration.

In the report, the DEP sought to engage the industry by focusing on the economic aspects of the loss:

> Natural gas is . . . valuable. Gas wells are essentially a facility designed to extract that value in a controlled manner. This concept of "controlled" is at the heart of the commonwealth's Stray Gas Migration concerns. Control of natural gas released from its geologic reservoir is required for the well operator to maximize the economic benefits of their investment. If the release is uncontrolled, the gas is lost. Not only is its value lost but uncontrolled gas can create extreme hazards to people and property. Once control of the gas is lost, the operator not only faces costs to regain control but also potentially expensive and litigious remediation costs.

And then the DEP delicately sought advice:

> The commonwealth recognizes and appreciates that we need to deal with the legacy wells and that the ability to plug those wells comes from fees paid by the modern industry. But the natural gas industry must also recognize and appreciate that modern day wells need to be designed, constructed and maintained to prevent the uncontrolled release of gas from the geologic reservoir into the surrounding community.
>
> [W]e have come to the Technical Advisory Board to ask for its technical advice. We know that controversy may ensue, but we are confident that the idea of prevention of health, safety, and property impacts is a fundamental requirement of the production of natural gas. We are equally confident that once Pennsylvania's Natural Gas Industry gives thoughtful consideration to the issue, it will arrive at the same conclusion.

The industry reaction to the proposal was unenthusiastic. "We all understand the goal of course," said Stephen Rhoads, the president of the Pennsylvania Oil and Gas Association. He questioned the technical soundness of the proposed casing and cementing requirements and of the well monitoring plan. "We need to make sure they accomplish what they need to without causing harm, financial harm, to the industry." Komoroski had a similar reaction. The draft, he said, "wasn't particularly well developed by those with sufficient knowledge of such materials and techniques." He called for "further evaluation of whether additional regulatory changes are needed or will be helpful." Board members also raised various concerns, including the impact the casing regulations would have on the shallow gas wells in the western part of the state, some of which produced gas from wells that didn't use any casing at all. This technique, "open hole completion," would be banned under the new rules. While taking stock of the industry feedback, Watson asked DEP Secretary Hanger for time to evaluate the proposed rules.

In summer 2009, as it became clear that Marcellus problems were becoming the single biggest environmental issue facing the Rendell administration, John Hanger, who reported directly to the governor, made the first of several six-hour round-trip drives with his aides from his office on the top floor of the Rachel Carson State Office Building in the state capital to Dimock to meet with Victoria and the other residents for a firsthand accounting of the problems. Victoria coordinated these early visits, and she did not invite the media or Brady Russell because she wanted to avoid the partisan voices of advocacy and the self-consciousness and pandering sometimes fostered by media coverage. Hangar met with the Carters, Norma, and the Sautners and heard their stories and frustrations with Cabot. He observed the vent protruding from Norma's well, and the ineffective filters and tanks in the Sautner's basement, and Victoria and Jimmy's unfinished house, where they kept crates of bottled water in what was supposed to be the wine closet.

Hanger assured Victoria and the Carter Road residents that he had lawyers, engineers, and administrators dedicated to resolving the problem. In November 2009, the agency issued a consent order negotiated with Cabot documenting methane contamination in the aquifer and related problems stemming from three defective gas wells, as well as multiple spills, illegal discharges, and records violations. The agreement called for a $120,000 fine and gave Cabot five months to provide solutions, including "whole house potable water and/or mitigation devices" for thirteen homes with thermogenic methane in their wells, along with plans to "permanently restore or replace" affected water supplies. The contract also called for the company to provide a list of complaints from other residents that the company had addressed independently in the 9-square-mile area around Carter Road. The agreement was held up as an example of regulatory toughness by the DEP Press Office and as a corporate achievement by Cabot. Company officials felt "very good" about the agreement, Komoroski said, which ensured the company is "conducting its drilling and well completion activities in complete accord with the department's interests."

By the end of 2009, Governor Rendell had begun his push (against formidable pro-drilling resistance) for a severance tax to pay for more enforcement efforts, the DEP opened an office in Williamsport with additional staff dedicated to Marcellus oversight, and the industry and various governmental review boards were considering the DEP's proposal to upgrade casing regulations to prevent methane migration. As these measures were reported in the news, Victoria could see they looked like progress to the rest of the world; but she found them to be superficial. She was sure the mounting problems in Dimock were an unheeded warning, as were signs of similar problems in Bradford County and other places with intensifying shale gas operations. Despite these signs, the state was tentatively poking along with a trial-and-error approach to shale gas regulation that Victoria referred to as "Rigs before Regs," and she was mystified that more elected officials and residents of Pennsylvania could not see this folly. "This plan is being written while the industry continues to drill," she wrote to her state senator, Gene Yaw, a Republican and drilling advocate who stridently opposed Rendell's calls for taxes and regulations. "What other industry gets to write the rules as it goes along at this pace, and then figure out what to do about the negative consequences?" She dropped the letter off at the post office, but she was done waiting for a response, from Yaw or anybody else.

6. THE DIVISION

Vera Scroggins, the independent-media videographer and advocate, joined a small assemblage of reporters and photographers on the Carters' front lawn and unpacked her video equipment. News was about to happen. It was a mild day in late November 2009, a year after Ken Ely first gave Victoria a tour of Cabot operations on his property. Since then, in addition to the six gas wells on Ken's land, Cabot had drilled fifty-seven others within 9 square miles and had obtained permits for sixty more. The Pennsylvania DEP had logged more than twenty spills and accidents related to these operations, issued seven citations, and documented three faulty wells blamed for polluting the aquifer. Now, eight months after Ken Ely resorted to a quarry-stone blockade to slow Cabot down, his surviving family members and neighbors were about to make their own stand against the company. This one would be in front of the media.

Victoria, Norma, Ron and Jeannie Carter, the Sautners, and Pat Farnelli sat in folding chairs in front of the Gesford 7 well pad next to Ron and Jeannie's porch. They looked like they had been gathered for a snapshot at a family reunion, except nobody was smiling. Behind them stood Scott and Bill Ely and members of a total of fifteen families. Several grim-faced lawyers and clerks, dressed in wool suits, scarves and dress coats, stood conspicuously amid the casually dressed group.

The cameras rolled. Leslie Lewis, an attorney with Jacob Fuchsberg law firm from New York City, began summarizing the lawsuit against Cabot, which her firm had filed in federal court the previous day on behalf of fifteen households with forty-six family members residing on and around Carter Road. She referred to the group of plaintiffs as "The Carter Road 15." As a result of Cabot's negligence, Lewis continued, placing her hands on the back of Norma's chair, these residents "have suffered contamination of their water, loss of property value, permanent damage to their property, loss of use and enjoyment of their lives, loss of the only thing these people had for the next generation in this community."

The press conference was just the type of public relations contrivance that Victoria had once wished to avoid. But by now she had come to accept such events as necessary tools for the kind of fight she was engaged in. Reform was driven by political pressure, and political pressure was leveraged through media exposure. That lesson had been reinforced a few weeks earlier, after she finally agreed to be interviewed by Laura Legere of the *Scranton Times Tribune*. The resulting article ran on the front page of the October 26 edition, distributed throughout northeastern Pennsylvania. The headline was "Dimock Residents Thirst for Gas Well Fix," and it ran with a picture of Victoria standing in her unfinished great room and another of Norma in her trailer embracing a great-grandchild. The article recounted the explosion of Norma's well, the spills and other violations, and the unresolved water problems that Victoria and other residents had faced for ten months. It summarized Cabot's response, delivered through Ken Komoroski: an effort was "underway" to determine "whether or not it would be appropriate to either provide them water or a water treatment system." Komoroski was quoted as saying that a determination had been delayed by the response of the company to a series of chemical spills at one of its well sites and by a subsequent stop-work order from the DEP in September. Legere's article also quoted DEP Secretary John Hanger, who cited new rules under consideration to make companies more accountable for methane migration. Hanger added, in the context of Cabot's operations in Dimock, that he was "not fully satisfied" with the response to complaints under the current system and suggested more regulation was needed. Several hours after the Scranton daily paper was delivered to doorsteps and posted online, Victoria, the Carters, and Norma each received a call from a Cabot official. Water deliveries were to be arranged shortly.

Some of the plaintiffs were worse off than others financially. The gas wells from which Norma and Pat were allocated royalties were both defective and nonfunctional. Their families were among the most needy, and the gas that was supposed to provide royalty income for them was, according to the DEP analysis of the problem, instead going into their water. In response to the DEP consent order to fix the methane problems, Cabot plugged the Gesford 3, Gesford 9, and Baker 1 wells. Norma Fiorentino and the Farnellis had suffered the ill effects of drilling with none of the expected returns; and the lawsuit was their chance for compensation.

Victoria and Jimmy were receiving royalty payments, but would have gladly returned every penny, Victoria told me, to have their confidence in the water quality restored and to see a regulatory presence established that would control the drilling "free for all" erupting all around them. Absent that, a trial was a chance to bring public attention to the problems on and around Carter Road, and Victoria was ready to wait for however long it might take for that to happen. She was also prepared to endure whatever pressure the company might apply in the meantime; at least, that is how she felt on that particular November morning. "I've gone to every congressman, representative . . . DEP, Cabot, anyone I could think of to bring this issue to the forefront," she explained to Legere at the press conference. "We couldn't get help anywhere. . . . We're not greedy people. We just want some justice for something terribly wrong that happened here."

Still, Victoria knew the lawsuit would escalate the conflict. There would be even more money on the line; moreover, there were egos and reputations at stake. Cabot was resolved to discredit the claims brought against it. Until now, Cabot executives had sent their responses through Ken Komoroski, and he had presented the company position with unwavering politeness and strategic vagueness. Engaged now in a lawsuit, no one could count on Cabot to continue to be polite.

Parties on either side of the lawsuit began collecting evidence from the other to support their cases. In this process, called discovery, lawyers from Cabot came to the Switzers' new house, escorting heavy machinery into her yard. Based on the rules of this legal contest, Victoria and Jimmy were unable to intervene when Cabot contractors brought in drilling pipes, swinging them from a crane in front of their plate glass windows. From dawn to dusk, the workers used this equipment to collect geological samples that would support their employer's interpretation of what did or did not happen to the Switzers' water well. Victoria was not surprised to see this kind of effort on the part of Cabot; the company was seemingly able to summon unlimited legal and political resources to support its case. She was unprepared, however, for the reaction from her own neighbors to her complaints against Cabot.

The lawsuit was one of the first related to Marcellus Shale production to be filed against drilling companies in northeastern Pennsylvania, and it triggered a new wave of media interest. A few weeks after the press conference, Victoria was featured as the lead to a story in the *New York Times* titled "Dark Side of a Natural Gas Boom." In the meantime, members of the Carter Road 15, whose names had never appeared in the newspaper outside of a birth or marriage notice, were quickly becoming featured sources in national stories about the policy debate over on-shore drilling. Reporters from *Vanity Fair*, *Discover Magazine, Forbes*, CBS's *60 Minutes*, CNN, *NBC Nightly News,* and others were visiting the Carter Road homes to interview, film, and photograph residents and learn about their experiences with shale gas drilling. Networks in Quebec—where exploration was beginning in the Utica Shale—also recorded stories of early expectations and eventual disillusionment in Dimock.

Each story added momentum to the debate. Web sites, blogs, and Listservs conveyed facts, argument, propaganda, and rumor. There were venues, both traditional and electronic, for every agenda and every group imaginable. The Central Broome Landowners Association, one of the coalitions represented by Peter Hosey, the Texas lease lawyer, posted a page on its website urging members to counter the "negative publicity" coming from the media accounts of events in Dimock and other places by sending elected officials letters with this language:

> Unfortunately, the anti-drilling crowd seeks to prevent us landowners from recovering the wealth beneath our lands in the guise of sustaining the environment. Through misinformation and fear mongering they are giving the impression gas drilling will create all kinds of horrific events in an effort to turn public officials against developing this valuable resource. They use fabricated "facts" rather than science to instill fear in the uninformed population.
>
> They suggest they are the only protectors of the environment and the "greedy landowners" are only interested in the money, with total disregard for the environment. Nothing could be further from the truth. We harvest

fruits and vegetables from our orchards and gardens, and our farming neighbors raise crops and produce milk from our lands.

A reader of one of Legere's stories in the *Scranton Times Tribune* chronicling the hardships and frustration of the Carter Road residents responded with this post on the newspaper's forum:

> There are two things that have been documented about the owners of these gas drilling companies. First, most of them have 3 eyes, dark green in color, with a built in laser beam gun in one. They wink at you with one, rather than tell you the truth; they stare you down with two when you ask why your well just blew up; and they will zap you with the laser if you get in the way of their quest for obscene profits at the expense of your well being. Second, they don't live in the areas where they drill for gas. Three green eyes aside, they are not stupid. But don't worry. When the gas travels up your sewer line and blows you, while still sitting on the toilet, out the window, they will buy you a new toilet. That's the least they can *do*.

SENATORS REACT

The Marcellus debate was tied to a set of broader policy considerations on which the energy future of the nation depended, and those were woven into storylines of global warming, the wars in Iraq and Afghanistan, and the failing economy. In March 2010, the fledgling Obama administration opened large portions of the Arctic Ocean, the Gulf of Mexico, and the Atlantic Ocean to offshore drilling just as the 111th Congress honed a bipartisan climate bill that would cap carbon emissions while encouraging the nuclear power and natural gas industries. Whereas the political underpinnings of the cap-and-trade proposal was an abstraction to many Americans, the shale gas industry suddenly cropping up in their backyards was not. Gas exploration and development in New York and Pennsylvania, once the niche for small and mid-size independent drilling operators, suddenly factored into the long-term strategic plans of the world's largest companies. In December 2009, Exxon Mobil entered the shale gas business with a $31 billion purchase of XTO energy (which controlled rights to the land of Dewey Decker's coalition in New York). Five months later, Royal Dutch Shell—the largest European oil company—announced it was buying East Resources for $4.7 billion to "grow and upgrade" the company holdings of shale gas in North America, according to a statement from Peter Voser, the CEO. (The purchase included rights to Jackie Root's land in Pennsylvania.) In addition to the Marcellus and Barnett in Texas, new shale gas reserves being proven in Louisiana, Arkansas, Texas, and Oklahoma resulted in a 50 percent increase in the number of rigs operating domestically in 2010 over the previous year, leading a rate of domestic gas exploration unseen in more than forty years.

Proponents portrayed the gas boom as an evolution away from coal to cleaner energy sources—a "bridge fuel." The Exxon and Shell purchases were interpreted as signals to investors that the most influential industry decision makers supported this view. "There is broad recognition that we will be living in a carbon-constrained world," Chris Tucker, a spokesman for Energy in Depth, told me after the Exxon deal. "This is the direction that everybody is moving in. It's completely changing the entire dynamic. The reports written before the shale boom can be thrown out."

For industry supporters who touted domestic gas production as the road to energy independence and economic revival, "drill-baby-drill" was the rallying cry. For opponents, Norma's well and the images of flammable water became icons of corporate exploitation and greed and were brought into the mainstream with the premier of Josh Fox's *Gasland* on HBO in 2010. Victoria had mixed feelings about the attention focused on her community. It was stressful and polarizing, and the complete opposite of the idyllic visions she once had of her retirement to the country. And while she was glad the Carter Road case study brought weight to the discussion of policy reform in the context of Marcellus development, she remained vexed about the apparent indifference to her plight and that of her neighbors expressed by her own elected officials. The public support of Marcellus development by Pennsylvania State Senator Gene Yaw and his resistance to any attempts to tax and regulate the industry came without so much as an acknowledgment of the problems in Dimock. Victoria had written formal letters, sent emails, and made phone calls, all to urge the senator to come and tour Dimock and speak to the affected residents.

"I've read the newspapers and I know from them about the lawsuit," Yaw told a newspaper reporter who raised the issue. "I have no interest in their trying to get me to take sides." Victoria responded, to the same inquiring newspaper reporter, that the senator had ignored her pleas to visit her community before the lawsuit was filed. The senator might be interested, she added, in the status of the consent order the DEP had filed, on behalf of the state, to address the problems in Dimock. She sensed from her meetings with DEP officials, including Hanger, that there was a growing rift between the agency and the company over the interpretation of language in the consent order to "permanently restore or replace . . . affected water supplies." DEP officials said the case of damage to the aquifer caused by Cabot's operations was clear under Pennsylvania law; but as events unfolded, the company was forming its own interpretation of what was an acceptable response to the problem, as well as what the Commonwealth of Pennsylvania could hold Cabot accountable for. These differences could become a matter of legal precedent transcending the situation in Dimock and affect the ability of the state to control the industry operating anywhere in Pennsylvania. She wondered why this wasn't worthy of the senator's interest. "At $10 million a square mile, you would think we would be living the high life here," Victoria told me. "We can't take a shower, we can't drink the water and our senators won't talk to us. What's wrong with this picture?"

Although Victoria's efforts were aimed at elected officials in Pennsylvania, they found their mark with lawmakers in New York. A week after Victoria was featured in the *New York Times*, a dark-colored sedan arrived in front of the Carters' trailer. Antoine M. Thompson, chair of the New York State Senate Environmental Conservation Committee, stepped out along with four aides. They were greeted by Ron Carter and Victoria, while Jeannie contained her dog Brandee, who was wild with excitement. Ron ushered the guests into the family's small living room, where the Sautners, Pat Farnelli, and Doug Heitsman were gathered at the table under a picture window looking out over the Gesford 7 pad. Introductions were made, and Jeannie passed a plate of pastries and poured coffee as the senator explained the interest of New York in the events related to gas exploration in Susquehanna County.

Thompson was something of a political prodigy. He had risen from the rank of a legislative staffer for the Buffalo Common Council, where he had helped craft policy directives for affirmative action and minority business development before serving several terms as a councilman. At the time of his visit to Dimock, he was a second-term state senator and not yet forty years old. His trip to Dimock followed his participation in a guided tour, earlier that day, with Chesapeake officials of that company's operations in Towanda. Chesapeake had begun offering these tours for the media and elected officials to boost its image in an area where it had critical long-term interests in acquiring and maintaining rights to the land and to convince policymakers that gas companies were in fact good corporate neighbors. At the Towanda site, escorted by Chesapeake representatives, Thompson had seen a shipshape operation, a pristine workplace packed with high-tech equipment. Walking among industrious and good-natured roughnecks, Thompson had viewed tanks containing fluids and waste that were arranged systematically around the 5-acre site. These were the preferred alternative to open pits, he was told. The drill pad was covered by an impermeable barrier—a $750,000 investment—and surrounded by a berm to contain spills if they should happen. These measures were embraced by the industry as good business, he was told. Further, this good business would reduce U.S. dependence on foreign oil and do so in a way that was "both environmentally sensitive and economically rewarding."

The Chesapeake tour was impressive, and it was easy to see how prospects of this type of economic development generated excitement. Thompson learned that the Carter Road residents he was sitting with had once felt that way, too. He heard each of their stories. The Sautners displayed a jug of polluted water. Pat related how Cabot had decided water from her well was safe, as long as it was vented, even though wells around her were polluted. The family had initially resumed using the well water, but found that it was making them sick, she said, and now she had to buy bottled water for the needs of seven people. Thompson also heard how operators routinely spread flowback on the road in the name of dust control, even during wet spells. The residents told him about Norma's well exploding and about vents periodically blowing off others. Doug Heistman related details about the spills on his property, which Cabot hadn't told him about until after he heard about them from other sources and news accounts.

Victoria then recounted how it had all started, with a visit from "a kindly elderly gentleman—a landman." She described how, after being told there might be "a well or two out there," she and Jimmy had agreed to a lease for $25 per acre. "We were misinformed, uninformed and naive," she continued. The transaction "translated into $90 million dollars from a 9-square-mile corporate gas field that is our home." Cabot had drilled sixty-three wells and, "despite what has gone terribly wrong here," the company was planning eighty-three more. "And they will keep going after that. The gas company is not going to leave until it has exhausted the land and extracted the last isotope of gas. Not only do they control the gas, they control the water. And if we are bad, and speak out against them, we do not get our water. This is not free enterprise at work under a democratic system. This is an occupation. . . . We're all tired of being told that 'we are sorry for your inconvenience.' Not having change for parking is an inconvenience. Losing your water is not an inconvenience."

Victoria very much had the young senator's attention. When she was done, Thompson recounted how the modern-day environmental movement had been spawned by events at Love Canal, in his own district. A group of ordinary citizens in the La Salle neighborhood of Niagara Falls forced Occidental Petroleum, the parent company of Hooker Chemical, to clean up 21,000 tons of toxic waste buried under residential properties. Through perseverance and a belief in the system, the residents of Love Canal eventually had brought about sweeping federal action. That change came in the form of the Superfund law of 1980, which marked the beginning of the modern regulatory era. He added that Victoria should consider running for office.

Two hours later, Victoria left the meeting glad that their efforts were helping shape policy in New York but discouraged that there was, in her view, little concern in Pennsylvania. "These energy companies are not used to playing by the rules," Victoria warned Thompson as they headed to their cars. "They have convinced our senators here that they can't do this with rules. If there are too many rules, they will take their rigs back to Texas and Oklahoma, and we will miss our golden opportunity."

ROUGH SAILING

The rules under which the industry would operate in New York were still being determined in fall 2009. Administrators at the New York DEC released a draft of the SGEIS on September 30 and, in response to requests by advocates to facilitate public participation in the policy review, scheduled a series of hearings throughout the state, including one in Broome County on November 13.

Lines began forming outside the Chenango Valley High School an hour before the doors opened at 5:30 p.m. By that time, about one hundred people waited to fill out a card that would allow each to speak in the order in which he or she arrived. There were people picketing by the door with signs that read "Don't Frack on Me" and "You Can't Drink Gas or Money." Some wore costumes—one was a barrel of toxic waste; another, a gas company executive billionaire with money coming out of his hat. The pro-drilling contingent was also represented, most visibly by people who wore t-shirts that read "Pass Gas, It's a Movement." Environmental conservation officers wearing ranger's hats and bearing sidearms stood attentively at various entrances and milled about the lobby as hundreds of people began filling the auditorium to capacity.

Before the meeting, the deputy commissioner in charge of the SGEIS review, Stuart Gruskin, was busily engaged with clusters of agency people, politicians, and advocates. All of them were looking to offer their views or ask him questions. Walter Hang worked his way within Gruskin's field of vision and waited for him to break away. It was a week after the article featuring Hang's assessment of the problems related to natural gas production had run in upstate New York newspapers. The article, which had emphasized that problems in New York State went unreported or underreported, had struck a nerve with staffers at the Mineral Resources Division and the repercussions resonated up the agency hierarchy, eventually prompting a letter from Commissioner Peter Grannis. The letter, directed to legislators in Elmira and Binghamton, defended the agency's record and dismissed Hang's work as skewed.

Stu (as Gruskin was known to Walter Hang) found a graceful exit from his conversation with one group and turned to shake the hand of his former NYPIRG mentor. The two exchanged pleasantries before the conversation turned to Hang's public criticisms of the agency. "You made it look like we were hiding things," Stu complained. "We have nothing to hide."

"Your agency keeps telling everybody that there's never been any problems," Walter replied. "There have been all kinds of problems, and you don't even know of them yourself, because the system allows the industry to cover its tracks. There are no disclosure requirements, and when problems crop up, they are dismissed or never made public."

"There have been a lot of wells drilled, and the reality is there have not been a lot of problems," Stu said. He stopped there. He didn't want to begin a debate that he knew would go nowhere, especially with Walter; and, moreover, his role in this forum was to listen. "That's why we are here," he continued. "To vet all this—it's not always going to be smooth sailing. That's the nature of policymaking. You know it as well as I do."

"It's going to be very rough sailing," Hang promised.

Stu reflected on this conversation months later, and on his long relationship with Walter. The issues surrounding how the development of the Marcellus was unfolding charted a division between him and Walter and also a conflict within himself. On the one hand, Stu believed strongly in the ability of the public to sort out and resolve contentious issues given a sufficiently open public process; and he did not shy away from the sometimes bruising debates from which he knew sound policy was derived. He had been called to this policy process as an undergraduate, when he had served on the NYPIRG student board (when Walter was a staffer) and when he had studied rhetoric before going to law school. "Walter's voice was very important to the public debate," he told me, "and people like Walter need to be heard." On the other hand, he feared the Marcellus debate was being "spun out of control" by both industry opponents and critics. "The industry had done a miserable job of presenting itself and the issues from the start," he said. "It had been minimizing the environmental consequences and over-generalizing the economic gain." The environmental lobby, on the other hand, "was guilty of hysteria that was vague and misleading. Nobody called them on it, and I think they were at the risk of marginalizing themselves."

After his brief conversation with Walter, Stuart headed to the microphone in front of the auditorium, which was now filled with more than 1,000 people, most seated and many standing. Gruskin thanked the audience for attending, noted the importance of the event, and advised speakers that their comments would have the most influence if they related specifically to aspects of the proposal rather than their general feelings about drilling. He wasn't surprised when this advice was ignored. In what sounded and looked like a pep rally, speakers took their turns, offered impassioned praise or criticism of the drilling industry and its plans to set up shop in Broome County, and drew hearty applause and hoots of derision for their digressions from Gruskin's agenda. More than 170 people signed up to speak during the allotted 5-minute periods. Three hours later, fewer than forty of them had gotten a chance to present their views.

The first draft of the SGEIS had been released on September 30, 2009, six weeks prior to the hearing at the school, and fourteen months after the New York DEC began the review process. The five-hundred-page document reviewed the history of gas exploration in New York and the regulations and permitting process that had been developed to date, including the 1992 GEIS (to which the SGEIS was an amendment). The draft of the supplemental document explained the differences between traditional and unconventional drilling, as well as the risks of methane migration and water pollution associated with each. It recommended a series of steps related to permitting shale gas development near principal aquifers, including upgrading the casing requirements for well bores and mandating on-site inspections during cementing. It also proposed the following specific regulations: requiring the monitoring of private water wells within 1,000–2,000 feet of drill rigs; requiring companies to disclose fracking compounds and concentrations to the DEC; and requiring a series of more rigorous checklists and paperwork to track water withdrawals, waste disposal, and technical compliance for well construction. In detail and scope, there was a lot to grapple with in the draft SGEIS. Measured by the breadth of public interest that it generated, the draft was among the biggest environmental policy reviews in state history. Amid a flood of written responses to the plan, the thirty-day comment period was extended to ninety days. In that three-month period, the agency received more than 14,000 formal comments, and it had to address them all before the plan could become final.

The first draft was generally embraced by drilling proponents, who urged the state to finalize it with relatively minimal suggestions for changes. Summing up the position of advocates for drilling, Brad Gill concluded that the Mineral Resources Division had struck "a balance that continues to protect New York's environment and allows responsible exploration for natural gas" while ensuring the industry's "outstanding record of environmental and operational safety in New York." He added in his statement for the press, "We as an industry welcome a high environmental bar, but time is now of an essence." Barb Fiala, representing the interests of all the landowner coalitions in Broome County, also called for a speedy wrapping up of the process. Without referring to them by name, Fiala called out members of her own party, including Binghamton Mayor Matt Ryan and state Assemblywoman Donna Lupardo, both of whom urged a slow and cautious approach, "I have to say I am dismayed and frankly, a little embarrassed, for those who have made the argument that starting over on the draft Supplemental Generic Environmental Impact Statement (SGEIS) or holding off the permitting process indefinitely is the prudent path." Chris Tucker, the spokesman for Energy in Depth, praised the initial draft of the SGEIS for addressing the very concerns that had prevented the industry from moving forward. "It seems that the state set the conditions so that actual production could proceed, even under tightly defined parameters," he said. "That's why we support this and they [opponents] do not. There is no way to split the baby on this."

The range of criticism directed at the draft SGEIS was broad, coming from both environmental groups and stakeholders representing municipal and federal government agencies. These groups were in no hurry to see the plan put into effect, and they detailed the shortcomings of what, they noted, was a first draft. An overriding concern was how the proposed additional measures for overseeing drilling operations and enforcing the rules could be funded as the state was forced to cut back on staffing in the deepening recession. After the release of the report, local government officials throughout the Southern Tier—including some in Fiala's administration—told me they feared that enforcement measures, such as water testing and erosion control, would be left to local governments, which had no say in permitting the wells. The Tompkins County Council of Governments characterized the review as incomplete and insufficient on at least a dozen points. In response, the council called for a ban on any chemicals in fracking suspected to be carcinogenic, mutagenic, or endocrine disrupters; more comprehensive testing to protect watersheds, as well as private wells; measures to ensure that companies, rather than governments, pay for mitigation and protective measures; and consideration of the impact on sectors of the economy that depend on "our existing environment, clean water, and viewsheds . . . higher education, high-tech spin-off industry, grape growing and wine production, agriculture, and tourism." Moreover, the council noted that the "failure [of the SGEIS] to address cumulative impacts on the landscape and on communities is a fatal flaw that undermines everything else in the Document."

Union leaders, representing some workers who would be responsible for overseeing shale gas development, issued a similarly harsh response. New York State Public Employees Federation Division 169, the union of environmental conservation employees, criticized the SGEIS for failing to identify how enforcement efforts would be handled by state agencies that were already understaffed and overextended. The employees' union also claimed that the document lacked a fair assessment of the cumulative impacts of drilling, including "what to do with the recapture, treatment and disposal of flowback—sewage treatment plants must be identified and modernized to handle radiation, dissolved solids and chemicals." For its part, the NYC-DEP itemized several dozen failures of the document, including its failure "to examine impacts of activities that are integral to natural gas drilling, such as wastewater treatment and disposal, among other things." The NYC-DEP also issued its own technical analysis of the potential impacts from a projected 3,000–6,000 Marcellus wells, which would "industrialize" the 1,900-square-mile area of the Croton and Catskill/Delaware watersheds. These impacts were summed up by Acting NYC-DEP Commissioner Steven W. Lawitts: "Seven million truck trips, one million tons of concentrated chemicals and millions of gallons of waste water that are necessities and byproducts of current extraction methods pose a substantial threat to the water supply. . . . The known and unknown impacts associated with drilling simply cannot be justified." City of Syracuse officials issued a similar assessment of dangers that fracking posed to its water supply, which is primarily drawn unfiltered from Skaneateles Lake, and concluded, "no amount of regulations can safely protect this unique water supply from the risks of hydrofracturing."

Officials from the New York State Department of Health noted, for the record, that an analysis of wastewater from the Marcellus drilling operations had found levels of radium-226 and related alpha and beta radiation up to 10,000 times higher than drinking water standards. Concluding that this "could be a health concern," Department of Health officials urged the DEC to require more testing to identify the drilling sites that pose radiation risks and ensure "hot" drilling waste is handled and disposed of properly. "The issues raised are not trivial but are also not insurmountable," the memo concluded. "Many can be addressed using common engineering controls and industry best practices."

The New York Farm Bureau responded that the SGEIS "is not protective enough of New York State's agriculture, environment and people." The agency recommended the SGEIS should include "a comprehensive assessment of the impacts on the environment and human health by numerous gas wells, develop a practical plan for the disposal of all waste water, include a comprehensive list of safer alternatives to currently used fracking chemicals, and prohibit gas or oil companies from using water from aquifers." Having traveled the state to talk with grassroots stakeholders, Lindsay Wickham, the farm bureau field advisor, wasn't at all surprised by the critical reaction to the state's first attempt to overhaul policy. "People were lining up to paddle this baby before it was born," he told me. "But that's part of the process. When this finally gets done, it will be worth the wait. It will happen on terms established by residents and landowners, and not the gas companies."

Long after the SGEIS had been publicly critiqued for the first time, I learned how the skeptical crowd packing the school auditorium at the Farm Bureau meeting in Greene in the summer of 2008—attended by Gruskin and Judith Enck—had affected the course of shale gas development in New York. Before that meeting, the changes to adapt the state's policy to accommodate shale gas development "had been discussed in an academic way," Gruskin told me. "Once you get into a room full of 600 or 700 people, it takes it out of the academic setting into the real." Judith Enck had been a fundamental figure in recharting the DEC's course from the outlook depicted in presentations by the Mineral Resources Division in summer 2008—"We have been doing fine so far. . . . No problems"—to embracing a sweeping policy reform that gave

New York the distinction of being the only state in the country to pause the rush toward shale gas exploration. At the time, Enck's career was still taking off, and she wasn't inclined to jeopardize it by discussing with me her role in politically sensitive administrative policy; she accordingly declined to answer my questions about exactly how the de facto moratorium behind the governor's order had come about. I found the answer to be implicit in the public record, and my conclusions were supported by my later conversations with people who worked inside Governor Paterson's inner circle.

The public record shows that it was Enck who stepped in to clarify the message by the Mineral Resources Division to the community in the summer of 2008. Up to that time, skepticism and doubt had been standard responses to the agency's presentations. This was evident when Enck drove from the executive office in Albany to the small high school in Greene that summer and assured the angry crowd that "The DEC is going back and doing its homework," and then prophetically added, "I'm sure you will hold our feet to the fire and make sure it gets done." What insiders told me was really not a surprise—Enck, along with Gruskin, and DEC Commissioner Peter Grannis, charted the plans for the governor's executive order to suspend the permitting for high-volume fracturing until the state policy on drilling permits could be revisited. In so doing, Enck had effectively hit the pause button on the process.

In November 2009, it was announced that Enck had been appointed by President Obama to head the EPA's Region II office, which included New York and New Jersey. A month after she began the job, on the next-to-last day before the public comment period on the SGEIS expired, Enck's Region II office issued its own comments on the SGEIS, signed by John Filippelli, a senior EPA planner. The letter, released to the media, pointed out that, although the federal EPA lacked direct jurisdiction over hydraulic fracturing, it did oversee other water protection measures that pertained to the handling and disposal of flowback. It also pointed out that federal EPA officials had a special stake in protecting New York City's watershed and urged the state DEC to heed the city's concerns. The letter added, "just because fewer people rely on upstate sources, does not imply that these supplies are not worthy of protection." In addition: "We have concerns about potential impacts to human health and the environment that we believe warrant further scientific and regulatory analysis. Of particular concern . . . are issues involving water supply, water quality, wastewater treatment options, local and regional air quality, management of naturally occurring radioactive materials disturbed during drilling, cumulative environmental impacts, and the New York City watershed." Those comments echoed, essentially, the points that Enck had summed up during her trip to Greene a year and a half earlier when she was Paterson's top environmental advisor: "The thing that stands out the most is the need to have a comprehensive mechanism to protect groundwater and surface water."

The tone of Filippelli's letter to the DEC was as noteworthy as its content. It signaled a departure from the manner of the EPA under the Bush administration and anticipated the renewed interest of the agency in the safety of hydraulic fracturing.

EXPECTATIONS

By the middle of 2010, the Marcellus development in Pennsylvania was meeting the expectations of its most enthusiastic proponents and its most ardent critics. Between July 1, 2009, and June 30, 2010, 632 Marcellus wells produced 180 billion cubic feet of gas, more than doubling the annual state production from all other sources prior to shale gas exploration. The last six months of 2010 showed an even steeper production curve—256 billion cubic feet of gas from 1,147 wells. Even with gas prices suppressed by a weak economy and a glut in the market, exponential production gains were forecast through 2012, when some expected the state to pass 1 trillion cubic feet on its way to becoming one of the world's top producers. It was no longer the game of the small independents that had traditionally worked the gas fields of Appalachia. The low prices were offset by economy of scale, enabling well-capitalized multinational companies—the Walmarts of the gas industry—to harvest and warehouse gas with the expectation that growing supplies and stable prices would encourage new markets and eventually displace coal in power plants and gasoline in vehicles. At least, that was the vision held by some.

There was another line of thinking that held that full-scale shale gas exploration would produce diminishing returns rather than economy of scale. Although shale gas wells had proven prolific over the short term, there was uncertainty over how long they would produce. Critics pointed to evidence that suggests the life span of these wells may be shorter than expected and that production gains would therefore require not only rapid development of new wells but an aggressive maintenance schedule to restimulate old wells. This, in turn, would create more disposal problems and raise costs to keep gas flowing as production volumes for each well fell. According to this line of thinking, the investment prospects of shale gas extraction had been inflated by overstated returns and understated costs, and the industry was headed for a bust on Wall Street in much the way the dot-com and housing bubbles had burst in the 1990s and early 2000s.

That possibility of inflated shale gas economics was the theme of a front-page story that ran in the *New York Times* on June 26, 2011. The article cited "hundreds of industry emails and internal documents, and analysis of data from thousands of wells." Deborah Rogers, a member of the Advisory Committee of the Federal Reserve Bank of Dallas and a former stockbroker with Merrill Lynch, stated in one of those emails about the Barnett Shale: "These wells are depleting so quickly that the operators are in an expensive game of catch-up. . . . This could have profound consequences for our local economy." Industry proponents characterized these fears as unfounded and exaggerated. Even if the price of gas were to remain indefinitely low—which was unlikely—the shale gas economics would be supported by manufacturing efficiencies and technological advancements, and by new and growing markets for energy that is cheap, abundant, and cleaner than coal.

"Shale gas supply is only going to increase," Steven Dixon, executive vice president of Chesapeake, told the *Times*.

A year prior to the *New York Times* investigation, Anthony Ingraffea began a series of public lectures using industry figures and science to show why, in his view, shale gas operations are a net social and environmental loss. In 2010, Ingraffea gave more than fifty lectures, and he had no trouble filling community halls and library conference rooms in upstate New York with rapt audiences. His presentations, like those of Terry Engelder, were backed by convincing credentials. His scholarship at Cornell University deals with rock fracturing mechanics, on which the science of hydraulic fracturing is built.

Tony, as he is known to his friends and colleagues, is a man of modest height and erect bearing, with wavy white hair and keen dark eyes. He is typically dressed in a coat and tie for these lectures, but employs a speaking style that is informal, entertaining, and blunt. He has been an avid fly fisherman since his boyhood days, and an opening slide of one of his PowerPoint presentation featured a trophy-size rainbow trout from one of his excursions to Inman Creek, south of Ithaca. "I love New York and I would love to keep it the way it is," he told a group in the Vestal Public Library in the spring of 2010. "That ain't gonna happen. So the next thing we should do is figure out how to keep it as close to the way it is, for as long as we can, without harming each other economically or environmentally." He then launched into the meat of this talk, which showed how Engelder's projections of Marcellus development could not be achieved without "massive and intensive industrialization" of the region. Effectively extracting the gas would require eight wells per square mile, and 48,000 wells in the seven New York counties overlying the most promising Marcellus reserves. "What have you heard from your county legislators? Maybe 2,000 or 3,000 wells?" he asked the crowd of fifty or so people in the library meeting room. "That's BS."

Victoria Switzer was in the audience, and after Tony finished his two-hour talk, she introduced herself and recounted her experiences. "You've hit this right on the head," she told him. "Anybody who wants to see for themselves can come to Dimock. I'll give them a tour." He said he would take her up on that offer.

When Tony first began making these presentations, he told me, he received a call from Engelder, accusing him of "yellow journalism." Terry wanted to know why his former research associate was so bent on proclaiming the dangers of shale gas, which Terry saw as an exaggeration. "Terry's a consensus builder, and it bothered him that not everybody saw this the way he did," Ingraffea recalled. Yes, Terry was troubled about the events in Dimock, which he read about in the papers and in emails sent to him directly by Victoria Switzer, who thought he needed, in her words, "a reality check" about the promises of shale gas development. Engelder held that the industry was fallible and that it should be held accountable when things go wrong; but that didn't diminish his belief in the overall good that could come from full-scale development of the Marcellus. Along these lines, Engelder had suggested that he and Ingraffea could publicly come to terms about their differences. He proposed they form a speaking panel to collectively provide information to community forums and the media. This panel, which Engelder called the "Gang of Five," would also include Switzer; Tom Murphy, the Penn State Extension representative who coordinated outreach programs for landowners; and Seamus McGraw, who was writing a book about the shale gas rush in northern Pennsylvania based on the stories of Dimock residents and his own experiences as a landowner.

Terry's proposal never materialized, at least not in that format. There were too many differences, and too many incidents happening—in Dimock and elsewhere—that pulled the sides further apart.

In March 2010, a pit containing 400,000 gallons of wastewater from an Atlas Energy drilling operation ignited in balls of flame and black smoke visible for miles from the site in Washington County, Pennsylvania. The fire, fueled by a mixture of hydrocarbons that condense from natural gas, also consumed a tank containing fracking fluid and burned the plastic pit liner. On June 3, 2010, a combustible mix of gas and fracking fluid erupted from a well operated by EOG Resources in Clearfield County, Pennsylvania, on the southern end of the Allegheny region, and shot a geyser 75 feet into the air. The torrent continued unchecked for sixteen hours while a team of specialists traveled from Texas to address the blowout. By the time it was brought under control, 35,000 gallons of toxic wastewater and brine had washed over this well site at a remote hunting ground. At the same location, prior to the blowout, inspectors found evidence that waste fluids seeping from one or more pits had degraded Alex Branch (a wild trout stream) and polluted drinking water at a nearby hunting camp with barium levels four times greater than federal drinking water limits.

Five days after the blowout in Clearfield County, a Marcellus well in Marshall County, West Virginia, exploded when crews drilled through a coal seam and ignited a pocket of methane. Homes near the well, 55 miles southwest of Pittsburgh, were evacuated and seven workers were burned.

Just as coalmining communities learn to live with threats of cave-ins and explosions, drilling communities live with "blowouts"—the industry term for what happens when pressure from a well exceeds the measures to control it. In cartoons, a blowout is depicted as a gusher; and in the old days, it was a sign everybody watched and hoped for. Although they were often dangerous, wasteful, and damaging, gushers meant somebody had struck it rich. Modern-day gas wells are generally deeper and more sophisticated; the volumes are also greater and the pressures more extreme. An entire field of engineering is devoted to providing controls and backups, including drilling mud, designed to exacting standards for weight and viscosity and pumped into the well bore to balance the pressure from above with the pressure from below, and blowout preventers, heavy valves that shut off the bore in the event of an emergency.

Even with the best safeguards, accidents happen. The blowouts in June 2010 were the first in the Marcellus. Although they drew banner press coverage due to interest from a growing number of mainstream stakeholders in northeastern shale gas development, they were nothing new to the industry; and they served as reminders of the difficulty of extracting energy of any kind from the earth, along with hazards that had to be managed and the limits on the technology designed to mitigate them. In the first decade of the development of the Barnett Shale, the Texas Railroad Commission, which regulates the industry in that state, had logged more than 230 blowouts of oil and gas wells, causing 25 fires, 28 injuries, and 4 deaths. A 2006 blowout of a Fort Worth shale gas well, similar to those being drilled in the Marcellus, had killed a rig worker who had mistakenly opened a valve. The Fort Worth explosion prompted the city to enact an ordinance doubling the 300-foot zone separating wells from occupied structures.

The Marcellus blowouts may have been dismissed (along with the hundreds of similar industry accidents that had preceded them in well fields throughout the country) as an unavoidable cost—collateral damage—of meeting the energy demands of the country, if not for an even greater spectacle drawing worldwide attention to drilling hazards. On April 20, 2010, in the Gulf of Mexico, crews began battling one of the single largest U.S. environmental disasters—and it was caused by a blowout. British Petroleum's Deep Water Horizon well was hemorrhaging oil from the ocean floor and providing an unmatched public display of the perils related to energy extraction. It began when methane gas from the well shot up the drill column and over the platform, causing an explosion that killed eleven workers. The rig, engulfed in flames, sank two days later. Industry officials and regulators were at a loss after repeated attempts to cap the well in the Macondo Prospect failed. Over the course of the next three months, 185 million gallons of crude poured from the hole in the ocean floor, shutting down fisheries, soiling beaches, and placing unknown long-term burdens on the gulf ecosystem.

VICTORIA'S TOURS

Institutional landowners—governments, churches, schools, and nonprofit agencies—were among the front-line stakeholders in the Marcellus controversy, and the types of accidents covered in daily news reports added to the complexity of the decisions they faced. In addition to issues of solvency and land stewardship, there were matters related to their reputations and the mind-sets of constituents and directors, who sometimes fell on opposite sides of the issues. This challenge applied to Tony Ingraffea's employer, Cornell University. In making decisions about leasing 11,000 acres owned by the university, the board of directors and the president had to take into account the views of faculty and students, many of whom had been active in, or sympathetic to, antidrilling protests. With this, the administration had to balance the potential for tens of millions of dollars in income, along with administrative and legal dynamics, ranging from research objectives to the possibility of compulsory integration. Accordingly, Provost Kent Fuchs appointed an advisory committee of faculty, students, and administrators to evaluate the issues and make recommendations to guide university policy. "Any decision Cornell makes regarding gas development on its lands could impact its mission," a committee report noted. "Such a decision will affect its reputation as a leader in sustainability efforts, its relationships with surrounding communities, and its ability to recruit the best students, faculty and staff who, at least in part, want to come to Cornell due to the quality of life Ithaca and the surrounding area provide."

In spring 2010, Victoria Switzer, microphone in hand, stood in the front of a bus loaded with members of this Cornell committee. They had already taken the tour given by Chesapeake at a well site in Bradford County showcasing the technology and safeguards of the industry; and now, at Ingraffea's urging, they were getting another perspective. The bus drove down Carter Road, passing trucks, pipelines, and well pads in various stages of construction or reclamation, and Victoria told her story and the story of her neighbors who felt they had been deceived by the landmen when they signed away rights to their land. Now, Victoria explained dryly, these residents look forward to Friday as "water delivery day." She pointed out the "uglification of Pennsylvania that started in Dimock" and added that the biggest impact—the damaged aquifer—was impossible to see or photograph in a drive-by tour. "We're not as dramatic as what happened in the gulf," she said. "We're a slow poisoning."

Inevitably, there were other conflicts between gas companies and residents as the shale gas boom progressed through Pennsylvania in 2010, but the collection of circumstances in Dimock—perhaps because of Norma's well or Victoria's dream house or Cabot's growing insistence on distancing itself from the problems—continued to draw international attention. "It is unfortunate the word 'Dimock' is known around the world—literally," Hanger complained to a reporter at a town hall meeting in Williamsport as he recounted how he was asked about the circumstances of the Carter Road residents by a Russian television crew. Hanger and his staff, in interviews and testimony, referred to the Dimock methane problem as "a black eye" for the industry and the agency. "It's been an enormous headache. If Cabot doesn't get this message, the company has got an amazing hearing problem," Hanger said.

The DEP public relations headache over Dimock only got worse. A month after the Cornell bus tour, Robert F. Kennedy Jr. led an entourage of high-profile advocates to Dimock. Among other roles, Kennedy is the chief prosecuting attorney for Riverkeeper, senior attorney for the Natural Resources Defense Council, and president of the Waterkeeper Alliance. He was joined by Mike Richter, a Riverkeeper board member and former all-star goalie with the New York Rangers; Mark Ruffalo, actor and resident of the Catskills; and an assortment of administrators, board members, and staffers from the Natural Resources Defense Council, Catskill Mountainkeeper, and Damascus Citizens for Sustainability—all regional environmental groups with interests in protecting the Delaware and Susquehanna River watersheds. They met at the Switzer house, which was still under construction but now inhabitable. Victoria hosted this group, as she had the contingent from Cornell. She introduced the luminaries to the Carter Road folks and gave the visitors a tour of the well sites.

The Kennedy meeting represented a victory for the Carter Road residents, if only a small one. The cause of the Carter Road 15 was not a popular one among many Susquehanna County residents, including innkeepers, landlords, construction contractors, and restaurateurs who counted on the industry for their livelihood (some of whom also sat on, or were allied with, local government boards). Nor was the antidrilling message popular with some of Victoria's landowning neighbors, who were happy with the financial returns they were realizing from gas development. Kennedy's visit was a morale boost for residents who were feeling worn down and ostracized from their fight against Cabot, which seemed to drag on forever with no sign of resolution. Victoria had the satisfaction of hearing Kennedy proclaim that those representing the industry "all seem to be pathological liars. You can make deals with them and they're going to break the deals. You've seen that happen at the local level; I've seen it at the national level." Still, Victoria, reflected later, what would this visit ultimately accomplish? It didn't appear to immediately change Kennedy's position—that coal was the worst enemy of a green planet and that natural gas was a viable alternative—nor did Kennedy's visit have a visible impact on the practical circumstances of the Carter Road residents. "They are in the ivory tower," Victoria told me of large advocacy groups in general. "I'm on the battlefield, and that battle is being fought now in the villages being staked out by gas companies. I'm going village by village to help where I can. Dimock is already lost." In summer 2010, she accepted an invitation to be featured on a panel that included Josh Fox and Tony Ingraffea at the premier of *Gasland* in Binghamton. After that event, which was attended by more than five hundred enthusiastic supporters of her cause, she began accepting speaking engagements at gas forums in neighboring townships in New York and Pennsylvania; and she told her audiences what she had told Kennedy: "Don't leave here thinking that Dimock is a freak show. There are many Dimocks out there, and they are not getting their stories told."

Legislators in New York, meanwhile, were clearly hearing Victoria's message. On August 3, 2010, they passed a bill—introduced by State Senator Thompson—that would ban hydraulic fracturing in New York until the following May. The bill would have no practical influence on Marcellus development because permitting could not begin until the New York DEC finalized its draft of the SGEIS, and that was not due until well after the legislative ban was scheduled to expire. But the bill carried significant political clout that could impact the broader discussion in New York. It showed that state lawmakers were not ready to embrace shale gas development and they could summon votes to resist it. Accordingly, industry leaders implored Governor Paterson to veto it.

"This bill is a job killer, an upstate business killer and potentially an industry killer," Brad Gill said in a statement issued by IOGA New York. "The governor must be made to understand the vast unintended consequences and act quickly to reject this needless legislation. I remind the governor that hydraulic fracturing has been used successfully and safely on water, gas and oil wells for 60 years in New York, and that drinking water has not been adversely affected. Members of the Legislature have based their votes on a ridiculously false assumption that hydraulic fracturing fluid will make its way into groundwater and surface water."

The bill also drew plenty of rhetorical flourishes from lawmakers, and even those who voted against it were eager to make some political hay by holding up Pennsylvania and the DEP as examples of how not to handle Marcellus development. Thompson noted, in front of the cameras, that Pennsylvania "was so thirsty to get this development opportunity" that the government had neglected to inspect the wells or protect landowners. State Senator Libous, on the other side of the political fence from Thompson, still managed to get his shots in at their shared rhetorical target. "Shame on the state of Pennsylvania," said Libous, who voted against Thompson's bill. "Shame on their Department of Environmental Protection . . . because they screwed up badly. They didn't keep an eye on those who were drilling. They didn't keep an eye on environmental factors on behalf of the citizens of that state."

DEP Secretary Hanger (who once told me he had considered a career in politics) did not shy away from the political rough and tumble of the public arena. As cameras rolled and reporters took down his response, he accused New York of hypocrisy. "If they are so ashamed of what's gone on here perhaps they should stop buying Pennsylvania gas." He added, "If New York demands to have no impacts from drilling, then they better have a moratorium that extends not just through May 2011, but forever. You cannot have drilling, even done well, and get zero impact." The secretary also used the opportunity to push (unsuccessfully) the Pennsylvania legislature to adopt an extraction tax on shale gas. "Right now we're getting taken to the cleaners by New Yorkers who are getting tax-free Pennsylvania gas and then complaining about it."

UNDER SURVEILLANCE

In the wake of spills and blowouts in the summer of 2010, Cabot stepped up its own public relations efforts. The company hired George Stark, a public affairs specialist, to take over as a community liaison for Komoroski; and one of Stark's first tasks was to organize a free community picnic on the school grounds in Montrose. Stark has a runner's build, with a precise haircut, rimless glasses, angular features, and deliberate body language that together project a confident manner. Unlike Komoroski, who had a background in law and engineering, public relations was Stark's career specialty. He had twenty years of experience in that field with Columbia Gas before taking the Cabot job.

The picnic, held on a mid-August day, was part trade show featuring Cabot and Cabot contractors, and part local fair with several acres of displays, picnic tables, and large white tents. Halliburton staff in red polo shirts and khaki shorts barbecued pork, chicken, and corn. Children jumped in inflatable structures and amused themselves with *Guitar Hero* and other video game displays, while men and boys ogled rows of heavy equipment and trucks in showroom condition. Inside one of the large tents, a woman handed out samples of fool's gold, Marcellus Shale fragments, and colorful rocks. A crowd gathered around a man standing over a cut-away model of a well bore as he explained casing guidelines and industry best practices. After interviewing a few passersby, I got a feeling that most were there for the food and entertainment. The crowd seemed curious and open-minded. Most had yet to experience drilling first hand, and a relatively small percentage seemed to have passionate feelings for or against drilling.

It was impossible not to notice the uniformed state troopers on chestnut horses moving through the crowds. I learned they were part of a security detail that included four troopers and four Montrose School police officers. All the men had the physique of body builders and wore wide holsters with side-arms, handcuffs, and clubs. When Julie Sautner made her way up the walk-way, in an anti-fracking t-shirt, they were waiting for her. Sautner had her hands full with a file of brochures and a plastic milk jug filled with a brownish liquid. She wore her hair in a bun, and she stuck out her lower lip and blew away the stray lock of hair that occasionally drifted across her forehead. She was looking for Victoria Switzer, Virginia Cody, and Vera Scroggins—with whom she had arranged to set up their own informational table—when she was stopped by one of the patrols and ordered to hand over the jug.

"It's my water," she snapped, pulling away. "Maybe I'm bringing it to drink, because Cabot says it's okay." The man held her arm and forced the jug from her fingers. He emptied it in a port-a-potty, and then crushed the container and discarded it in a recycling bin.

Vera Scroggins appeared a few minutes later and, after learning of the incident, joined Julie in enthusiastically lecturing the police about First Amendment rights. The man who seemed to be coordinating the security detail listened politely and appeared to be sympathetic. He kept reiterating his point that the jug violated the open-container rules. Vera insisted residents be given a spot to set up their own displays and distribute their own material. Victoria arrived and, with the calm and firmness of a teacher handling a class conflict, explained that the women were not seeking confrontation. They simply wanted to be allowed to participate, in an orderly and peaceful manner, in a public informational event held on public property. After several minutes of back and forth with the women, conferences with each other, and phone calls to unseen parties, the police directed the women to a corner of the property near the walkway. There the women advocates stood and chatted with each other next to a folding table with brochures that highlighted the dangers of drilling and made the case for stricter oversight. They offered these brochures to people arriving amid the scent of barbequed pork and chicken. Some people accepted them absently. Others politely declined with a wave of a hand or ignored them altogether.

The women's plans for this protest, if it could be called that, had been listed on a bulletin issued to local law enforcement agencies and energy companies by the Pennsylvania Office of Homeland Security that detailed possible terrorist activity targeting key commercial facilities and energy infrastructure. Pennsylvania Intelligence Bulletin No. 123, August 11, 2010, under the heading "Adversarial Protest Strategizing" read:

> The Gas Drilling Awareness Coalition has released a special communication harshly ridiculing and criticizing this (Cabot's) event. However, the activists themselves are far from agreed as to a course of action. One activist said that he or she was "compiling a fact sheet against Cabot . . . to make up hundreds of flyers and stick them under windshield wipers during the event and hand them out if people want to see the 'real facts'." In addition, the activist "would like to try to get a group of people together that want to ask Cabot questions that put them on the spot and try to get people to understand the truth."

In light of this, the bulletin warned, "Local officials and law enforcement should maintain situational awareness regarding the potential for some confrontations between anti-drilling environmentalist militants and gas drilling company employees."

After receiving a copy of this and other briefings, leaked through a friend of a friend, Virginia Cody, one of Victoria's "warrior women" allies on the Susquehanna Gas Forum, was astounded to recognize her conversations with other citizen activists listed in the same intelligence network tracking Al Qaeda. Thinking it was a farce, she posted the notice on the group's Internet forum. She promptly received an email from Pennsylvania Homeland Security Director Jim Powers, who had apparently mistaken Cody as a drilling supporter trying to faithfully disseminate information.

The public domain was not a good place for departmental security briefings intended for industry and police, Powers warned Cody. "Please assist us in keeping the information provided in the PIB [Pennsylvania Intelligence Bulletin] to those having a valid need-to-know; it should only be disseminated via closed communications systems. Thanks for your support. We want to continue providing this support to the Marcellus Shale Formation natural gas stakeholders while not feeding those groups fomenting dissent against those same companies."

Cody was stunned. "I have the right to foment dissent over anything I want," she later complained as she shared Power's email with an elected official. "It's the Constitution!" After Cody went public with the revelation that agents were spying on citizens, lawmakers began to investigate, and a series of hearings in the state Senate revealed that the intelligence gathered on the "middle-aged women in comfortable shoes," as Cody characterized them, was part of a broader assignment commissioned to the Institute of Terrorism Research and Response (ITRR). The firm had a $125,000 contract with Homeland Security to gather information on terrorism, which it apparently found wherever it looked. The list of events with the potential for terrorist activity, according to bulletins issued by ITRR, included a forestry industry conference, a screening of *Gasland*, several municipal zoning hearings regarding Marcellus Shale gas field development, a political demonstration in favor of a tax on gas extraction, an animal-rights protest at a rodeo in Montgomery County, "Burn the Confederate Flag Day," Jewish holy days, and the Muslim holy month of Ramadan. Also on the watch list were gay pride parades, deportation protests in Philadelphia, antiwar groups, and Tea Party activists.

"My voice was loud enough to rouse the attention of my government, but not in the way I had hoped," Cody testified. "Instead of paying attention to what I had to tell them, my government sicced the FBI, the Office of Homeland Security and the Institute of Terrorism Research and Response (ITRR) after me. My government branded me an extremist, or possibly an eco-terrorist."

Coverage of the story and the subsequent state Senate hearings—chaired by Lisa Baker, a state senator from Luzerne County and chair of the Veterans Affairs and Emergency Preparedness Committee—exposed the issue and prompted an avalanche of criticism from liberal and conservative groups, along with ample comparisons to McCarthyism and Big Brother. "If someone cannot tell the difference between a citizen activist and a terrorist, what sort of protection are taxpayers paying for?" asked Senator Baker.

In pursuing an investigation, Baker related the "appalling revelation" of the list in a letter to Governor Ed Rendell and her belief that it would damage the already "exceedingly low" public confidence in government. "This will have a chilling effect on public activism at a time when it is critical for diverse views to be part of the debate," Baker wrote. "This sort of involvement should be commended, not retaliated against." In late summer 2010, Rendell canceled the $125,000 state contract with the ITRR and appeared before cameras to apologize for the intelligence operation, which he said had happened without his knowledge or consent. Days later, Powers resigned.

ENVIRONMENTAL PROTECTION AGENCY HEARING

In late 2009, Congress passed a $32 billion Interior and Environment Appropriations Bill that directed the EPA to use a portion of its allotted funding for a peer-reviewed study of the relationship between hydraulic fracturing and drinking water, "using a credible approach that relies on the best available science." To prepare for that endeavor, which was facilitated by Maurice Hinchey, the EPA scheduled a series of public meetings in summer 2010, including one in Binghamton.

The Binghamton hearing on September 13 drew about 1,500 people to the Binghamton Forum, a community theater in the heart of downtown. Outside the building, hundreds of protestors representing both sides of the issue, many with signs, drums, pamphlets, or other props, lined up literally facing each other behind makeshift barriers on opposite ends of the street. The industry supporters were mostly local and regional landowners led by Dan Fitzsimmons. The opposing contingent included people from New York City, Ithaca, and Dimock, including Julie Sautner, who carried another jug of brown water. It was a month after the Cabot picnic and three weeks before news of the Homeland Security scandal broke. Josh Fox, standing in front of the protestors, gave a live interview on location for CNN, with a rebuttal via satellite from Chris Tucker of Energy in Depth in Washington, D.C. Police presence was heavy, and cops energetically enforced the rules prohibiting protesters or apparent protestors from loitering in the area—set up like a demilitarized zone—directly in front of the forum.

Victoria made her way between traffic barriers set up to contain the rallies, passed under the marquee of the forum, and through the glass doors. In the lobby, she waited in line to sign in and retrieve a numbered bracelet that allowed her a two-minute turn at one of two podiums facing a stage where Judith Enck and other EPA officials sat behind folding tables covered with pleated white cloth. A screen above them showed a large digital timer that began counting down whenever a new speaker was called to the microphone.

Maurice Hinchey was first. Minutes before he was called, he walked down one of the theater aisles leading to a podium in front of the stage. He continued past the podium and disappeared behind a curtained exit. When the moderator called him to the microphone, Hinchey appeared at the back of the stage and approached the officials, to the apparent surprise of the moderator. He took the microphone and began to speak, disregarding the two-minute rule, "We cannot and must not move forward with hydraulic fracturing absent an independent, scientific analysis, supported by empirical data, of the risks that hydraulic fracturing can pose to water supplies or air quality."

Over the course of two days, several hundred people had a chance at the microphone. Their testimony—intended to collect input on the technical aspects of the study—often digressed into impassioned speeches for and against drilling.

> Aaron Price: "High-volume hydraulic fracturing builds wealth, saves jobs, and gives hope . . ."
> Josh Fox: "Stop this practice now. There is a tremendous amount of suffering going on across the nation while this study is taking place . . ."
> Brad Gill: "Despite claims to the contrary, there hasn't been one case of groundwater being contaminated by the hydraulic fracturing process . . ."

On Monday evening, speaker 213, Victoria Switzer, was called to the podium. She condensed the essentials of her story into two minutes. She was a Dimock resident whose water had been ruined by methane migration. The DEP had confirmed that, but she suspected there was something more; and now she had proof. She recounted tests from Farnham and Associates that documented the presence of solvents in her water. "No fracking fluids ever found? Fact: I have fracking fluids in my drinking water. EPA, do your job," she said. "Please demand accountability. I offer you a case study: myself, Dimock."

MORE THAN METHANE

There was no indisputable empirical understanding of how the bedrock aquifer from which the Carter Road residents drew water might be linked to Steven's Creek on Heitsman's property to the west, or to any of a number of other drilling and fracking operations in the area that may have been the pathway for solvents to enter the aquifer. Nor was there any consistent and reliable disclosure of the fracking products being injected into the ground at hundreds of other sites around the region. There was also another missing piece of information: under the regulations in place up through 2010, Cabot was not required to report water complaints to the DEP. In an undetermined number of cases, the company found ways to avoid public disclosure. The events in Dimock demonstrated a disconnect between the official version of the problem and the actual number of wells affected. The documents available through the DEP press office point to the eighteen water supplies that the agency ordered Cabot to replace. But the company had addressed problems in at least thirty-six water wells, many without the involvement of the agency.

Dan Farnham found the anecdotal reports of water wells going bad along the Burdick Creek hollow troubling. Farnham was owner of Farnham and Associates, an engineering firm that specialized in commercial wastewater issues. Water testing and analysis happened to be his expertise, and drilling operations were one of his niches. Farnham's company had tested more than 2,000 water wells in northeastern and north central Pennsylvania, and he knew of similar problems in Bradford County, where fifty or more homes were getting replacement water supplies provided by gas companies. Farnham had a unique stake in the issue as a resident of nearby Hop Bottom (a hamlet south of Dimock), a Little League and soccer coach, and father of two. To understand a problem defined largely by anecdotal information, applying common sense was step one. "When your water is yellow, there is two inches of foam on the top and brown scum on the bottom, you don't need a Ph.D. to show impact," he told me. Applying science was step two. As a community service, he told me, he began routinely testing samples from half a dozen suspect wells along Burdick Creek in spring and summer 2010 where the DEP had documented dangerous levels of methane. He found a pattern. Hydrocarbon solvents—including ethylbenzene, toluene, and xylene—were showing up periodically after heavy rains in the wells of people who also had methane problems, results he found "very interesting and very disturbing at the same time." (Testing by Farnham also found elevated levels of metals in water supplies near Chesapeake Energy drilling operations in Bradford County wells.) "I'm not here to argue with the gas company," he said. "My objective is just to illustrate that something's going on here and it needs to be investigated."

Cabot denied the pollution was related to drilling. "Cabot never used those chemicals," Stark told me. I asked him about Farnham's findings and pointed out that the chemicals detected by Farnham were the breakdown products of solvents found in LGC-35 CBM, the solvent that had spilled into Steven's Creek from the Heitsman property. Stark raised the possibility of other sources, unrelated to drilling, including a fuel truck that had crashed and spilled about 500 gallons of home heating oil into Burdick Creek near the intersection of Carter Road and Route 3023 in February 2010, and an auto-repair shop near that same area. He didn't mention the 2,500 gallons of diesel spilled by Cabot contractors in five different accidents in the Burdick Creek watershed. Without a complete data set of contamination cases and more thorough hydrological models, the problem of empirically tracing the solvents to their source became enormously complicated; and that task would be left to lawyers and the courts.

IN THE PULPIT

In late September 2010, two weeks after Victoria made her plea to the EPA, a secular event held in a Baptist church in Dimock came to define the legal and political course of the Dimock case study. It was a public meeting between Carter Road residents and Pennsylvania DEP Secretary John Hanger. The meeting, originally planned for outside Ron and Jeannie's trailer where lawyers had assembled the Carter Road 15 to announce their lawsuit the previous fall, had been moved to the church at the last minute because of a driving rainstorm.

Ron and Jeannie Carter took their seats in the front pew of the tiny white church where they had been married and waited for Hanger. The eight pews arching around the altar were filled to capacity. In addition to Ron and Jeannie, there were about sixty other people, including members of the fifteen households with contaminated water and advocates and attorneys supporting their cause. Bill and Scott Ely and Victoria and Jimmy Switzer were there, and so too were Norma, the Sautners, and Pat Farnelli. The aisles were filled with banks of cameras, independent media, and reporters, including Vera Scroggins and Josh Fox. The gathering overflowed into a coatroom and an adjoining vestibule.

At promptly 10 a.m., Hanger appeared from an anteroom behind the altar and stood at the pulpit. A trim, middle-age man, he had a politician's bearing and fine sandy brown hair parted on the side; he wore a tailored dark suit, a light-blue shirt, and an expertly knotted silk tie. "We are here to right a wrong," he announced, his voice resonating through the small white church. The Dimock conflict had become the landmark controversy in his tenure as the state's top environmental administrator. Now, on September 30, 2010, nearly two years after Norma's well exploded and two and a half years after the first diesel spill, he summed up the problem: "This all shows it is much better to do the job right the first time than to deal with a mess when you don't do the job right the first time." He reviewed the DEP's tests that showed methane from gas wells was migrating into the aquifer, and the terms the DEP had negotiated with Cabot to fix the problem. "Since April, we have extended extraordinary opportunities for Cabot to meet the schedule. We had thirteen meetings or calls with Cabot, almost weekly since April. The heart of the matter is this: We have had people here in Pennsylvania without safe drinking water for close to two years. That is totally, totally unacceptable. We have given Cabot every opportunity. We have entered into agreements that they have signed, and their attorneys have signed," he continued, referring to the various versions of the consent order that had been signed since November 4, 2009. However, the company took the position that these orders were not binding. "They have said they signed this under duress," Hanger announced. "Mr. Dinges is the CEO of a major corporation and he has many resources at his command, . . . He is not a ninety-year-old widow sitting in a nursing home incompetent to sign a document. Even without counsel, he is fully qualified to sign a document. And he had counsel. That argument tells you all you need to know about how Cabot has proceeded."

The answer to the problem, at this point, Hanger stated, was to build a water line from Montrose. The project would cost more than $11 million, and the DEP would make sure that Cabot funded it. "We will move this project forward. We will do everything we can to hold Cabot responsible for this. . . . We will fight with Cabot if that's what Cabot wants to do. We are not going to wait for long legal proceedings."

A voice piped up from the back of the room. "Excuse me, but Cabot is represented here." It was Vera Scroggins, and she was pointing to George Stark, who stood against the back wall with his hands folded across his chest. "Perhaps it's a good time for Cabot to address these issues," Vera asserted.

"Thank you for the opportunity," Stark said. "I came to listen."

Somebody from the audience then complained to Hanger that Cabot officials had been coming on to private property with armed guards. Hanger responded, "I would like to say to everybody involved: Put the guns away. Cabot has to follow the law and you have to follow the law. Emotions are high. This has to be done in a peaceful way. We're in a church, and I think it is an appropriate place for that message."

After Hanger's announcement, Stark exited, followed by a cluster of reporters seeking his response to the water-line proposal. "Was Cabot accountable for the problem or not?" one asked.

There were "questions to the validity" of the consent order, Stark replied, "based on the duress that was taking place when those documents were signed." He continued, "We think today, there is overwhelming, scientific proof, and historical evidence, that we are not the source of this methane migration." He pointed out that the company had worked with other families to address water problems and that, with these families, "things are going very well." The company could refurbish the Carter Road water wells, he said, but litigation was preventing that.

Then came the question: Was Cabot using guns, and if so, why? "There have been instances where Cabot has run into armed residents," Stark replied. "We had to install and to employ the use of an armed guard, every once in a while, when we run into an armed resident. It is not our wholesale practice by any means to use armed guards."

ENOUGH ALREADY

The water problems in Dimock bothered some residents more than others. Among those least bothered were owners of tracts drawing royalty payments from multiple wells, and people with jobs and incomes related to the gas business. They included the Teels, Victoria's neighbors to the south, and the Lewises, across the street. When the Teels' water well went bad, the family didn't mind using bottled water for drinking and cooking as they waited for Cabot to redevelop their damaged well and install filters. Lewis was concerned about the diesel spill on his property, but later his confidence was restored by the company's tests that showed his water was unaffected. The people who sided with Cabot were at first disinclined to grant interviews. But as the public profile of the conflict grew, they, like Victoria, began going public. "Everything is back to normal," Ron Teel told one investigating television crew inquiring about the quality of their water in October 2010. "The water seems good. They put a couple of filters on, Cabot did. Everything is working great right now. Everything is fine."

By this time, Don Lockhart's lunch counter had become a regular gathering spot for industry people. Roughnecks, executives, and landmen were often lined up at the counter and seated in the dining room among local residents. The consensus there was that the Carter Road plaintiffs were holding on to a problem that could be easily fixed in hopes of leveraging a legal claim. "You're not getting any common sense. All you hear about is 'these poor people,'" Don Lockhart told the press. At the lunch counter, he kept a petition asking the state infrastructure investment authority to deny approval for the water-line project. There was talk that the people on Carter Road were malcontents. They were exploiting the industry, and it was this type of attitude that would end up driving the industry away, along with jobs for Pennsylvania and royalties for landowners. I heard similar assessments from patrons at Lockharts and other establishments that catered to the industry. Some people said they believed the Carter Road plaintiffs had framed Cabot by contaminating their own wells; and I heard from one person who believed, or who wanted me to believe, that the explosion of Norma's well was actually caused by a hidden meth lab.

The waterline plan, from the industry's perspective, would establish a costly precedent for methane cases, and Cabot CEO Dan Dinges had no intention of going along with it. Upon receiving the order from the DEP, he began a public relations offensive that included the claim that Pennsylvania regulators had fabricated the explosion of Norma's well as part of a broad conspiracy against the company. "You continue to trumpet this fictional incident in an attempt to stir up public opposition and distrust toward Cabot and its activities in the region," Dinges asserted in a letter to John Hanger on September 28, 2010. Dinges then took his case to the public in a full-page advertisement, laid out in the form of an open letter to the Citizens of Pennsylvania, that ran in regional newspapers in New York and Pennsylvania in late September. The order for Cabot to build and finance the pipeline, Dinges stated in the ad, "seems unreasonable, unprecedented and unfair. No private business model would support such an investment." Dinges noted that Cabot was providing potable water to affected homes and "has attempted to work toward a long-term solution in the form of individual water treatment systems" even though "Cabot does not believe it caused these conditions and intends to fight these allegations through its scientific findings." The company, he continued, "wants to continue as a good corporate citizen" and "to provide jobs, opportunity and royalty revenue to our fellow citizens for years to come." He closed by urging residents "to view Cabot's concerns fairly and seek the facts." With that, the advertisement directed its audience to the company website to learn more.

Dinge's message, and the implication that the pipeline order would drive the company and related business away, was picked up by industry supporters who began organizing a group called "Enough Already" to oppose the waterline project. The effort was coordinated by, among others, Bill Aileo, a local attorney who claimed the water pipeline would, in addition to discouraging gas development, generate strip malls between Montrose and Dimock that would degrade the beauty of the rural community. Enough Already members printed and posted fliers, featuring a blue waterline pipe crossed out in red, at businesses, on public bulletin boards, and along the proposed route from Montrose to Dimock. Cabot began featuring some of these pipeline opponents on its website. There, on a page called Susquehanna Community Education, among images depicting community picnics, library donations, local scholarships, and Christmas charities, was a link to a video showing Victoria's neighbor Harold Lewis drinking water from a garden hose running in front of his barn, and dismissing the proposal to provide a public water supply as a waste of money. "Go with Cabot," he advised. "They are willing to drill [water] wells for free. Try that. Or the water treatment system. I think that's a real good idea." The video featured several other residents, including Char Myrick, who suggested the Montrose water, containing "chlorine and that sort of thing," was inferior. "What I find really difficult," she said to the camera, "is when we have neighbors who have a water problem. No questions asked, Cabot is willing to come in and fix the water—in some cases drill a new well, in some cases put in a water treatment system like I have. When they are willing to do that to no cost to the homeowner, no cost to the taxpayer, and for these people to go, 'uhh, I don't want that.' It's like, why don't you want good water? I don't understand that."

On October 21, members of Enough Already held a town meeting at the Elk Lake School auditorium. The featured speaker was Bob Watson, the emeritus professor of petroleum and natural gas engineering at Penn State, chairman of the Technical Advisory Board that advises the DEP, and co-author of industry-dubbed "Penn State report." Cabot had commissioned Watson (for $25,000) to review its operations in Dimock in light of the DEP order. After doing this, Watson produced a report that itemized the company's procedures at various wells identified in the consent order. The report concluded that Cabot "has used and is using procedures for drilling, casing, and cementing" that meet or exceed Pennsylvania requirements and industry standards and that no Cabot wells were "currently causing . . . methane migration."

The auditorium was filled with both local stakeholders, and drilling opponents and industry supporters who had traveled to the meeting from out of town—including some I recognized from the advocacy groups based in New York City representing interests of stakeholders in the Catskills. Watson, who prefers a colorful, offbeat approach to his lectures, introduced himself and explained his credentials as a teacher and academic. He then added, "I am a redneck. I take great pride in that. I am an environmental wacko. I take great pride in that. I am also anti-war. Politically, I believe I am registered as an agnostic." He drew some chuckles with that. He drew groans when he added, "I'm not Michael Vick. I don't have a dog in this fight." After the warm-up, consisting of several more stories and anecdotes, he got to the topic. Methane migration has been a long-standing problem in the state, he said. Much of it is related to natural occurrences, and some of it is related to abandoned wells. In World War II, he explained, some steel well pipes were pulled from the ground for material to support the war effort. Regarding the modern day Marcellus rush, he explained how the DEP "got caught absolutely flat footed by new developments" and added that the industry was far ahead of the regulations with its current practices. He summarized his findings that showed Cabot practices in Dimock to be above par, and then he shifted to the long-term prospects related to the industry in general. "There is every reason to believe the Utica Shale will be just as profitable as the Marcellus Shale," he said. "So if you are looking for jobs, you are looking at an industry that will be around here for a hundred years."

During the question and answer period that followed, moderated by Aileo, a showdown began between proponents and critics. The moderator said he didn't understand a questioner's reference to "the Halliburton Loophole. . . . I don't know what that is." Some of the drilling opponents, in frustration, began shouting answers from the audience. Aileo had warned the crowd that it had to be respectful, and with these outbursts, he ended the meeting. With this, the out-of-town protesters stood and began chanting, "Make Cabot pay!" Victoria, who wore a blue ribbon to convey her position but carried no sign, worked her way through the crowd to the panel and apologized for the outbursts. She respected the forum and appreciated the dialogue, she told them. The raucous did not represent the way she wanted to handle things.

PARTING SHOTS

John Hanger traveled from his office in Harrisburg to Dimock for the last time on November 4, 2010, two days after Tom Corbett was elected governor of Pennsylvania on a platform opposing taxes and regulatory burdens on industry. Hanger's SUV exited Interstate 81 at Wilkes-Barre and headed north through back roads threading the Endless Mountains. It passed tankers and flatbeds laden with pipes traveling to and from remote well sites scattered across the countryside and sometimes visible from the road—illuminated by klieg lights against the pale November twilight. The truck traffic caused no bottlenecks, at least not on this route at this time of day. Less than three hours after Hanger's party set out from the capital, shortly after nightfall, they approached Dimock on the route that ran along the Burdick Creek hollow, with the Teel and Lewis farms on the west side of the road and Ely's Mountain, the Switzer home, and the branch off to Carter Road on the east. They had traveled the last few miles in darkness, but now they found the section of road approaching the turnoff into the Switzers' driveway ablaze in artificial lighting—the same high-intensity stadium lights used at the drill rigs.

Directly opposite Switzers' home, a two-story makeshift billboard on the side of the Lewises' barn was illuminated by one of the lights. It had the words "Water Pipeline" inside a crossed out circle. The front yard sloped down toward the road and the Switzer property and Burdick Creek on the other side. About two dozen people, holding signs that said "No Water Line," had been lined up in the Lewis yard facing the Switzers' property for several hours as they waited for Hanger's arrival. On the other side of the road, Craig and Julie Sautner stood in the Switzers' driveway. Having caught wind of the protest, they had arrived early with a sign of their own: "Make Cabot Pay." Hanger's vehicle parked in the Switzers' driveway, and the secretary stepped out with an aide and Scott Perry, chief of the DEP Oil and Gas Bureau. Hanger crossed the road to address the protesters. He was followed by his staff and the Sautners, who tried to shoulder their way into a circle of people that closed around the DEP chief. There was some jostling, and somebody yelled, "I'm warning you, stay out of my face!"

Hanger spun around, looking for his staff. Perry, who has the build of a body guard, got between the drilling proponents and the Sautners, who retreated to the edge of the property. Some more protestors showed up. There were demands for people to leave, and an argument broke out on the roadside over what was public and what was private property.

"Look, I'm not going to participate if there is any violence," Hanger warned, raising his voice to the pipeline protesters, who in turn pointed at the Sautners, complaining that they were the instigators. Hanger, dressed smartly in a suit and tie, was tense with the expectation that the conflict would degenerate into a brawl. But he was determined to hear the protesters out and to have his say, as well. He listened to Ron Teel and others tell him the state had no right to conduct the business of a public water line out of the public eye.

The project would have to go through the appropriate channels, and those included more than the DEP and a collection of residents who were intent on "making Cabot pay." Hanger knew that complaint well. His answer by now was well rehearsed, but in the heat of the moment his voice was taut: "The case that Cabot has affected people's water in Dimock is overwhelming. We gave them two years to fix this problem. That's too long. These people need safe water …"

From her house, Victoria could not avoid seeing the lights from across the street, and the sign that Harold had built several weeks earlier was directly in her field of vision every time she left the driveway. Now she heard the shouts and commotion of the protesters while Jimmy stood outside the door and took stock of the situation. "I don't think you should go over there," Victoria warned Jimmy. "No good will come of that."

After responding to the protesters, Hanger walked down the Switzers' driveway with his staff and entered their home. "Just your neighbors, exercising their rights," he remarked to the group—the Carters, Norma Fiorentino, Pat Farnelli, and others. There was no need for elaboration. Hanger turned his attention to the latest analysis of DEP data, printed on sheets that Perry passed around the table, and explained the agency's assessment that their water source was permanently damaged by unsafe levels of methane.

Hanger knew the agency's order to install a water line from Montrose to Dimock would not likely hold up under Governor Corbett's administration, and the victory of the pro-drilling candidate had given Dinges leverage. In his final weeks, Hanger's staff accepted a compromise with Cabot. In exchange for the DEP dropping a lawsuit to enforce its order for the $11.8 million pipeline, Cabot would pay a total of $4.1 million to nineteen families with affected wells as determined by the DEP analysis (which included some families not involved in the lawsuit against Cabot). The settlement also required Cabot to install systems to fix the problem in each of the homes, and reimburse the DEP $500,000 for the investigation. Although Hanger believed the aquifer that fed the Carter Road wells was permanently ruined, the settlement allowed Cabot to remedy the problem with "whole house gas mitigation systems" on affected properties, in addition to compensating landowners with payments totaling twice the assessed value of their properties, with a minimum of $50,000.

By late December 2010, the executive suite of the DEP on the top floor of the Rachel Carson State Office building was mostly empty. With the exception of a secretary and receptionist, John Hanger's staff had been dismissed or had left, and he, too, was packing his boxes.

The 2010 elections, in addition to influencing the outcome of regulatory policy in Pennsylvania, had fortified conservative Republican representation in Congress. With control of the House of Representatives, Republicans would be in a strong position to challenge the Obama administration on regulatory and tax issues. Days after the election, Karl Rove, the Republican strategist and former adviser to President George W. Bush, addressed an audience at an energy industry conference at the David L. Lawrence Convention Center in Pittsburgh, while several hundred anti-drilling protesters marched outside. The Republican victories would curb legislative threats to the drilling industry and prevent any efforts to bring high-volume hydraulic fracturing under federal control, Rove said. "Climate is gone," he added, referring to efforts to legislate impacts on global warming.

In light of these events, I thought Hanger, now leaving office and fresh from the fight, would be in a good position to reflect candidly on a state's regulatory effectiveness over industry. As we sat at a table in his corner office overlooking the Susquehanna River, I asked him how much discretion the DEP had in enforcing environmental laws. He replied that environmental issues, as with finance regulation, tend to reflect the priorities of any given administration. "The state police doesn't approach its traffic enforcement any differently when the governor changes. That's not the case with drilling. I wish I could tell you laws are professionally and independently enforced. There is a real difference with each administration."

I asked him to assess the most significant and most inflated dangers associated with shale gas extraction. The disclosure of fracking formulas, he said, was a secondary issue to methane migration. Regulatory changes under the Rendell administration to upgrade well casing standards and improve accountability "should have been done a decade ago," he added. More changes are necessary to prevent long-term problems, including legislation that gives the Pennsylvania DEP more enforcement power. "Right now the department really has questionable authority to deny permits," Hanger explained. In his view, methane migration, in particular, remained a threat, especially from wells that companies abandon after they become unprofitable. As with the legacy left by abandoned coalmines, 1,600 oil and gas wells in Pennsylvania are potential contributors to long-term problems because it's too easy for companies to walk away from them. "There's no money to pay for plugging them because we didn't require the gas companies decades ago to post a reasonable bond," Hanger said. "The industry has resisted any changes to that law."

I asked Hanger where natural gas fit into his vision of green energy. It was better than coal, he said, the mining of which destroys mountaintops and the burning of which produces CO_2 gasses and mercury that accumulates in the food chain. In his view, a surge in the development and demand for alternatives, including solar panels, is imminent, due mostly to China's burgeoning interest in the industry. "The baby has been born," he said. "China is intent on building solar panels for the world. Those changes will come and those changes will be rapid—less than five years."

Despite his battle with Cabot and call for more accountability for abandoned wells, Hanger remained a supporter of natural gas development. People need to accept, he added, that all development, extraction industries in particular, come with an environmental impact. "You need to walk the walk," he said. "If you are not willing to have gas drilling, please don't go home to your gas stove or your gas heat."

As I left his office, I wondered how close we really were to fueling our cities with solar panels. For the length of my three-hour drive to Dimock, I saw capital investment everywhere—drill rigs, trucks, and pipelines—that suggested, not very. Eventually, I passed over Meshoppen Creek and passed the road to Ken Ely's Mountain. I turned onto Route 3023 and drove by the Lewis property, which still had the No Pipeline sign facing across the street at the Switzers'. I took the branch on to Carter Road. It was cold and dark, and the Carters' porch was vacant, illuminated faintly by light from the kitchen window. It was suppertime, so I decided not to stop. When I had last spoken with Ron that fall, he was distressed by accusatory letters to the editor directed at the Carter Road plaintiffs and talk that they had fabricated or exaggerated problems. Moreover, he appeared worn down from the controversy and media exposure that, like gas development in Dimock, seemed endless.

I drove west to New Milford and merged on to the interstate heading north. With no view beyond the headlights, the road rose and fell and rose again, and within minutes I was crossing into New York State, where the outcome over the Marcellus controversy was yet to be decided.

7. SUPERIOR FORCES

Dewey Decker had no regrets about becoming a gas millionaire, except for maybe that he missed his cows. After signing over his mineral rights to XTO for $2.77 million, the leader of the Deposit Coalition in New York sold off the dairy herd and replaced them with beef cattle, which required less work and suited his idea of retirement. But unlike the docile Holsteins that were accustomed to intimate handling on a daily basis, beef cattle tend to be independent, ornery, and "just as happy to run you right over as not," he informed me. Sentimental differences aside, Dewey's love for his livestock was tied to a larger set of principles: Whether dairy or beef, his cattle were part of the country's gross domestic product that would increase in value under his stewardship. To him, agriculture, like mineral extraction, timbering, and manufacturing, was a fundamental element of national independence. In Decker's view, the United States could operate these industries more responsibly and effectively than the nations from which it imported goods in such great quantities, as he liked to remind me when I came to see him in my Toyota hatchback. "Nobody wants to have water wells blow up," he said. "But compare gas production in this country to Russia or China, and what they do to the poor people over there."

I spoke with Dewey in August 2011, more than three years since he leased the mineral rights on his 1,150 acres in eastern Broome County—and there wasn't a drilling rig in sight. The stake marking the site of a future well pad on his land was obscured by a tangle of wild berry bushes and milkweed. Dewey had his money, but he wanted more production from his land—that, and the 15 percent royalties that he and his neighbors would gain after the crews began fracking and the gas began flowing. He wanted the coalitions that had formed throughout the Southern Tier, led by Dan Fitzsimmons and Jim Worden to get their money in exchange for their energy. Dewey was ready for the Marcellus era to arrive and transform New York.

REVISING THE SGEIS

On a sunny Saturday in early July 2011, Dewey opened his mailbox and found a three-page letter from XTO. The first two pages recounted New York Governor David Paterson's executive order of July 2008 that mandated a review process to update the state's permitting guidelines, and the resulting "de facto moratorium" on high-volume hydraulic fracturing. XTO could technically attempt to bypass this hurdle, the letter explained, by undergoing a "site-specific review" available for companies seeking permits outside the generic application process. This step, however, "would require the same public comment and technical analysis as the supplemental GEIS and is thus pointless," the letter stated. "In other words, at least until the DEC's process is completed, the development of horizontal shale wells utilizing high volume hydraulic fracturing is at a standstill." The regulatory uncertainty that was preventing permits constituted a "force majeure event," the letter concluded. *Force majeure* is a French term meaning "superior force." and pertains to an unforeseen and uncontrollable event such as a natural disaster, act of God, or a labor strike that prevents parties from fulfilling contracts. A force majeure clause is standard in gas industry leases and invoking it is a strategy for companies attempting to extend leases, without additional compensation to landowners. Therefore, Dewey's lease would be extended "so long as XTO is prevented from conducting drilling or reworking operations on or from producing oil or gas" from his property. The letter in Dewey's hands was one of thousands of force majeure letters being sent to New York landowners, including those who had leased mineral rights prior to the shale gas boom to companies pursuing pay-zones that could have been developed with conventional methods.

Dewey was all too aware of the events that had led to the standstill and the invocation of force majeure. With Dimock as a high-profile example just across the state line, a significant part of the New York public had become suspicious of the immediate impact of fracking, the cumulative impact of the waste it produced, and the degree of transparency of the entire process. By encouraging large showings at public hearings and a logjam of criticism through the public comment process, drilling opposition had been able to parlay the SGEIS review into a kind of referendum over the industry's near-term future in New York. Not surprisingly, there were conflicts within the Paterson administration regarding how to proceed.

According to insiders, Larry Schwartz, secretary and top advisor to Paterson, was pushing DEC Commissioner Pete Grannis and Deputy Commissioner Stuart Gruskin to resolve permitting issues so shale gas drilling could begin. Schwartz saw shale gas as an essential economic asset for upstate counties in light of the Deposit Deal with XTO and Barb Fiala's advocacy in Broome County; and he pushed Grannis to expedite the review. Grannis pushed back, complaining that Schwartz was micromanaging his agency. Tensions between the two men came to a head in October 2010, when Grannis wrote a memo to the Budget Office detailing the weakness of the DEC as it faced the shale gas expansion and other challenges. The budget called for cutting 209 more DEC jobs by the end of 2010, leaving the agency with 2,926 staffers, a 23 percent reduction from the 2008 level. The agency was "in the weakest position that it has been since it was created 40 years ago," the memo warned. "The public would be shocked to learn how thin we are in many areas." The correspondence was leaked to the *Albany Times Union*, where it was featured in a front-page article on October 19. Two days later, Schwartz notified Grannis via e-mail that he was fired, "Either you co-operate with regards to your resignation or a release will go out saying you have resigned by the end of the day. . . . All the calls that you are having people make (to the governor) are not going to change the decision. You can either do this in a cooperative fashion or a hostile fashion. That is up to you." Grannis chose to be fired rather than resign, and in subsequent interviews he criticized the direction of the administration under Schwartz's influence and publically referred to his adversary as a "hatchet man" and "thug."

When Andrew Cuomo succeeded Paterson as governor two months later, Cuomo hired Joseph Martens as commissioner. Martens was formerly president of the Open Space Institute, an agency dedicated to land preservation, and his appointment to head the DEC was lauded by the same environmental groups that had condemned the firing of Grannis. At the same time, in what could be best explained as an effort to balance political interests and personalities in his inner circle, Cuomo kept Schwartz as a senior level advisor, and later promoted him to secretary, the top advisory position he held under Paterson.

In contrast to Paterson, Cuomo began publically pushing the process along. The governor set July 1, 2011, as a firm deadline for the release of a revised SGEIS, which had grown from 1,000 to 1,500 pages. The executive summary of the SGEIS noted that, since the first draft of the document a year and a half prior, the DEC had gained "a more detailed understanding of the potential impacts" of shale gas development from events in Pennsylvania, reports and studies, and the public's response. The new guidelines would limit open pits, increase casing requirements, and regulate water withdrawals and waste disposal. One of the most significant changes, though, was that shale gas development would be banned in unfiltered watersheds that feed large populations, including New York City and Syracuse, along with "primary aquifers" in various other areas. That decision would keep about 15 percent of the Marcellus in New York off limits. It would also open the door for lawsuits from affected property owners who claimed that the ban infringed on their rights to develop their land. While there was immediate talk of the legal ramifications of such a ruling, it couldn't be tested in court until permits were issued (or denied); and permits wouldn't be issued until the SGEIS was complete and final. That would require another round of public hearings, and environmental groups were not going to make it any easier this time.

Walter Hang characterized the DEC's ban of high-volume fracking in various watersheds as "an admission that the practice is simply unsafe" and he advised his considerable following on the Internet to "get ready for action. . . . A decision has clearly been made to allow Marcellus Shale horizontal hydrofracturing to begin in New York. That effort can and must be stopped. Nothing else matters." Hang's demand for a complete regulatory overhaul dealing with all gas wells, rather than a supplement to existing permitting guidelines, was a stipulation embraced by a broader environmental movement in New York. Deborah Goldberg, an attorney for Earthjustice, told me advocates were ready to sue the DEC if the agency began issuing permits for high volume fracking before the rules were codified in regulations, rather than approved simply as permitting guidelines per the SGEIS review. "Our concern is this is a very powerful industry and the DEC is a very understaffed agency," she said. "They will always be trying to negotiate waivers, deals, and special exemptions that are impossible for the environmental community to monitor if they have to review every permit and look for every waiver that might be granted based on 'special conditions' that might apply. Regulations are more transparent and enforceable." Such a strategy, if successful, would also be time consuming, perhaps delaying shale gas development in New York for several more years.

Dewey was not one to proselytize, but he didn't mind sharing his view with people who inquired: All the resistance was not only unnecessary, but unpatriotic. There were many cases in Pennsylvania and elsewhere where drilling turned out just fine and benefited communities. In other cases, naturally occurring methane was the cause of contaminated water, not drilling. "These environmental groups won't be happy until there's no drilling at all. If we were more like Pennsylvania, and let them come in here with less restrictions, everybody would be leased," he said. More regulation would only result in delay—it would not contribute to safety and it certainly would not help landowners with mineral rights under lease.

The industry, likewise, characterized New York's revised SGEIS as restrictive and discouraging. The policy proposal, according to Brad Gill of IOGA New York, represented "unjustified, excessive and inequitable rules, regulations, requirements, mitigation measures, permit conditions and access restrictions" that would make New York uncompetitive with other states in attracting shale gas development.

As the complexities and conflicts of shale gas development grew, New York and Pennsylvania each formed panels to advise policymakers. In his first term, Corbett charged the Marcellus Shale Advisory Commission to "develop a comprehensive, strategic proposal for the responsible and environmentally sound development of Marcellus Shale." The thirty-member panel, headed by Lieutenant Governor Jim Cawley, included heads of the Departments of Environmental Protection, Agriculture, Conservation and Natural Resources, Transportation, and Corbett's top energy executive, Patrick Henderson. Although local governments and environmental interests were represented, about a third of the panel's members represented drilling companies, a ratio that drew fire from both drilling opponents and Corbett's political rivals. In mid-July, the commission issued ninety-six policy recommendations, ranging from streamlining permitting operations to creating an impact fee (as opposed to a flat tax) to direct a share of drilling revenues to the communities most affected by operations. (Corbett used the threat of a veto to stymie a previous legislative push for such a fee.) The impact the report had on Corbett's agenda, if any, wasn't immediately clear. Months after the recommendations were issued in July 2011, the governor remained uncommitted to any of them. "I'm looking to see that which I agree with and there's maybe one or two things I don't agree with, and things I don't know if I agree with or not that I have to get some more information on," he told one reporter.

To help guide the state policies in New York, DEC Commissioner Martens formed the High-Volume Hydraulic Fracturing Advisory Panel in July 2011. The eighteen-member panel, selected by Martens, initially included Senator Tom Libous, Assemblywoman Donna Lupardo, and Robert F. Kennedy, Jr. As chief lawyer for the Natural Resources Defense Council, Kennedy was once an icon of mainstream environmental support for natural gas as a bridge fuel away from coal to clean burning alternatives. Counter to Victoria Switzer's expectation, however, he changed his views after visiting Dimock in light of what he characterized as the industry's "reckless conduct," concerns over underrated greenhouse gas impacts associated with shale gas extraction, and the amount of capital that shale gas development would divert from sustainable energy endeavors. The group was rounded out by representatives of academic and business communities, with two members representing industry: Brad Gill, of IOGA NY, and Mark K. Boling, executive vice president and general counsel for Southwestern Energy, a Houston company that had been working with the Environmental Defense Fund on a set of model standards for safe drilling that could be used as a guide for state governments. The panel was asked to provide guidance on staffing levels and resources needed to monitor and enforce hydrofracturing operations and manage impacts to local governments and communities. At one of its first meetings, panel members reviewed an internal DEC report that showed the agency would need an additional 226 staff members—roughly the same number lost in the last round of cuts under Paterson. Adequate staffing would require $25.3 million in salaries, non-personnel costs, and equipment.

While New York State was charting its direction, municipal governments representing largely anti-drilling citizen bases, like those in Tompkins County, began to take matters into their own hands by listing fracking and drilling as incompatible with land use plans. That approach, known as "home rule," would soon lead to a landmark legal challenge to the state's authority to site gas wells in towns that did not want them. The controversy reinvigorated town hall debates as local governments held their own hearings about the merits of gas development. By the summer of that year, when Dewey received the letter from XTO blaming the state for stopping shale gas development on his property, at least eleven New York towns had joined Buffalo in enacting fracking bans, and seven New York counties banned fracking on county land. Another twenty towns either had bans under formal consideration or strong movements pushing for them.

Sanford was decidedly not one of those towns. Dewey and the town board, eager to do what they could to support drilling efforts, passed resolutions urging state authorities to back off regulations. One resolution asked the DEC not to limit drilling in the New York City watershed; another urged the Public Service Commission to approve an application for gathering lines and a compressor station connection to the Millennium Pipeline within the town borders to encourage companies "to commence drilling in the town rather than more distant locations." One drilling proponent, who with her husband had used the XTO gas money for a beagle kennel and a new barn with an addition to house falcons, told me she would "personally help crews dig the first six feet" of a well on her property. This pro-fracking position appeared to represent the majority in the Town of Sanford, but I had also spoken to Sanford town residents who feared the consequences of drilling and who felt powerless as they were in the minority both in terms of land holdings and political representation.

The de facto moratorium and the gas companies' strategies to turn it to their advantage were a few of many forces that would collectively shape the outcome of shale gas development in New York. Many of these issues—the right of property owners to develop their mineral resources in areas where drilling was banned, the right of local governments to prohibit drilling, and the right of people to be protected from its hazards through regulations— would play out in courts on a schedule independent of political and bureaucratic calendars. Others, involving taxes, fees, and enforcement, would be decided by elected officials and bureaucrats. The relevance of all these influences, however, depended on a single overriding force: The current and future demand for gas.

CONTESTING THE FLOW

The centerpiece of the natural gas industry's display at the 2011 Chemung County Fair was a 575-horsepower Chevy Camaro, gunmetal grey with cobalt blue stripes. It was parked amid booths set up by IOGA NY and Chesapeake Energy at Gate 2, near the midway. Above the car there was a banner which read: "A Natural Gas Powered Muscle Car: A Modern Fuel for an American Classic." Ashur Terwilliger, the Chemung County Farm Bureau president and a director of the fair, circled the vehicle and whistled approvingly. Ashur had arranged the display with the American Natural Gas Alliance, a trade group, and he liked what he was seeing. According to the specifications, the car could, from a standstill, knock off a quarter mile in less than twelve seconds, reaching 119.5 miles per hour in the process. It was fueled by compressed natural gas (CNG) stored in tanks near the back of the trunk. "I tell you what," Ashur said. "I wish to hell we had a plant here to make this stuff. Why shouldn't farmers want to run their tractors off gas from under their own land? By God, I'm sure interested in that."

In terms of greenhouse gases, particulate matter, mercury, and volatile organic compounds, natural gas vehicles have the cleanest tailpipe emissions of all fossil fuels. CNG cost over 40 percent less than diesel fuel on an energy equivalent basis and was expected to cost 50 percent less by 2035. Despite these advantages, the idea of CNG vehicles remained a foreign one at the advent of the shale boom in the United States. In December 2010, there were about 12.6 million natural gas vehicles in the world, more than half of them in Pakistan, Iran, Argentina, Brazil, and India. The United States, a place with more vehicles per capita than anywhere in the world, had only about 112,000 CNG vehicles. Compressed natural gas was cheap compared to gasoline or diesel, but CNG vehicles—lacking a mass market—were more expensive than conventional vehicles to produce. Their biggest limitation, though, was fueling. There were fewer than 1,000 CNG fueling stations in the United States, and only about half of those were public—a scarcity that tended to discourage CNG vehicle development beyond a niche domain for transit fleets and commercial trucks commonly run on diesel. These vehicles were easy targets for air quality improvements, operated within a limited geography, and were refueled at a central location.

The answer to Ashur's question about running tractors on natural gas went well beyond the technical merits and drawbacks of any particular fuel, however. It involved influential people looking for competitive advantages in global energy markets. In early July 2011, Chesapeake Energy President Aubrey McClendon announced to shareholders that the Utica shale—sitting below the Marcellus with a footprint extending even further into upstate New York—was nothing less than one of the largest hydrocarbon finds in U.S. history. Based on fifteen test wells and "geological and proprietary data," Chesapeake determined that, in addition to methane in the eastern part of the play, the Utica is rich with natural gas liquids (NGLs) further west.

These include ethane, propane, and butane, which are all valuable both as fuels and petrochemical feedstock for an array of manufacturing processes and products, including packaging, textiles, fertilizer, coatings, and adhesives. The westernmost part of the formation, protruding into Ohio, also holds oil reserves that, like natural gas and NGL, could be produced only by high-volume fracking. All told, according to McClendon's assessment, the Utica was capable of producing 25 billion barrels of oil-equivalent energy (almost ten times the amount exported annually from Saudi Arabia) and the company was looking for partners to develop it. Chesapeake had "quietly and efficiently" built the largest leasehold in the play—1.25 million net acres—that McClendon estimated would immediately increase the company's worth to shareholders by $15 billion to $20 billion. The announcement, in late July 2011, was followed by exuberant forecasts with a familiar ring. Ohio Governor John R. Kasich called the discovery the "beginning of a new and extended positive chapter in Ohio's economy," with a focus on both "job creation and environmental stewardship."

At that time, investors from U.S. and abroad were banking on U.S. shale, and judging by the size and experience of the companies involved, it looked like the smart money was on shale gas plays across the country. The purchase of promising sections of the Marcellus by Exxon Mobil, the world's largest energy company, was followed by similar buy-ins from other global giants, including Royal Dutch Shell, based in Britain, and Reliance Industries, based in India. And to Dewey Decker's chagrin, China National Offshore Oil Corporation (Cnooc), one of the three big Chinese state-run oil companies, bought a one-third interest in Chesapeake's shale projects in Colorado and Wyoming. China was the only country with shale reserves estimated to be bigger than that of the United States; and the Chinese government—several decades behind the United States in terms of developing the capital, infrastructure, and technology to produce it—was reported to be gaining experience in the U.S. oil patch. These companies joined others that were already well invested, including: Range Resources; Cabot; Chesapeake; Chesapeake's Norwegian partner, Statoil ASA (formerly StatoilHydro); Norse, also of Norway; and Talisman of Canada. An industry survey in late 2010 calculated that, since 2008, investors had spent $16.5 billion on the Marcellus alone, with expectations of another $11 billion investment in 2011. Now, according to McClendon's forecast, the industry could be expected to spend $10 billion a year for twenty years to drill 25,000 wells into the Utica. All told, investments in shale gas underlying the mid-Atlantic and northeastern regions would reach well into the hundreds of billions of dollars as the Marcellus play matured, and the Utica began, with hundreds of billions also being invested in other U.S. shale plays.

This international rush into America's shale boom was accompanied by renewed questions about its economic sustainability. In August 2011 readers of *Bloomberg* were informed that the federal government "will slash its estimate of undiscovered Marcellus Shale natural gas by as much as 80 percent after an updated assessment by government geologists." The assessment came from the USGS, which estimated the Marcellus to hold 84 trillion cubic feet rather than the 410 trillion cubic feet estimated by the U.S. Energy Information Administration (EIA). Production from initial wells coming on line through 2011 remained strong, but the revised numbers underscored talk that Marcellus wells might taper faster than expected.

Uncertainty over the longevity and, consequently, the value of shale gas reserves was part of a growing set of public relations problems for the industry. A week prior to the USGS announcement, New York Attorney General Eric Schneiderman began an investigation into whether companies accurately represented the profitability of their wells in the Marcellus. Range Resources, Cabot, and Goodrich Petroleum Corporation received subpoenas for documents detailing formulas used to project how long the wells can produce gas without additional hydraulic fracturing. The Securities and Exchange Commission was also investigating how energy companies calculate and publicly disclose the performance of their shale gas wells.

Both industry analysts and government officials downplayed the significance of disparities in federal agencies' calculations of Marcellus reserves. Philip Budzik, an operations research analyst with the EIA, told *Bloomberg* that the differences showed, at most, that there was still relatively little comprehensive information to draw on regarding estimates. "Layer on that the fact that over the next 30 years, the technology is going to evolve," Budzik said. "It will have an impact on our projections. We all need to keep a sense of proportion here." The same report offered this perspective from Kevin Book, managing director at ClearView Energy Partners, a policy analysis firm: "One fifth of a big number is still a big number. . . . It shouldn't tell you anything about your conclusions. It should tell you what you need to know about estimates: They get revised." The proof, as more than one industry representative told me, ultimately would play out with drilling results; and as of 2011, they were looking good. So good, in fact, that a market glut was inevitable without an easing of production or a surge in demand. With all the capital in play, an easing of production was not a likely option. In other words, a big push to produce America's shale gas would necessitate a big push to sell it.

In the summer of 2011, as the CNG Camaro made its way to various fairs in Pennsylvania and New York, McClendon, working with other industry heads, outlined plans for such a market push. Chesapeake would divert 1 to 2 percent of its forecasted annual drilling budget away from production and put the money toward projects to stimulate demand. This amounted to some $1 billion over ten years to support infrastructure, such as fueling stations for natural gas-powered vehicles, in order to, in the words used by the company's press office, "reach a tipping point" where manufacturers "will have sufficient confidence to increase their production" of natural gas–powered vehicles. The investment would also encourage other "end-use technologies" including the conversion of methane into a liquid fuel that could be blended with diesel or gasoline, or even as a replacement product for both of those fuels. To provide a model, the company was converting its own fleet of 5,000 trucks to run on natural gas.

Chesapeake had some powerful allies motivated by similar goals, including the influential energy baron, T. Boone Pickens. Pickens had achieved both fortune and fame as an oil wildcatter and later as founder of Mesa Petroleum, which had grown into one of the world's largest independent oil companies. As the shale gas boom became established in Texas, Pickens also formed Clean Energy Fuels, a company that owns and operates natural-gas filling stations and production plants to supply them. When Marcellus prospecting heated up in July 2008, the eighty-year-old Pickens began a campaign to "wean the country off foreign oil" through aggressive development of natural gas. Pickens, by his own account, spent $82 million over three years promoting these initiatives in "The Pickens Plan." The plan was delivered to mainstream America through television, newspaper, Internet, and magazine news coverage and advertising that touted natural gas as clean, abundant, and cheap. America was depicted as the "Saudi Arabia of natural gas."

By 2011, a manifestation of the Pickens Plan found support in Congress. House Resolution 1380, commonly known as the Natural Gas Act, would provide a series of tax breaks over a period of five years to trucking companies, vehicle owners, vehicle manufacturers, and fueling station owners. These tax breaks were geared to encourage the transition from gasoline and diesel to natural gas. That summer, 108 Democrats and 75 Republicans signed onto the bill, which also received encouraging signals from the Obama administration. Meanwhile, President Obama unveiled an energy plan of his own that summer—Blueprint for a Secure Energy Future—which recognized the importance of shale gas development, including a component to support global efforts to displace oil with natural gas.

HR 1380 looked like a legislative winner, until it was knocked off track by an opposing political force, nurtured in conservative circles and backed by another tycoon with a fortune staked to energy: Charles Koch. Koch and his brother, David, control Koch Industries, a $100 billion privately held conglomerate with major holdings in cattle, timber, and oil. The company is also vested in petrochemical companies, the profitability of which depends on cheap natural gas supplies. The Pickens Plan posed dual threats to Koch. First, it encouraged higher prices with the stimulation of natural gas demand and, second, it encroached on the oil market on which Koch-owned refineries and pipelines depended. Charles Koch went on the attack, casting HR 1380 as an unfair and unhealthy meddling with free markets, a "well-intentioned but misguided suggestion . . . promoted, in large part, by those seeking to profit politically, rather than by competing in a market where consumers vote with their wallets." Koch articulated a view supported by editorial writers of oil trade and financial market publications, and conservative political action groups, including the American Conservative Union, and Americans for Prosperity (a group cofounded by David Koch).

The Koch brothers were not Pickens' only commercial antagonists. Using natural gas as both a fuel and feedstock, Dow Chemical Company had a hand in manufacturing 3,300 different products, from paints to personal care items, and the company stood to lose if government-subsidized initiatives encroached on its supply of natural gas. Petrochemical engineering is an energy-intensive business. Dow alone uses the oil equivalent of 850,000 barrels per day (approximately the daily energy use of Australia) mostly in the form of naphtha, natural gas, and natural gas liquids. As HR 1380 circulated in Washington offices, Dow presented its own position paper to the Committee on Energy and Natural Resources in the U.S. Senate. It flatly summarized the downside of attempts to reduce dependency on foreign oil and coal: "The potential exists for demand [of natural gas] to outstrip supply, assuming that fuel switching from coal to gas continues to accelerate and factoring in the proposals by some to displace 25 percent of our oil imports with natural gas." While Pickens's pitch relied heavily on the foreign oil-dependency theme, the petrochemical industry countered with the promise of jobs, a promise that never lost its appeal given the economic problems of the day. Access to the vast shale gas reserves that made available abundant cheap feedstock and fuel to supply the petrochemical industry, the Dow report concluded, "will provide an opportunity for more than 400,000 jobs—good jobs." This was the future promised by the natural gas boom, "barring ill-conceived policies that restrict access to this supply."

The fight pitting Pickens and Chesapeake against the Koch brothers and Dow Chemical was, in fact, a fight over the rate and flow of shale gas, and it had everything to do with the question posed by Ashur: "Why shouldn't farmers want to run their tractors off gas from under their own land?" Chesapeake and Pickens, in the supply side of the business, stood to gain from a broader market. Moreover, Dow and Koch, on the consumptive side of the equation, stood to lose.

A MATTER OF FAITH

Market forces of supply and demand for natural gas, dynamic and unpredictable, had everything to do with the physical landscapes over shale gas reserves and the wellbeing of their inhabitants. Whether full-scale shale gas production would be boom or bust was the question of the day and possibly of the century, and speculation over the answer continued to drive public opinion and media coverage. On January 14, 2011, Terry Engelder, the geologist and researcher from Penn State, and Tony Ingraffea, the Cornell engineer and fracturing mechanics scholar, drove to LaPorte, Pennsylvania, to debate the issue in a public forum. The event, at a midway point between Cornell and Penn State, had grown out of a challenge by Ingraffea to Engelder and others who had criticized the Cornell professor's public attacks on the industry. At the time, Ingraffea and Robert Howarth, a Cornell ecologist and biochemist, were finishing a paper challenging the hypothesis that natural gas was a clean alternative to coal. Their peer-reviewed methodology took into account something that was generally unaccounted for: fugitive emissions from methane released from shale formations during hydraulic fracturing. Methane is a potent greenhouse gas, and with this factored into the equation, Howarth and Ingraffea found shale gas extraction was a greater global warming threat than coal mining. Engelder, who had seen the preliminary results in a white paper written by Howarth, characterized Howarth's finding as "disingenuous" and "simply not true" in an opinion piece titled a "Gusher of Hogwash," published in the *Philadelphia Inquirer* in spring 2010.

Terry began with a lecture spiced with the folksy humor and facetious references to popular culture that made him popular with students. Unlike most of his presentations, however, the one in LaPorte was delivered to an audience that was skeptical of his message and eager for an accounting of problems. Accordingly, much of his talk was devoted to explaining what happened in Dimock, and why these problems represented growing pains both inevitable and necessary for the betterment of the country. He alluded to John F. Kennedy's call to "Ask not what your country can do for you—ask what you can do for your country"—and added this: "The people in Dimock have already done that, in spades. It's their sacrifice, hopefully, that is necessary if the gas industry continues to evolve and help other people." Within this framework, Engelder wove empirical information relating to economics, geology, well casing schematics, and his interpretation of a federal EPA study of pollution near gas wells on the Wind River Indian Reservation in Pavillion, Wyoming. In Terry's opinion, the problem was probably not caused by hydraulic fracturing, but leaks from surface impoundments were suspect. He returned often to the main theme: the importance of "doing it right," and his expectations that the industry would eventually get it right, despite early mistakes. "It takes the industry a considerable length of time to go through the process of learning how to do it right. Clearly Dimock was a case study of how not to do it. It's presented some challenges that we learned from and hopefully the industry will move forward from that." He was willing to stake his reputation on it. He showed a slide of a tombstone that read "Here lies Terry Engelder. He was full of hogwash." "With this slide, I've laid down the gauntlet to industry," he explained. "That is, I would hope the industry finds the leaks and fixes them, otherwise this will be my legacy."

Ingraffea then took the stage and said he would not be speaking long because he would rather use his time for a direct and spontaneous debate with Engelder. Using Terry's estimates as a starting point, Ingraffea calculated that recovering nearly 500 trillion cubic feet of gas from counties overlying the Marcellus in Pennsylvania, New York, Maryland, Ohio, and West Virginia would require 400,000 wells—170,000 in Pennsylvania. For the purposes of the discussion, he used a more conservative estimate for Pennsylvania: 60,000 wells. At the time, just over 2,000 wells had been drilled, and less than a third of them fracked. All the problems Pennsylvania had been experiencing were but a hint of what was to come, Ingraffea explained. "You're only 800 frack jobs into a 60,000 frack job experience!" Ingraffea hit the remote and another image filled the large screen above the stage. It was an overhead view of Taughannock Gorge in Ithaca, New York, where Terry had taken me the summer before to explain, in the context of Pearl Sheldon's work, his revelations that the cracks in the rock were themselves a product of gas. Now Tony, citing Terry's work, was making the same point: Devonian shale formations are typically highly fractured and rich with gas. What could be seen at various outcrops represented hundreds of thousands of square miles of gas-bearing rock stretching under the northeastern states. When cleaved by hydraulic forces at any given point, the natural system of cracks provided an optimal "flow-network" of gas, not all of which could be captured by man-made systems. "That brings me to another point, about scale," Tony continued. "In conventional wells you poke holes in the ground and hope you hit the place where gas is. With shales, the gas is everywhere. There is variability from well to well, but if you drill, you're going to get gas. But now you have to ask the questions, if gas is everywhere, where should the industry drill . . . ? Everywhere." Even if the industry could someday "get it right" and eliminate all spills and leaks, Ingraffea continued, the global impact from burning nonrenewable energy would be a net loss. The idea of using natural gas as a bridge to some brighter sustainable energy future was a carefully crafted industry myth. "When are we going to stop kicking the can down the road to our kids and grandkids and suck it up, and solve the damn problem now?"

After their presentations, the two scholars stood next to each other in front of the stage as people lined up at the podium in the aisle. Early in his presentation Terry described himself and Tony as "identical twins" in their research interests as applied to gas production. Their physical appearances and demeanors, however, differed as much as their viewpoints. Terry, striving for a conciliatory approach delivered through a mix of self-deprecation and humor, wore blue jeans, hiking shoes, and a Nordic sweater with a blue ribbon—presented to him by Victoria—pinned to his chest in sympathy to the Dimock residents. Tony, dressed in his customary blue blazer and paisley tie, had been forceful and blunt.

Now they considered a question from a young woman in glasses and a T-shirt that advertised an upcoming "Walk for Water" demonstration. It was a simple question that cut to the heart of the shale gas debate: "What do you base your beliefs on?" The question was directed to both Ingraffea and Engelder. Both men had grounded their views in science and arrived at contrasting beliefs.

"A belief is something we can't even talk to one another about because our perception of this is very different," Terry began. He paused and looked over at Tony, who stared at the floor fixedly, his chin in his hand. Terry reached toward the shoulder of his former research partner with a little jab, but stopped short as Tony continued to look down. Terry turned back to the questioner and continued, "What we hope—I think this is fair—by the two of us appearing together, the industry is goaded into action and if the industry is not able to, so to speak, clean its act up then the penalty they pay I suspect is a much stronger set of regulations at the federal level. And we've debated a great deal about whether or not the state regulatory agencies are capable of handling the job. I would hope so, but that has yet to play out."

Ingraffea accepted the microphone from Terry. "This is not a religion to me," he said. "From a science point of view, on a global scale, I think that going down the road of burning much more gas, most of it from unconventional sources, without a concomitant decrease in burning fossil fuel from other sources, is taking us where nobody in this room wants to go."

Victoria Switzer, four rows back and within plain view of both presenters, was tired of hearing Dimock and its people being labeled as a learning experience. As she listened to Terry's talk, she saw the essence of the problem more clearly than ever. Industry was messing up and its supporters were taking it on blind faith that it would get things right, and they expected everybody else to believe it. It was a faith she once shared when she organized community meetings with elected officials, but any shred of trust Switzer had in the industry vanished when Cabot CEO Dan Dinges accused the Pennsylvania DEP of fabricating the story of Norma's well as part of a smear campaign against the company.

RALLYING CRIES

The number of faulty well casings cited by the DEP continued to climb in 2011, even after Pennsylvania's new standards took effect in February. During the first nine months, the department issued eighty-nine citations for faulty casing and cementing practices—seven more than for all of 2010—at Marcellus wells throughout Pennsylvania operated by Range Resources, Cabot, Chesapeake, Chief Oil and Gas, Hess, Exco Resources, Williams Production, and XTO Energy.

The industry continued to characterize problems as irrelevant and exaggerated while opponents held them up as harbingers of disastrous consequences to come. Politicization of the issues intensified as the 2012 election year approached.

On September 7, 2011, about 1,500 industry leaders and proponents gathered at the Philadelphia Convention Center for Shale Gas Insights, a two-day conference headlined by Aubrey McClendon, the Chesapeake CEO; Tom Ridge, the former governor of Pennsylvania and Homeland Security secretary, who was completing his one-year $900,000 contract as an industry spokesman; Pennsylvania governor Tom Corbett; and former governor Ed Rendell. At the same time, a like number of activists gathered in the streets of Philadelphia's Center City for a two-day event of their own, dubbed Shale Gas Outrage, featuring protests, rallies, marches, concerts, and demonstrations. Activists chanting "It's our water, it's our right" marched through Center City toward the industry conference at the convention center, waving signs and toting banners: "We All Live Down Stream," "They Said It Was Safe in the Gulf, Too," "Fight Back Now."

Inside the convention center, Ridge had kicked off the conference that morning with a speech equating shale gas development with national security. He warned about the dangers of dependence on oil from politically unstable countries, and called to make natural gas a featured part of the country's energy policy. As McClendon began his lunchtime speech, a parade of protesters began arriving in front of the convention center windows. The Chesapeake CEO lampooned the crowd as "eco-extremist," spreading a message of "unfettered fear mongering" that was the enemy of progress. "What a glorious vision of the future: It's cold, it's dark and we're all hungry."

Chanting protesters continued to fill the block in front the convention center, facing a stage with large speakers. Josh Fox, who had recently won an Emmy for *Gasland,* took the stage and addressed the cheering crowd. "We are running out of fossil fuels," he announced. "Instead of doing the logical thing, the rational thing, the scientific thing, the civilized thing, which is to move our society to renewable energy as fast as possible . . . the energy industry has decided to go insane . . ."

Fox could not be heard inside the convention center, where McClendon told his audience that concerns over methane migration were overblown to begin with and now irrelevant in light of improved casing standards adopted by the industry. "Problem identified, problem solved," he said. "That's how we do it in the natural gas industry."

Nor could Fox hear McClendon. The filmmaker-turned-activist called for acts of civil disobedience to stop the industry, and recounted how liberated he felt when he was handcuffed and led into a paddy wagon at a recent protest in Washington, D.C. To loud cheers, he called for gas supporters to do the same if regulators permitted drilling in the Delaware watershed, the source of Philadelphia's water. "It's time to shut them down," Fox yelled, his voice reverberating off downtown buildings. "We will blockade the well sites. It's time to get into the paddy wagon."

Julie and Craig Sautner were also at the rally. They took the stage and Craig recounted the story with Cabot: what the family's well water looked like before drilling began near their home on Carter Road, what it looked like after, and Cabot's refusal to accept blame for the contamination. "We got to do something about this," Craig shouted. There was a swell of cheers and applause. "It's about time they fixed that water. Three years is long enough! We want action!"

The Sautners would get action. Although it would take time and there was no way to foresee the outcome, Cabot would make sure that the Sautner household, where all the troubles began in September 2008, would be permanently eliminated from its list of problems.

The rally in Philadelphia was one of many in 2011 and 2012. Some were in Pennsylvania where activists refused to give in to the advancing industry, but the soul of the anti-fracking movement was nurtured in New York, where the battle over shale gas was yet to be decided.

Walter Hang, for his part, made good on his promise to Stu Gruskin for "rough sailing" with the DEC's review to develop permitting guidelines for fracking. Drawing on his experience and contacts as an organizer with NYPIRG, Hang began coordinating rallies at the New York State capitol shortly after Cuomo began his first term. The first event, on March 24, 2011, featured several busloads of activists and public speakers from Binghamton, New York City, and Ithaca who gathered in the vaulted stone and marble halls of the capitol. Their robust and prolonged chants of "no fracking way" prompted those in nearby offices to close doors in an attempt to dampen the noise. The protests grew in size, sophistication, and potency after that and became an unavoidable presence at the governor's annual state of the state addresses in 2012 and 2013.

While Hang's focus on New York remained trained on the SGEIS, the fight grew well beyond the mechanics of a policy instrument for granting fracking permits in New York State. A contingent of pop stars, intellectuals, and scholars—some in concert with Hang and many working independently or with their own groups—rallied followers to take up the anti-fracking cause as a statement against fossil fuel consumption and for renewable energy. Tony Ingraffea helped found Physicians, Scientists and Engineers for Healthy Energy to educate the public about the science of sustainable energy and impacts from carbon fuels. Sandra Steingraber, an author and ecologist often compared to Rachel Carson, joined the early rallies as a public speaker and later became one of the movement's most renowned voices. Steingraber, a scholar in residence at Ithaca College, won a $100,000 cash prize from the Heinz Foundation for her memoir, *Raising Elijah*, about raising her son in an era she finds rife with corporate recklessness and government negligence. With the award, she helped found New Yorkers Against Fracking, an umbrella group that organized local grassroots endeavors statewide. She later served time in jail for participating in a human blockade to a propane storage facility proposed at the south end of Seneca Lake. Natalie Merchant hosted a sold-out show—"Dear Governor Cuomo"—in Albany's premier performing arts center in the Empire State Plaza, which featured pop stars, folk singers, and movie actors portraying shale gas development as a pending disaster. Robert F. Kennedy—former in-law and friend to the governor—joined Alec Baldwin in denouncing fracking on talk radio shows in New York City. Pete Seeger, at the time well into his nineties, joined a rally of thousands of protesters and performers crowding a quarter-mile concourse between the capitol and the convention center where Cuomo was giving his 2013 state of the state address. To get to the speech, lawmakers and officials had to pass the gauntlet of shouting protestors. Yoko Ono and Sean Lennon founded a group called Artists Against Fracking to focus on "New York as a first step in a global effort" to ban fracking. The widow of John Lennon, the famous Beatle, and their son personally helped deliver more than two hundred thousand written comments, packed in dozens of boxes, to the DEC headquarters in Albany. The logjam of paperwork—organized by a multi-group campaign—prevented the agency from meeting a deadline to finalize a plan for codifying regulations to go with the SGEIS.

It was easy to envision, if Ken Ely were alive, advocacy groups recruiting him, as Josh Fox had the Sautners, as an example of a common citizen exploited by Big Energy. The anti-frackers' call to "shut them down" at the Philadelphia rally came two years after Ken Ely had tried to do just that on the quarry road that ran along the ridge of his property. He would have been a compelling addition to this effort, but also an unlikely one. "He wasn't against drilling," Ken's brother Bill once told me. "He was against Cabot messing up his land and water." Ely was, with few exceptions, not much different from other residents of Dimock, where the story of shale gas development could not be neatly extracted from the context of personal circumstances.

For the Carters, 2011 had been a dismal year, apart from their water problems. Ron's son Todd, who ran the family quarry, had died unexpectedly. Earlier that year, Jeannie had lost a brother and a sister to illnesses, and Ron had been in the hospital for acute bronchitis. Although it was difficult to see past Ron and Jeannie's stoicism in a given interview, they would not mind putting the controversy over the gas rush behind them. While the lawsuit filed by the Carter Road 15 dragged on with no guarantee and no end in sight, the settlement offer brokered by the DEP in lieu of the water line would immediately compensate landowners with payments worth twice the market value of their respective properties. For the Carter's 73 acres, that would amount to $344,560—enough to build a home across the street to accommodate all four generations of their family during holidays. But the uncertainty about the water nagged at Ron. He knew technology existed to purify sewage water to drinkable standards, but he wondered how much that would cost and whether they could trust it. His voice grew hard when he talked about this. The deal between Cabot and the DEP was "made in the dark" without input from the residents, he told me. It would allow Cabot to "wash its hands of the problem." There were complications involving lawyers. So he held off, day by day, but he didn't know for how much longer.

The year had been no better for Norma Fiorentino. She had taken a fall that put her in the hospital for a spell. Her daughter's cancer had advanced beyond treatment, and she was spending what time she could with her. Her house was expensive to heat and needed repair, and she was not looking forward to spending another winter there. On the coffee table by her sofa, she kept a picture of a three-bedroom home on Elk Lake she was determined to buy—one-story, furnished, and with a hot tub. That was her hope, that is, if all this lawyering would produce some results while she was still around. Norma's share of the settlement offer brokered by the DEP would be $229,000 for her 3-acre homestead. Between that, and the pending lawsuit, she was convinced some money would be coming her way during her lifetime. If not, then maybe it would improve the lives of her children and grandchildren, she told me matter-of-factly.

The Switzers, who had invested their life savings in the construction of their dream home, would receive the smallest of the sums. After more than four years, the magnificent dwelling was finished in the fall of 2011, save some painting and interior decorating; yet Cabot had based the settlement offer on an assessment from 2007, when the Switzers signed the gas lease. At that time the only structure on the 7-acre property was an old trailer. The figure, times two, came to $162,000.

After nearly four years of shale gas development, some residents had been able to cash out and move on without worrying about the outcome of lawsuits. Harold Lewis, who had hosted the protest against the water pipeline across the street from the Switzers, sold his 60-acre farm for $780,000, including gas royalties, to a New Jersey–based mission that rented the house to drilling crews. But for Switzer, moving wasn't an option. "Everything we have is in this home," she told me. "It embodies our courtship, our relationship, our dreams. It's hard to give up your dreams, especially when you waited half a century for them to happen."

The settlement money Cabot offered to each of the families—under the condition that they drop all other legal claims—would be available in escrow until the end of 2012, should they change their minds.

8. TURF WARS

O f all the people confronted with Cabot's take-it-or-leave-it settlement terms, Craig and Julie Sautner were among those most ready to leave it. Events unfolding in late 2011 and early 2012 only served to strengthen their resolve. Their water problems, the first to appear on Carter Road, triggered an emergency intervention by the federal EPA—a development that suddenly changed the dynamics of the Cabot lawsuit.

"We finally have somebody looking at this without being tied to Cabot's agenda," Julie Sautner told me. "Before it was two against one; DEP and Cabot are both on the same team." It was January 24, 2012, roughly four months after Craig and Julie had roused support at the Philadelphia rally and a year and four months since Victoria Switzer had appeared at the EPA hearing in Binghamton to urge officials to "do their job" and include Dimock in their national study of fracking's impact on groundwater. That was now happening. Julie sat on her couch clutching Emmi, the family's overprotective chihuahua. Craig was at his job as a cable splicer for a phone company where he had worked for more than forty years. As we talked, a cacophony of barking periodically rose from a backyard kennel as EPA technicians and contractors entered and exited a basement door. The workers carried tubs of lab paraphernalia—rubber gloves, hoses, buckets, coolers, beakers, and test tubes packaged in clear sterile wrap—from three vehicles parked along the roadside in front of the house. Next to the vehicles' open tailgates were folding tables under canopies with even more equipment that spoke to the rigors of the investigation underway.

Julie and I chatted about the significance of the federal investigation transpiring in her basement. It was preceded three years and some months earlier, following the Sautner's original complaint, by an investigation by Cabot and then by the Pennsylvania DEP. During my first trip to Carter Road in 2008, Julie told me that she had confidence in Cabot to make things right—"good people" and "cooperative" were her words to characterize the company reps. But her faith in Cabot was displaced over the years by distrust and then by anger. Things came to a head one day in September 2010 when Cabot environmental contractors showed up on the Sautner property accompanied by an off-duty police officer. The Sautners told them to leave and each party later complained to the DEP that the armed presence of the other party posed a threat. After that, Cabot agreed to stop going on the Sautner's land altogether. Tensions did not improve when, the day after Tom Corbett was elected governor, the DEP withdrew its order for Cabot to build a pipeline to bring fresh water to Carter Road residents. A year later, the DEP allowed Cabot to discontinue water deliveries to affected homes.

"Don't get me started," Julie responded when I turned the conversation to Cabot. I sensed an affected lightness to her tone, but she wasn't smiling. "We lived with this for more than three years, and that damned company says it's our problem. They've ruined our water and installed that contraption in our basement. It's supposed to be a filter—yeah, right. And it's our problem?"

The $11.8 million water pipeline from Montrose was off the table, but the $4.6 million settlement offer stood until the end of the year. Cabot determined the market value of the Sautner's three-bedroom ranch with a two-car garage on nearly 4 acres of land to be $132,860. Multiplied by two—per the settlement offer—this would amount to $265,720. It might be enough to allow the Sautners to move from the area and start over, but being displaced was not Julie and Craig's idea of a happy ending.

The Sautner house had become a public reference point in the anti-fracking battle. Julie and Craig had turned their front yard into a gallery of anti-fracking posters on stakes. In the gardens they arranged an assortment of plastic jugs—representative of the vessels they used to carry polluted water to rallies or to receive clean water to drink. The setting became a backdrop for occasional media events. Just weeks prior to the EPA's arrival, Josh Fox and actor Mark Ruffalo led a busload of activists from New York City to deliver fresh water to Carter Road residents and, as part of the program, held a rally in the Sautner's sloping gravel driveway. The Sautner's approach—blunt,

sustained, and public—became a threat that registered in the industry's public relations offices and a target for counterattacks from not only Cabot, but
also from a dedicated alliance of gas supporters, both local and national, determined to discredit them. Now, the EPA's involvement represented the
Sautner's hope and the industry's fear: affirmation of the anti-fracking cause
would begin on Carter Road, punitive measures would follow, and the door
to federal regulatory reform would open.

"AN URGENT MATTER"

As a precursor to the activity at 1101 Carter Road on this bright January
morning, a delegation of EPA staff had been compiling information throughout the previous fall that included a review of records kept by the DEP and
Cabot, and notes from meetings with residents living within the 9-square
mile area in Dimock where water problems persisted. Public statements by
the agency in early 2012 about plans to intercede were vague and mixed,
reflecting the political volatility of the issue and conflict within the agency's
hierarchy as to how to proceed. Officials told the Sautners the agency would
sanction and fund the delivery of fresh water to certain homes. A day later,
after the media jumped on the story, the agency announced that news of water
deliveries was premature; the EPA was "actively filling information gaps and
determining the next steps."

Behind the scenes, however, rank-and-file agency staff had acquired evidence of water contamination at the Carters' and several other homes that
they found to be both credible and troubling. Karen D. Johnson, head of the
EPA's regional groundwater and enforcement branch, summarized the findings in an email to Scott Perry of the Pennsylvania DEP on November 25,
2011: "It is clear from the volumes of data that DEP has collected . . . that
there are ground water effects on the private wells, and that in some cases the
treatment systems installed at least by Cabot in Dimock, are not capable of
removing all the contaminants." Contaminants, Johnson noted, included gasses, ethylene glycol, volatile organic compounds such as toluene and xylene,
and radionuclides.

Agency staff saw this as a chance to move the science forward, but the
politics were nettlesome. Tom Corbett, like many governors of oil and
gas states, saw federal regulatory involvement in his state's business as
an unwelcome intrusion, and Corbett was counting on Michael Krancer,
the man he hired to head the DEP, to address it. Krancer has extensive

experience as a partner with Blank Rome, a law firm that specializes in counsel to energy companies regarding regulatory matters, and as a head judge on Pennsylvania's Environmental Hearing Board. He is on record as a fossil fuel proponent and a climate change skeptic. Most significantly, like the governor who appointed him, he is an outspoken promoter of the benefits of domestic petroleum production and is seen as an antagonist by those against it.

Krancer—an American history buff who has participated as a Union soldier in Civil War reenactments—embraces a style that is intense and pointed, but not without a sense of humor. In one congressional hearing he challenged Robert Howarth, Tony Ingraffea's colleague at Cornell University and an advocate for federal controls on fracking, to a duel. In characteristically blunt fashion, Krancer once asked Twitter followers—in the context of the Keystone Pipeline controversy—whether "we want to be a nation of wusses" about building petroleum infrastructure. So it was no surprise when, in November 2011, Krancer dismissed the litigants on Carter Road as malcontents and opportunists. "The real issue here is not safety," he stated in one public response. "It's about a very vocal minority of Dimock residents." These residents "continue to demand that taxpayers should foot the bill for a nearly $12 million public water line . . . to serve about a dozen homes," he added, sidestepping the fact that the order called on Cabot, not taxpayers, to fund the project.

Krancer's response to complaints from the Carter Road residents was the opposite of that of John Hanger, his predecessor in the corner office on the top floor of the Rachel Carson State Office Building. But the more Krancer dismissed these complaints, the more persistent they became. Now they were amplified through a series of warnings from the EPA. On November 30, just days after Johnson wrote to Perry, Jon Capacasa, director of the EPA's Water Protection Division, followed up with his own missive: "Extraordinary volumes of data collected by PADEP and the multiple orders issued to Cabot" showed "this has been a difficult issue to resolve, with multiple drinking water wells being impacted from drilling activities." Chief among those issues, Capacasa wrote, were doubts about the effectiveness of the filtration systems Cabot installed to address complaints. In addition to methane pollution, six homes that accepted treatment showed "periodic" levels of toluene, ethylene glycol, certain metals, and radionuclides that the systems were not designed to handle. "These are common constituents of hydrofracking fluids, flowback and produced waters," Capacasa noted. He urged the DEP to "consider whatever mechanisms" the agency might have to continue water supplies to homes where problems persisted.

Shawn Garvin, EPA's regional administrator, had already called Krancer to let him know the agency staff was collecting data on Dimock's water and they intended to follow up. Krancer knew Garvin well; their professional circles and backgrounds tended to overlap, and they shared a passion for hockey in general and the Philadelphia Flyers in particular. But Krancer's amicable relationship with the federal bureaucrat did not mollify his irritation with the EPA's interest in Dimock. "They were looking for any hook they could find to get involved with this," he later told me. "Frankly, I thought it was a joke."

Krancer was known among his staff and colleagues as the "epistolary secretary" for his fondness of administering directives and policy positions through letters. During his tenure with the DEP he wrote at least a dozen letters to Garvin and Lisa Jackson, head of the EPA, regarding the federal government's role in state matters ranging from mining permits to storm water discharges to the Chesapeake Bay watershed. In response to the emails from Capacasa and Johnson, Krancer began composing the body of a letter he would later send to Jackson about the Cabot investigation. Krancer correctly anticipated the media would have keen interest in this correspondence. He kept it simple. "We realize and recognize that EPA is very new to all of this and the EPA's understanding of the facts and science behind this activity is rudimentary," he wrote. "Fortunately, Pennsylvania is not new to all of this and we have a long history of experience at overseeing and regulating oil and natural gas extraction activities in our state, including hydraulic fracturing." To a reporter, Krancer later reflected: "It's amazing to me sometimes how stupid the EPA thinks we became . . . the day Corbett was sworn into office."

Capacasa, meanwhile, did not need Krancer's approval, nor did he wait for it. In what he characterized as "an urgent matter to the agency," Capacasa called on scientists at the federal Agency for Toxic Substances Disease Registry (ATSDR) to review all of the data from Dimock and flag any health threats "as soon as possible." The reply came three weeks later, amid the year-end holiday rush, signed by Charles Edge, an ATSDR health scientist, and his two superiors. The available documentation showed "a possible chronic health threat based on the prolonged use of water from at least some of these wells." They recommended that water in the Sautner home and at least three others remain offline, and that these and other homes in the area be tested with the "highest quality sampling methodology possible, including appropriate quality assurance samples." With this toxicological evaluation, the Dimock case met the standard of "nationally significant and/or precedent-setting"—a qualification that allowed the EPA to intervene in Dimock under the authority of Superfund.

Formerly known as the Comprehensive Environmental Response, Compensation, and Liability Act, Superfund was born in 1980 to address environmental emergencies that went beyond the wherewithal or interest of local and state governments. The act is commonly associated with an upstate New York community that became an enduring icon of the mid-twentieth-century environmental movement: Love Canal. The 70-acre tract of single-family homes, swing sets, picket fences, and a school was built over an industrial dump in the city of Niagara Falls. It was condemned after children and their parents developed rashes and other illnesses blamed on toxic chemicals percolating up through the ground. The Love Canal story unfolded in the late 1970s, when there were virtually no regulations, let alone enforcement, of what industry did with toxic waste. Emblematic of corporate and civic environmental recklessness, Love Canal offered contemporary critics a point of comparison with gas industry practices, which remain exempt from federal hazardous waste handling laws enacted as part of Superfund.

By Love Canal standards, where homes were condemned and fences erected, the ensuing EPA action in Dimock appeared unexceptional. The water tanker that turned onto Carter Road on January 20, 2011, looked like thousands of other tankers that had been dieseling along the rural back roads to and from fracking sites since 2006. But this tanker was not headed to a fracking site. It slowed as it approached the gravel driveway that sloped down to the Sautner home, where a party of several dozen demonstrators, bundled against the cold, joined the Sautners with banners and signs. They were surrounded by another half dozen or so people with cameras. The group cheered at the sight of the stainless steel tanker with blue lettering, and the driver reciprocated with a short blast of his air horn.

Victoria Switzer was huddled in the group with the Sautners, the Carters, Josh Fox, and others. Her relationship with her pro-drilling neighbors had grown worse since the confrontation outside her home in the fall of 2010, and many of them would show up at public events to exercise their own right to free speech; they were here now, holding signs on a neighboring property. In a short speech that oscillated between hope and confidence, Victoria explained that her and Jimmy's house was not on the immediate list to get water, but it was on the list of more than sixty homes scheduled to be tested by the EPA. "The EPA—I hope they are here to protect us," she said. "Because I've had Cabot come in and test my water and say, 'nothing in it ma'am.' And I know that's not true. I'm not letting Cabot test my water and I'm not letting DEP test my water. As far as I'm concerned that's the same thing."

Two days later, the first EPA testing began at the Sautners. After interviewing Julie Sautner inside her living room, I circled the yard and followed a technician into the basement. There I found Richard Rupert, EPA's site coordinator of the investigation, studying the array of plumbing, tanks, and filters that Cabot had installed in an attempt to purify the Sautner's water. Before his time with the EPA, Rupert was a bioweapons specialist in Moscow for Sabre Technical Services, a government contractor. I was with a photographer, and he regarded our approach with the same appraising coolness with which he studied the plumbing.

While political forces shaped policy in the upper levels of the EPA, science appeared to plod painstakingly along in the Sautner's basement. Before meaningful samples could be collected, field tests had to be designed and executed to determine baselines for metals, total dissolved solids, and other reference points. Plumbing would have to be unhooked to eliminate variables that might be introduced by filters, pipes, or hoses. GPS systems documented the precise location of the water well from which samples were drawn. Rules for chain of custody and testing protocol had to be observed to the letter and fully witnessed and documented.

After brief introductions, we began discussing technical aspects of the investigation, while Rupert did his best to steer clear of anything suggesting politics. The Pennsylvania DEP and Cabot were welcome but not essential parties to the investigation, he told me. Each was likely to use its own laboratory contractor to test a given sample, and that "gives more assurances" to the outcome, he added. As technicians collected the water in a series of test tubes and packed them away in the cooler, I asked a few questions aimed at gauging the significance of the investigation. Rupert didn't offer much, but his simple reply was candid enough. "A lot of people all the way up the line are paying close attention to this one," he said.

THE CONVERSION

The investigation that began in the Sautner's basement would become a flashpoint in a war between state and federal agencies over jurisdictional control of the shale gas bonanza in Pennsylvania. At the same time, another battle over jurisdiction was unfolding in New York, this one between state and local governments. The "home rule movement" as it would become known, was distinct from the cause led by Sandra Steingraber, Walter Hang, Tony Ingraffea, and others. It began years before Yoko Ono delivered crates of written criticism of DEC policy and before chants of "no fracking way" echoed through the marbled halls of the state capitol; and its inspirational source was far upstream from professional politics, lobbies, and activism in Washington and Albany.

Cooperstown, New York, a village of some 1,800 residents, sits among lush hills on the south end of Otsego Lake in the northern Catskills. Here the water runs clear and deep, as the Susquehanna River begins its 464-mile journey through Appalachia to the Chesapeake Bay. Cooperstown is the governing seat of Otsego County, a locality of 61,000 people settled shortly after the Revolutionary War by William Cooper, a land speculator, frontiersman, politician, judge, and father to James Fenimore Cooper of literary fame. Over the centuries, the Cooperstown region, like other select areas of the Catskills and Hudson Valley, has thrived as both tourist destination and prime choice for vacation homes and land holdings for New York City elite. This area includes the Town of Middlefield, flanking the east side of Otsego Lake, and the Town of Otsego, on the west. Not far from the stately brownstone eateries and boutiques on Main Street Cooperstown one can find the Farmers Museum, the Fenimore Art Museum, Glimmerglass Opera, the New York State Historical Association, various craft breweries, and the National Baseball Hall of Fame.

Picturesque, culturally opulent, and historically prominent, Cooperstown was very much to the liking of James "Chip" Northrup, an oil and gas investor from Dallas, Texas. In 2006, he bought a renovated nineteenth-century colonial home with a view of the lake's junction with the Susquehanna. He was fifty-eight, and he and his wife Nancy had found the ideal refuge to escape the scorching Texas summers, not to mention the area fit his fondness for rowing, opera, museums, and local produce. At 6-foot-5, square-shouldered, slim-hipped, with sandy hair and a good-ol'-boy charm, Northrup invites easy comparison to a cowboy. He does little to dispel the impression, but his cowboyish appearance and a breezy manner of speaking, occasionally punctuated with "dude," belie an Ivy League background. While growing up in Dallas, he spent summers at a prep school in New England. He later attended Brown and Manchester universities before earning an MA in finance from Southern Methodist University and an MBA from Wharton.

Chip was not the only oilman drawn to Cooperstown's charms. Lou Allstadt once headed Mobil Oil's drilling operations in the Western Hemisphere and oversaw the company's merger with Exxon in 1999 to create the world's largest corporation. He, too, chose to spend his retirement near water sports, culture, and expansive natural beauty. Allstadt, a former rugby player and current paddling enthusiast and art collector, had summered with his family in a waterfront house halfway up the lake for decades. When he retired shortly after the merger, at age fifty-six, he bought an 1815 Federal-style townhouse in the village.

Educated, accomplished, and affluent, Allstadt and Northrup fit with Cooperstown's professional class but were a social rung below the area's bluebloods—people like Henry Spotswood Fenimore Cooper, scion of both the region's founding patriarch and James Fenimore Cooper, and Jane Forbes Clark, heiress of the Singer sewing machine empire. Cooper, founder and chairman of Otsego 2000, an agency dedicated to preserving the area's cultural and natural features, maintains a country estate on the outskirts of town and a residence in downtown Manhattan. Clark is connected through the Clark Foundation to the stewardship of 10,000 acres in the Otsego Lake watershed, including sundry historic manors and estates, a hospital, and various business interests. She also chairs the board of directors of the Baseball Hall of Fame, a sport which remains, along with horses, a lifelong passion. Unlike the residents of Dimock, Pennsylvania, and Deposit, New York, these Middlefield residents and many of their neighbors and peers were not inclined to feel their day of opportunity had arrived when a prospector came to town with the intention of drilling gas wells.

That prospector was Orville Cole. He represented Gastem USA, a subsidiary of a Canadian firm with some experience exploring the Utica Shale in the province of Quebec. It was early July 2008, shortly after the $110 million deal between XTO and farmers in the Town of Sanford triggered eager speculation throughout the Southern Tier and drove residents to town halls in search of information. Although only about 60 miles northeast of XTO's leasehold, Middlefield fell just outside the prime drilling fairway as defined in those early days. The Marcellus Shale under Middlefield was generally thought to be too shallow and too thin to be developed safely and effectively, and the Utica Shale, below that, was uncharted. Still, as with all frontiers, this territory drew interest from a small band of opportunists and risk takers— wildcatters in search of a big discovery and corresponding payoff.

About 34,000 acres of Otsego County were under lease at the time to Covalent Energy, a prospecting firm that had cheaply secured acreage over the Oneida Sandstone—a conventional play—prior to the shale boom. After the Marcellus breakthroughs in Appalachian communities to the south, Covalent partnered with Gastem to drill exploratory wells that could test both the

Marcellus and Utica shales while targeting the Oneida Sandstone. But first a water source was needed, and that brought Cole to the Village of Cooperstown board of trustees meeting on a summer evening in 2008.

Over the course of a thirty-year career, Orville Cole worked mostly as an independent operator. He's a muscular man, with full dark eyebrows and well-proportioned features, sometimes featuring neatly groomed facial hair that adds a debonair touch to his appearance. For this occasion he was clean-shaven and dressed in dark trousers, polished shoes, and a crisp dress shirt with an open collar. He was explaining to the trustees why it would be a good idea for the village to sell him water—without much luck. Jeff Katz, deputy mayor at the time, and other trustees had received phone calls and emails from residents alerting them to the proposal, which had appeared in a legal notice in the local paper. Residents worried that it would establish a precedent for water use by unfamiliar industries with unknown consequences. The board appeared unimpressed with Cole's assurances that he would use far less water than his permit application suggested. His offer to pay a premium for the water left them similarly unmoved. When Cole noted that the village had sold water for other purposes, Katz responded, "How it is used is a legitimate reason to deny."

Although Northup didn't know Cole, he identified with the business of buying leases, selling leases, selling other people's leases, and funding operations. Moreover, he relished public debate, the contest of persuasion. At the board meeting, he felt a remote kinship with Cole. "Someone was seriously looking for oil or gas, which, since I was a kid, I thought was a good thing," Northrup later reflected.

Northrup's youthful vision of the oil industry was shaped by the movie *Giant*. Its protagonist, Jett Rink, was based on the real-life wildcatter Glenn McCarthy. McCarthy's character, played by James Dean, started as a lowly ranch hand and laborer who, with dash, swagger, and raw-boned tenacity, became one of the wealthiest of Texas oil tycoons during the Depression era. McCarthy's story, told through book, movie, and magazine cover, was the forerunner of a genre popularizing the mythos of the go-for-broke-bad-boy Texas wildcatter. Chip was a boy when he saw *Giant*, and the experience was crystallized in his memory of walking up the aisle after the show imitating James Dean's character staking an iconic land claim. Fifty years later, Northrup was out of the oil and gas business, but he still felt a charge of excitement—in his words a "visceral reaction"—upon seeing what he immediately recognized as an oil field truck on a remote Otsego County road and a waxing curiosity about the circumstances that brought the industry, of all places, to his sanctum in upstate New York. That curiosity had brought him to the village board meeting, and he listened with rapt attention as Cole explained that his firm would have to truck water from Cortland if he could not get it from Cooperstown.

After a brief deliberation, the village trustees denied the water request. As the meeting broke up Northrup moseyed over and introduced himself to the wildcatter with the intention of offering some conciliatory words and perhaps making an acquaintance. He began with some industry small talk about the Devonian Shale. Cole, who had been solicitous and engaging with the village trustees moments earlier, now appeared impatient and uninterested. "That's confidential information," Cole responded coolly. Northrup shrugged. "So much for you, dude," he replied as Cole walked away. A Canadian oilman in upstate New York wasn't exactly a clear fit with Northrup's experiences in Texas and New Mexico, anyway.

The exchange with Cole was a precursor to what Northrup characterized as his "Damascene conversion" on fracking in New York. This conversion was representative of a much broader phenomenon taking shape in Otsego County at the same time as the leasing frenzy unfolded in the Town of Sanford, sixty miles to the southwest in Broome County. As various municipalities in Broome County welcomed the gas rush, residents in Otsego County prepared for a fight.

Shortly after his encounter with Cole, Northrup was invited to lunch with Henry S.F. Cooper. The invitation came through Nicole Dillingham, a litigator from San Francisco who had taken a seven-year leave in Paris with her husband, Gaylord Dillingham, and their twin daughters. In 2007, eager to get back to the States for her daughters' education, Dillingham settled the family in Cooperstown—which she considered "the Walden Pond of New York." Dillingham, who had a summer place in Cooperstown prior to moving there, had met Cooper at a dinner party, and the two quickly bonded, she said, over "threats of industrialization of the area." It was a bond that, along with Dillingham's other qualifications, led to her appointment as president of Otsego 2000.

On a late summer day in 2008, Northrup, accompanied by Dillingham, arrived at Red Creek Farm—a mid-nineteenth-century Italianate farmhouse, with rooms graced by portraits of ancestral Coopers and shelves of books. They were welcomed by current resident Henry Cooper, the ninth-generation descendent of the town's founding father—a studious-looking septuagenarian with a shock of curly dark hair and animated grey eyes. Over a lunch of chicken and seasonal produce, harvested and prepared by Suji, Cooper's personal assistant, they discussed the history and future of the area through the context of the Otsego 2000 mission.

Cooper, a staff writer for *The New Yorker* in the days of typewriters and cigarettes, values the precise use of language. He elaborated on a prevailing ethic manifest in the family's legacy and reflected in James Fenimore Cooper's work. When William Cooper first laid eyes on the primeval natural landscape around Otsego Lake, the American wilderness was seen by settlers as something "dangerous and difficult to overcome," Henry Cooper explained. By the time his son, James Fenimore Cooper, came of age in the early eighteenth century, it already bore the scars of unchecked exploitation—extensive logging, overfishing, and overhunting. The genius of his ancestor's writing—*The Leatherstocking Tales* being among the most famous examples—is embodied in the characters and settings that represent the tension between wilderness and cultivation, with plenty of lessons about the recklessness of settlers. James Fenimore Cooper, Henry Cooper continued, was a critic of clear cutting, the decimation of the carrier pigeon, and excessive netting of herring in the Susquehanna River.

Northrup, a fan of Garrett Hardin's "Tragedy of the Commons," found this genealogy of the land preservation movement fascinating, and Cooper was happy to expound, but he was also eager to tap into the Texan's experience on the matter of present-day shale gas development. At that point it was still very new: the anti-fracking movement had yet to coalesce although there had been local pockets of intense concern and there was little in the way of policy development pending the release of the SGEIS. Northrup turned the conversation to what he knew about Texas, Oklahoma, and Louisiana. These "grown-up" petroleum states, he explained, taxed gas extraction to fund independent agencies to oversee inspections, complaints, and enforcement. For the Lone Star State, that would be the Railroad Commission of Texas. New York has no such tax and no such agency. The DEC serves the dual and, in Northrup's view, dissonant roles of environmental protector and regulator of development. The environmental responsibilities of the agency are bound to be "co-opted and corrupted" by the permitting body, Northrup reported. "If you have autonomous oversight, then you have a fighting chance from an environmental standpoint. Otherwise, you might as well be in Venezuela."

Employing the tools of grassroots political action, Otsego 2000 had won many fights against power lines, windmills, and parking lots in an effort to keep the landscape consistent with the values of James Fenimore Cooper. Those had been local fights, Henry Cooper explained to me later. They employed an argument not against development, but that Cooperstown's character constituted a vital local resource dependent on the protection of its history and viewshed. The fight against drilling in the area would unfold along similar lines, and progress to something much bigger.

When the first draft of the SGEIS was released in September of 2009, there were many eager readers, and Northrup was one of them. In his home study Northrup, highlighter and pen in hand, delved into the 500-plus pages of sections, subsections, references, tables, charts, and appendices. During short breaks, he refreshed himself with glances out the window at "Glimmerglass"—the name James Fenimore Cooper gave Otsego Lake in his novels. Just to the left of Northup's view another feature in Cooper's writing, Council Rock, protruded solidly from the water's surface. It was a striking natural feature and, as legend has it, a critical point of governance for Mohawk chieftains seeking to negotiate fishing rights with various river tribes.

The document before Northrup was the New York State government's attempt to make the record straight and clear regarding management of a coveted resource in the twenty-first century. Yet Northup found it to be riddled with ignorance about the operations it meant to describe. The setback for a drilling rig from a water source was 150 feet, and Northrup wondered why not zero, because on the scale of shale industry operations they practically amounted to the same thing. He concluded the setback must be a precaution to prevent a roughneck from drowning if he fell off the platform. He mustered a thin smile, and then groaned when he read the DEC's justification:

> Significant surface spills at well pads which could contaminate surface water bodies, including municipal supplies, are most likely to occur during activities which are closely observed and controlled by personnel at the site. More people are present to monitor operations at the site during high-volume hydraulic fracturing and flowback operations than at any other time period in the life of the well pad. Therefore, any surface spills during these operations are likely to be quickly detected and addressed rather than continue undetected for a lengthy time period.

Northrup, through logic and experience, knew observing a spill is something altogether different from preventing its impacts. Although it was a year before the explosion of the Deepwater Horizon oil well would sensationally demonstrate that fact to the world, Northrup had been in the business long enough to recognize what he called "happy talk from an industry brochure," some of which apparently had been copied with scant alteration to the SGEIS. Other measures that "mitigate the risk of surface water contamination" from well pad operations—blow-out preventers, pit liners, and permits and plans to manage storm-water flooding and run-off at sites—were nothing more than standard practice for conventional wells, unchanged in the original version of the document from decades prior. The proposal permitted operators to keep waste onsite in open pits for up to a week, and to spread the briny discharge from wells on roads to keep dust down.

Northup glanced beyond Council Rock to forested hills sloping down to shimmering water. The document told him little about where the "closely observed" drilling and fracking operations would go, where the waste would go, or how the operation and build-out of pipelines, compressors, and treatment plants would cumulatively fit into a given landscape. It was, as its name aptly conveyed, a "generic" approach to permitting based on the industry's needs to get at the resource, without consideration for a region's culture, character, or tolerance for certain kinds of development.

Despite its ungainliness, the SGEIS became popular reading in Cooperstown that fall, and, like the favorite pick of a book club, it was routinely discussed and analyzed by a group of professionals and intellectuals who gathered at the Alex & Ika Restaurant on Main Street. Nicole Dillingham, a chief organizer of these meetings, would appear with sections of the SGEIS, clipped together and covered with sticky notes and highlighter, clutched in her long, slender arms. Other regulars included Harry Levine, chairman of the Otsego Land Trust, Ron Bishop, a biochemistry professor at nearby Oneonta University, and Adrian Kuzminski, a scholar and author on issues of economics, ecology, and philosophy. They represented the core of a larger group—many affiliated with agency and government boards given to preservation causes—who compared notes as they wrote their formal critiques of the SGEIS to be submitted as part of the state's review. The exercise continued through the fall, sometimes in a conference room at the Otsego 2000 offices, but as often online, on the phone, and, to the dismay of the spouses of those most resolutely engaged, at dinner parties and social events.

Along with Northrup, Lou Allstadt was a voice of authority at these gatherings. While Allstadt's background suggests an oilman, his persona is more corporate than cowboy. He is small and athletic, with a strong jawline, white hair parted neatly on one side, and steady blue eyes behind round spectacles. His thirty-one years with Mobil were preceded by military service as a merchant ship officer during the Vietnam War and studies at Columbia University, where he earned an MBA in logistics. These experiences suited his penchant for a corporate culture that valued order and predictability. He had not seen *Giant* in his youth or even thought much about the oil and gas industry until he visited a job fair while he was at Columbia. A few of the major oil companies were recruiting and Allstadt quickly became intrigued by the industry's abundant opportunities to apply his major field of study. He took a job with Mobil, where he expected to stay only "a couple years to see what it was like in a large corporation."

Though Allstadt was a sought-after source of information when Gastem began prospecting in Otsego County, he was at a loss to explain what might be expected from these wildcatters. His corporate background did not mesh with the ways of independents, which he found to be headstrong and reckless. During his days with Mobil, he served as a board member of the U.S. Oil and Gas Association, where he "tried to get independents and small companies to focus on larger issues," he told me. He found natural gas prospectors even more notorious than those in the oil industry for their disdain for regulations, and the Independent Oil and Gas Association of New York seemed to embody the most intractable approach of all. "Willingness to deal with some level of regulation increases as size of company increases," he said. "Reasonable level of regulation gives larger companies a framework—a kind of protection from somebody doing something stupid and pulling down the whole industry." Regulations, he said, had to be written "for the worst possible player" and the SGEIS was a distressing example of how they were not.

Neither Northup nor Allstadt at the time had heard of Walter Hang, the career activist in Ithaca. But, starting from vastly different perspectives and experiences, they all arrived at the same conclusion: The SGEIS was an extension of archaic policy that had allowed the gas industry in New York throughout its history to escape reasonable regulatory oversight. Yet there was a fundamental difference between Hang and the group from Cooperstown. Allstadt articulated it succinctly when I interviewed him one day on the porch of his home, overlooking a garden well-tended by professional help. Some people are driven by causes, he said, and if it were not fracking, it would be something else. "People around here are not those kind of people.

They are people who felt a threat." They were, in other words, not like Hang or Hang's followers—committed to a lifetime of activism and environmental causes. They were accidental activists, like Victoria Switzer and the folks on Carter Road. But the Otsego County group was better connected, better credentialed, and had far greater resources for the ensuing fight.

HOME RULE

On June 1, 2009, Rich Nyahay, Gastem's vice president for exploration, received a long-awaited permit from the New York DEC. It came with an apologetic note from Brian Baker, a DEC functionary: "Sorry this took a while—we're still getting a feel for these things."

The permit would allow the company to test the Utica Shale with a low-volume frack. Gastem had made the application thirteen months earlier, when landmen were furiously competing for leaseholds in counties to the south as natural gas prices peaked. By the time the permit arrived, Governor Paterson had issued the shale gas moratorium, the national economy had tanked, and DEC Commissioner Pete Grannis was facing staff cuts that would trigger the complaint that led to his firing. Still, hope was fresh among shale gas supporters that the SGEIS would soon be squared away and shale gas development would begin. Orville Cole was determined to get a head start in the race to prove the worth of the upstate pay zones. Though the state was not issuing permits for shale gas development with high-volume fracking, it did not forbid testing horizons with the kind of low-volume fracks used on conventional wells, and that was Cole's plan from the start.

First, he needed somebody familiar with Otsego County geology. For this he had hired Nyahay, who had previously spent twenty years as an academic geologist for the New York State Geological Survey and State Museum. Nyahay saw the job with Gastem as a chance to play a role in a breakthrough in the state's drilling history, an offer over which he was "salivating."

For this breakthrough, Nyahay had recommended Crumhorn Mountain, a 1,900-foot land feature fifteen miles south of Cooperstown in the Town of Maryland. On top of the mountain was a hemlock forest sheltering a beaver pond, a scout camp, and a gun club operated by a group called the Crumhorn Cowboys. Under the mountain was the Oneida Sandstone, sandwiched between the Marcellus and the Utica shales. Cole, with Nyahay's guidance, was banking on the sandstone, which could be produced without fracking, as an immediate commercial prospect. Depending on the show from the Utica and the Marcellus, and the future permitting status of high-volume fracking, he was hoping to position himself for a "triple play."

The site was named Ross 1. Drilling commenced early Monday morning, September 28, 2009, and continued nonstop through the week. Cole, in coveralls and a hard hat, moved energetically among the crew. He could be seen climbing up to the platform, directing logistics on the pad, or stepping into a contractor's trailer to study monitors. Two readings in particular drew his attention: "down-hole pressure" and "mud-flow rate," values necessary to gauge forces in the earth—gas, petroleum, or brine—against the stabilizing resistance of drilling mud pumped down the hole. After a few days, the monitors began to relay signs as the bit penetrated the Marcellus at a depth just below 2,000 feet and gas surged into the well through natural fractures and pushed against the mechanical controls above. The crew lowered doughnut shaped cylinders—"packers"—into the hole to seal off the well bore above the horizon and direct gas into the drill pipe. It flowed up the well and was flared through a valve at the top. Without fracking to open up the Marcellus, the gas kick was shortlived, as expected, and the drilling resumed. Gas surged up the well again as the bit cut through the porous Oneida Sandstone. After testing the flow rate, the crew sealed off that horizon and continued drilling. More days passed. At 4,950 feet the bit cut into the Utica Shale—the vast and unexplored rock that, in this neck of the woods, inspired visions of newfound wealth.

Seven weeks later, with the drilling completed and the cement casing poured and set, fracking crews arrived in the damp chill of a mid-November dawn, and the site was soon bustling again. Though Cole had given Northrup the brushoff at the Cooperstown Village board meeting, he remained trained on a positive PR push. His enterprise in the Otsego County backcountry was drawing the attention of state regulators, local officials, and some environmental groups, and Cole intended to turn Ross 1 into a showcase. He treated chosen delegations of public officials, property owners, emergency responders, agency people, and media to tours of various stages of operations. Among them was Roger Downs, conservation director for the Sierra Club's mid-Atlantic chapter. At the time the Sierra Club—the country's oldest and most venerated hub of grassroots environmentalism—supported natural gas development as a preferable alternative to coal, and the national agency was accepting eight-figure gifts from Aubrey McClendon, Chesapeake's CEO. Although the donations were yet to be disclosed, there was already dissension among various chapters as club members developed a more discerning understanding of the risks and tradeoffs of shale gas development. Downs represented a chapter where skepticism flourished as the public learned more about developments in Pennsylvania.

The water arrived in a convoy of tankers, along with truckloads of sand and tanks of various chemical additives, and Downs watched intently as these were coupled to an intricate network of hoses, valves, pumps, and a large chemical mixer. Parts of the pad were lined with crushed stone and all of it was bordered with silt and hay barriers to control run-off. Inspectors from the DEC and the Susquehanna River Basin Commission were present, critically evaluating operations as reporters took notes. Nyahay, a large, affable man with a booming voice, enthusiastically explained the various safeguards to the media and officials.

Downs found the operation both impressive and troubling. He sensed Cole and Nyahay were acting in good faith, and he found the operators' efforts to engage the community and regulators commendable. He remarked to one reporter that the operation was "a pretty awesome thing to see." He added, noting the diesel fumes and noise wafting down the valley, "It's probably the most inappropriate area to put a well." Mostly, Downs worried that the operation and others like it would leave the public with a perception that conflated operations of a single well with the cumulative impacts of full-on shale gas development. Ross 1 was the first well targeting shale in New York since the moratorium, and it was getting exorbitant attention. Yet it was tiny by shale standards. A single, fully productive Marcellus well would require between twenty and thirty times the amount of fracking fluid and produce between twenty and thirty times the amount of waste. Downs tried to project what the landscape would look like when the inspectors now so conspicuously on hand were spread thin tracking hundreds or thousands of wells. He wondered about the outcome when the press was no longer around, and drilling in an unspoiled tract of hemlock forest was no longer exceptional, but business as usual.

As he shadowed Cole with a group of reporters and officials, Downs realized a critical part of the operation was altogether absent in this showcase.

"How are you going to get the gas to market?" Downs asked Cole.

"No worries," Cole responded. "There'll be a pipeline." Much later, Downs wondered how Cole knew that.

Gastem announced to investors in the first-quarter report of 2010 that Ross 1 "exceeded expectations" and all three pay zones would be "the subject of extensive programs in the coming months." Those programs arrived in the Glimmerglass watershed that summer with the mechanical din of seismic trucks and the occasional concussive thud of dynamite detonations on leased property. The commotion alerted a portion of the Otsego County population that had not already heard or had dismissed the possibility that a burgeoning gas play was at their door.

Cole had obtained a permit and drilled a well to explore shale gas in New York—no small accomplishment given the political tenor that lead to the moratorium. In Nyahay, he had found a person who could choose a drill site based on a career's worth of study of physical landscapes thousands of feet underground. Yet for all the preparation, Cole and Gastem misjudged the forces at work above ground in Otsego County.

In November 2010, a stream of Otsego County residents filed into Templeton Hall, a 175-year-old community building on Pioneer Street in Cooperstown, for an event organized by Nicole Dillingham and Otsego 2000. Many of the two hundred or so attendees understood the gas industry from what they had learned from the movie *Gasland*. It had been screened to a full house in the Cooperstown High School in June of that year, and Josh Fox had been the guest of honor at a reception hosted by Dillingham and Henry Cooper at Red Creek Farm. There was plenty of buzz about the flaming water faucets and disillusioned leaseholders featured in Fox's account. News coverage of the public fight in Pennsylvania between John Hanger and Cabot, meanwhile, was affirming existing doubts about the wisdom of shale gas development. Now residents took their seats on rows of folding chairs facing a wooden lectern while others in the swelling crowd stood where space allowed. One observer noted "there was a lot of angst among the crowd," which included more than a few elected officials, some of whom had leased land with gas companies and now wondered what they had gotten themselves into. The buzz faded as Dillingham, stately and business-like, introduced Northrup. The tall, lean Texan strolled to the lectern and picked up the microphone. Matter-of-factly and in a pleasant drawl, he told a story that did nothing to ease that angst.

"When I moved up here from Texas, a few things really surprised me," he said. "How antiquated the New York regs are. How comically out of date they are—how far behind the technology." For the next forty minutes, he itemized why the shale gas industry born in Texas was a pending catastrophe in New York. He started with the mechanics: high volumes of toxic solution—literally blasted into a poorly understood, fractured, and possibly fragile subsurface at pressures exceeding 10,000 pounds per square inch. Yes, fracking had been around for a long time, he said, but not this kind of fracking. This was something new, and it had been developed in the "seismically inert environment" of the Barnett Shale in Texas. "And now you are bringing this Texas-sized industry to upstate New York, to areas where the faulting is not known or well understood, and applying the world's biggest frack."

The problems were not just underground. "Everything on the surface, in a driller's view, is in the way of what's under the surface," Northrup explained. In vast, flat, arid expanses of Texas, that tended to be less of a problem. In the lush and relatively highly populated watersheds of New York, it was very much a problem, at least for the people who lived over the desired targets.

The problems of mechanics, logistics, geology, and natural landscapes were compounded by the lack of a sound regulatory plan, Northrup explained. In Texas, taxes on production added a critical regulatory stabilizer to the mix by offering both economic incentives for state officials to "check the wells and meters" and the wherewithal to do that checking. To get a permit in Texas, he explained, operators show up with their leases, spacing units, and seismic records documented "so they can demonstrate they are not going to blow a fault clean to the surface." If things check out, they get a permit, and if something goes wrong, he said, state law does not leave it to the people issuing the permits to investigate. Problems are examined by an independent agency—"a true environmental watchdog. It might be a toothless mutt, but it's independent. Here in New York, those two things are mixed together. . . . If you want to find a way to corrupt a bunch of officials, the oil and gas industry is much better than the gambling industry."

A pall of silence hung over the room as Dillingham stepped forward to moderate questions. One hand shot up, and then dozens more. Questions at first tended to circle around a common point: What do we do about impending gas development? Northrup's experience told him shale gas development was inevitable once the economic wheels got turning because "there's no standing between a barrel of oil and a Texan." His answers, accordingly, focused on regulatory measures and policy: write congressmen to support the Frack Act, urge legislators to overhaul state policy or, at the very least, ensure tests documenting water quality are properly recorded prior to drilling to shore up any legal footing against the industry.

The discussion had turned to the importance of water tests when a young woman in the second row, Michelle Kennedy, stood up and introduced herself. She was in her early thirties, dressed in a black turtleneck sweater and grey pants, with full wavy brown locks pinned back from her face. With a no-nonsense manner, softened by a youthful face and becoming smile, she asked Northrup if he was suggesting that people protect themselves by gathering evidence they can use to sue the gas and oil companies when their water gets ruined. Northrup said yes and, after a pause, remarked that this sounded like recommending "getting a physical before you were run over by a truck."

"The town is better off by just not allowing it," Kennedy said. Heads turned away from Northrup and toward Kennedy. She had recently moved to the town with her two preschoolers after her husband, a physician, accepted an appointment at Bassett Hospital in Oneonta. Prior to this move, when researching Otsego County real estate, she noticed an abundance of properties with gas leases. This raised both her parental wariness and professional interest. Kennedy is a graduate of George Washington Law School with experience ranging from interpreting municipal law to litigating family abuse cases and she had been thinking about gas development even before Gastem showed up in the county. Now, faced with the likelihood of drilling in her new home, she drew on her law school training and strove to make the matter simple. A municipal government could set land use ordinances to preserve its character and values, she stated, and it could prohibit uses that did not conform—even drilling.

Kennedy's assessment would gain acceptance over time, but it was radical and somewhat confusing at that moment. The role of local government throughout New York, when raised at all in the fracking debate, had been in the context of road maintenance and questions about public safety. Drilling applications were filed with state, not local, officials. The state was the authority that approved spacing units and well permits. Previous community meetings in upstate New York—including those organized by the Farm Bureau featuring Chris Denton and Ashur Terwilliger—tended to focus either on the state regulatory process or on ways landowners could legally protect themselves, apart from state regulations, through leases and water tests and jurisprudence. It was never the intention of Denton and Terwilliger or the landowners who attended those early Farm Bureau meetings to stop the industry, but after the Templeton Hall meeting that became the goal of the grassroots movement taking shape in Otsego County.

"Lou [Allstadt] and I could explain the hazards of fracking, but we were clueless about how to respond politically," Northrup later observed. Kennedy, composed, attractive, and on point, had the answer: zoning.

Throughout the fall and into the winter, Kennedy continued researching the subject from her small law office on the third floor of the Key Bank Building on Main Street. The notion that state authority over shale gas development trumped municipal zoning was, at the time, based on a prevailing interpretation of the state's Oil, Gas, and Solution Mining Law, and on one sentence in particular: "The provisions of this article shall supersede all local laws or ordinances relating to the regulation of the oil, gas, and solution mining industries." Issues dealing with roads and property taxes were the only exceptions. That language, legislated in 1981, came to be known as the supersession clause. Natural resources, the argument went, could not be effectively extracted for the public's greater good under a patchwork of restrictions that varied from one town to the next. The oil and gas industry needed "regulatory uniformity and predictability," as one industry lawyer put it, to develop wells and infrastructure regionally.

Kennedy found case law dealing with this issue to be scarce. Although more than 50,000 wells had been drilled in New York, most of them were small ventures for homesteads and local businesses in the preregulatory era, sporadically successful and widely disjointed in space and time over the course of generations. Even the bigger commercial plays and small booms in western New York in the late twentieth century were geographically limited. Consequently, the matter of jurisdiction had never been rigorously challenged.

Now, with expectations of an impending shale gas juggernaut in New York, the intention of the supersession clause was suddenly getting a close look from municipal stakeholders. Kennedy drew on a small and developing body of work by other lawyers and scholars on the rights of local governments to ban operations through a concept known as home rule. Throughout late 2010 and early 2011, lawyers such as Todd Mathes, Michael Kenneally, John Lyons, and Helen and David Slottje were independently and collaboratively publishing opinions challenging the industry's interpretation of New York's oil and gas law. This school of thought did not dispute that the oversight of functions such as setbacks, well casing integrity, runoff control, and the like fell within the state's jurisdiction. But determining whether drilling fit into a given town's land use rules was, they argued, up to the town. Put another way, the state had control over the "how" of drilling, but not the "if."

While the case law was scant, one decision was consistently cited in these papers. In 1987, the Town of Carroll prohibited Frew Run Gravel from developing a gravel mine in a district where it did not conform to zoning. The gravel company, citing the supersession clause in state law, sued and won in lower courts. On appeal, the high court agreed that the supersession language "preempted all local laws relating to the extractive mining industry." But the court ruled that the town's zoning interest related "not to the extractive mining industry but to an entirely different subject matter and purpose: i.e., 'regulating the location, construction and use of buildings, structures, and the use of land in the Town . . .'" The court ruled in favor of the town. In this case, the court established that the language in the supersession clause allowed the state to regulate exclusively, but that regulation was something apart from zoning.

The industry would surely challenge whether and how this interpretation might apply to the oil and gas industry. For the zoning argument to prevail, land use ordinances would have to be explicit. Lawyers could argue the hypothetical, but local laws were a product of local governance, and that was up to the people.

Kennedy, working mostly pro bono but funded in part by outside donations, was retained by Middlefield Neighbors, a community group energized by several mothers of young children, like herself, who were unhappy about the industry's advances in their town. They embarked on door-to-door campaigns with flyers and pamphlets emblazoned with a fist thrust skyward and clutching a bunch of grass with dangling roots, with the message: "Middlefield, powered by people, not by Gas $$$." Middlefield Neighbors, on the east side of the lake, was the first of a number of campaigns that flourished in Otsego County. Kelly Branigan, a registered nurse and real estate broker, became another leading organizer in Middlefield. Julie Huntsman, a small-animal veterinarian with a budding career as a local politician, coordinated opposition in the Town of Otsego, on the west side of the lake. Harry Levine, chairman of Otsego Land Trust, organized efforts in the Town of Springfield, bordering the north side of the lake. They conducted surveys, organized petitions, and urged residents to attend planning board meetings.

Similar home rule campaigns were taking shape simultaneously and independently in other regions of upstate New York. Notable among them was a municipality of 14,500 people in the Finger Lakes region, where anti-fracking sentiment was particularly keen. The Town of Dryden, a bedroom community to Ithaca College and Cornell University, is 12 miles east of Ithaca in Tompkins County. Like the towns in Otsego County, conservation and environmentalism is woven deep into Dryden's identity and is reflected in town politics and private sectors. Dryden, too, sits on the unexplored fringes of New York's anticipated shale fairway and many people who live there want to keep it unexplored. Principal players in the ensuing fight included Helen and David Slottje, the husband and wife legal team who would do much to popularize the argument for home rule statewide; Mary Ann Sumner, the town supervisor who won a second term on a ticket to support an industry ban; and Anschutz Exploration, a drilling firm based in Colorado that held leases in the area.

With Middlefield and Dryden as the test cases, others looked on, and many—including half a dozen towns in Oneida County in the Norse Energy leasehold—passed moratoriums on shale gas development to give themselves time to consider bans. The industry, having already spent millions on lobbying for federal regulation workarounds, now faced the prospect that its stake in prospective drilling zones could hinge on a town council vote over local zoning and it geared up for a fight in the press and in the courts. Brad Gill, executive director of the Independent Oil and Gas Association of New York, said the bans would effectively drive the industry away from New York altogether due to the "patchwork of green lights and red lights that an operator would have to navigate." The bans lacked legal merit, industry lawyers would argue, and infringed on the rights of landowners and the companies they sought to do business with.

Middlefield town supervisor David Bliss was an early and enthusiastic advocate of Kennedy's home rule plan. After preliminary hearings at the planning board, Bliss presided at a town board hearing in early March 2011 regarding changes to the town's master plan and zoning ordinances. The changes would explicitly identify drilling and all related functions of petroleum development as prohibited industrial uses—the final step to codifying a ban. Kennedy arrived at the hearing early, but not early enough to get a seat. She shouldered her way through the lobby and into a meeting room crowded with 140 people. After some shuffling, she found a place to stand along the wall behind the board members. Given the stakes, she was not surprised to see Scott Kurkoski, the attorney hired by the landowner coalition organized by Dan Fitzsimmons in Broome County. People in the front were given the first chance to speak, and Kurkoski, who made the hour-and-a-half drive from his Binghamton office to deliver a special message to the board, watched and listened from the back.

After an hour of ardent speeches in favor of a ban, with few exceptions, Orville Cole got his turn. It was Cole's particular misfortune to follow Bassett Healthcare president William Streck, who announced that the hospital's board of trustees and medical staff had issued public warnings about fracking. When Streck finished, the crowd erupted in an ovation that continued as Cole took the portable microphone for his three minutes in front of the board. "I don't suppose I'm in line for as much applause tonight," he said when the clapping subsided. Following a silent pause, he pressed on, assuring the crowd that he heard their concerns. "There is a good bit of misinformation in this exchange." He explained that Gastem was planning conventional wells, "which have been drilled safely throughout the state for generations." This work, using low-volume fracks, was not prohibited by the state's moratorium, and he was asking the board to amend the proposed town ordinance so he could continue.

Kurkoski, a tall man in his late forties at the time, watched with no discernable expression as Cole's assertion about the safe history of fracking was immediately challenged by the next two speakers. A contract lawyer who has litigated his share of cases, Kurkoski draws on a style more dependent on directness than oratorical flare. He also has a strategic knack for finding the right client. That person, Jennifer Huntington, was next to speak.

Huntington runs Cooperstown Holstein Corporation, a family farm with 500 head of cattle—mostly registered Holsteins—and operations spread across 1,000 acres, 394 of which were leased to Covalant. With broad shoulders, short-cropped hair, round spectacles, and a degree in animal science from Cornell University, she has the appearance and manner of somebody equally comfortable birthing a calf or talking economic theory. Speaking clearly into the microphone, she explained that revenue from the gas lease was vital to her business. She was a firm believer in the public benefits of shale gas, but no enemy of renewable energy. Gas royalties, she explained, would fund a better oil seed press to make biodiesel from soybeans. She could also use the money to update the farm's methane digester, built thirty years ago, which uses manure to produce natural gas sold to the county nursing home. She depends on clean water—her dairy cows consumed 4 to 5 million gallons of water a year. "This land and my family are my life," Huntington said. "I'm not going to spoil something I need to make my living and for future generations to come." She added, without anger but a touch of defiance, "I truly believe this is coming and I have a constitutional right to explore gas drilling."

Then it was Kurkoski's turn. He informed the board that he represented Huntington and thirteen other leaseholders, some of them third-generation farmers, who would sue if necessary to uphold their rights. "You do not have the authority to do this," he warned. "These are issues to be addressed in Albany, and your obligation is to uphold state law."

Kennedy spoke soon after Kurkoski. She walked from the back wall to face the board and began her three-minute turn by embracing Kurkoski's challenge. "Fourteen members of our community are now being asked to fight the gas companies' fight," she declared, adding that Kurkoski had misrepresented the state's authority. "Others in the community also have rights, and it is the board's responsibility to protect those rights." Drilling would irrevocably change the character of their community, an invaluable asset and a product of careful and deliberate planning reflecting the preferences of both current residents and earlier generations. Volunteers had surveyed more than four hundred Middlefield residents, she reported, and 85 percent of them were opposed to drilling. If there were doubts about the integrity of that survey, they should be eliminated by the sheer number of people who spoke in favor of a ban at this and previous hearings.

That number was indisputably large, even in Kurkoski's estimation. "That was the most anti-drilling crowd I ever saw," he told a reporter after the meeting. But there was something else he felt compelled to point out. Since the shale gas rush began, he had been to countless gatherings in rural fire halls, school auditoriums, and church basements, he said, and this was the first time he recalled one being interrupted with an announcement that a Jaguar was blocking a crosswalk and must be moved. I later asked him to elaborate on this observation. "It's Cooperstown," he said. "It could have belonged to any number of wealthy people there." The hearings in Otsego County, in his view, represented a privileged class with the time and resources to exert an outsized voice in policy matters. The voices of working-class members in favor of drilling were underrepresented. "Many people against drilling are well-off and they lead comfortable lives. And they don't want to change that. But they live apart from the broader community."

Northrup, who owns a Jaguar but was in Texas at the time of the hearing, had heard this assessment before from well-heeled industry stakeholders and their PR agents, and it struck him as cloying and hypocritical. It was a stretch for him to see people who inherited thousands of acres of land—the Dewey Deckers and Jennifer Huntingtons of the world—as indigent. "Kurkoski's remark fits a pattern of casting the opposition as elitist, pitted against the poor farmers. Of course the real battle has always been and still is between gas companies that can buy any state legislature or governor, against the residents they can so easily exploit." Northrup did allow, however, that Kurkoski could have selected a better place than Middlefield—a place where land preservation represents a salient cultural ethic and economic feature—to pitch a fight over drilling. Elite or not, the municipality had an ample supply of well-educated people with relevant professional expertise and policy experience, and the ability to donate their time to the home rule cause.

In researching the matter of a ban, the town board left little to chance, procedural error, or political miscalculation. On June 14, after exhaustive research, vetted by multiple public hearings that conformed to the letter of the law, the Town of Middlefield passed a ban on drilling and all other aspects of gas production and transportation. With that, the political discussion over drilling in the Susquehanna headwaters was over, and the legal fight was about to begin. On September 15 Kurkoski made good on his threat to sue Middlefield, with Huntington as the plaintiff. By that time, Dryden had passed a similar measure, and on September 16 Tom West, an industry lawyer and lobbyist based in Albany, filed suit against the Town of Dryden on behalf of Anschutz. Both plaintiffs and defendants were preparing to take the fight to the state's highest court. The issue was shot through by divergent political ideals and conflicts regarding the nature of individual and community rights within the state constitution. Money and control were also riding on the outcome and the case would attract powerful allies.

TURNING POINT

Stakes in the home rule case rose unexpectedly on June 13, 2012, with news that the Cuomo administration was ready to release the final SGEIS and begin issuing permits for shale gas development. The report, by *New York Times* staff writer Danny Hakim, appeared on the front page and was sourced to a "senior official at the State Department of Environmental Conservation and others with knowledge of the administration's strategy." Shale gas development would begin with fifty wells or fewer in counties along the Pennsylvania border, more or less along the path of the Millennium pipeline, and build out from there. This area, which included XTO's leasehold in Deposit, encompassed both the anticipated sweet spot for Marcellus development and a region that was reported to be facing economic hardships. In Hakim's report, officials pitched the plan as a pilot project to establish both the public safety and economics of unconventional gas extraction. Watersheds supplying New York City and Syracuse, along with areas deemed to be environmentally sensitive, would remain off the table, as stipulated in the most recent draft of the SGEIS released the previous fall.

The leak—confirmed by documents that would later come to light—was reasonably assessed by Albany pundits as a "trial balloon" for Cuomo to gauge public reaction. It was not incidental that the towns targeted for drilling permits were the towns without bans. The governor himself declined to elaborate on the leak, or even confirm it, but he did cast the idea of permitting shale gas wells in the context of the "state's versus local governments' role" over matters of public interest. It's natural for local governments to resist state control over certain issues that affect localities, Cuomo said, the same way that states justifiably resist federal control. "The state takes the position of, 'Thank you federal government for your opinion, but we're going to do it our way,'" he said in an interview with columnist and radio host Fred Dicker days after the leak was reported. "Many times a town or a city will say, 'That's nice, I'm glad you think that Albany, but stick to Albany. . . . Government closest to home knows the best.'"

If it was the governor's intention to gauge public reaction to fracking in New York before committing to it, then the leak was an impressive success. While gas supporters hailed the project as an economic gift to struggling communities in the Southern Tier, the anti-frackers characterized areas targeted for development as "sacrifice zones" and the residents as "guinea pigs." The DEC's environmental review had deemed fracking too risky for watersheds that serve New York City and Syracuse, they reasoned, so it was immoral to think that it could be tolerated in other areas on a trial basis. "We see it as a matter of environmental justice," said Steingraber, the ecologist and author who in 2012 became a central figure in New York's anti-fracking movement. "A child who drinks water from an unfiltered well from an aquifer in Broome County has the same rights as a child in Manhattan who drinks from an unfiltered water supply that begins high in the Catskills."

Fracking protests, by this time, were becoming commonplace in Albany. Demonstrators—sometimes numbering in the dozens and sometimes in the hundreds—relentlessly shadowed Cuomo at scheduled public appearances. They stood on the fringes of these events for the most part, but occasionally interrupted with guerrilla-style tactics. During a policy summit on economic development hosted by Cuomo in midtown Manhattan, a magician and his assistant appeared out of the blue on the floor of the Sheraton convention room to demonstrate the "miracle of hydraulic fracturing" by showing "how the governor will turn water into brine." The crowd, which had been following a PowerPoint presentation on the economics of dairy farming, watched the magician, bantering on about his trick, pull a bottle of clear liquid from his suit jacket and pour it into a transparent cup. A security officer, slowed by linen-covered tables, many seated guests, and an effort to appear calm, eventually reached the man commanding attention in the center of the room. The magician, accepting his escorted exit as if part of the show, handed the cup of now-murky liquid to his "beautiful assistant." The young woman, in a sleeveless black dress, turned her back on a second security officer. In apparent indecision the man hesitated before reaching out a hand to escort her from the room. In that moment she downed the contents of the cup and, after weaving about dramatically, collapsed. The security officer either failed to catch her or decided against it, an outcome that heightened both the dramatic impact of the protest and the sense of discomfort for those watching.

While this and similar agitprops found their marks in the summer of 2012, the fracking movement was uncoordinated and divided between groups that favored Occupy-style guerrilla tactics and old-guard networking within established political channels. Mainstream environmental groups, including the NRDC, the Sierra Club, and the Environmental Defense Fund, were pushing for more assurances and stricter regulations for fracking, while the start-ups and left-leaning groups insisted on a ban. I spoke with leaders of the start-ups—including Steingraber—who found the institutional environmental lobbies too willing to compromise, and I spoke with members of the mainstream groups who found the grassroots groups unrealistic and impractical.

News of the fracking roll-out in the Southern Tier in the summer of 2012 proved to be galvanizing, however, and it precipitated an event on August 27 intended to leave no doubt about the political will, influence, and organizational wherewithal of New York's anti-fracking movement. The ensuing march on Albany—to be called "Don't Frack New York"—was organized and executed by groups favoring a ban and willing to promote a "pledge of resistance" to fracking, but it received behind-the-scenes support from mainstream groups. "Everyone could acknowledge that both political access and populist power were important, though seldom achievable by a singular entity," later recalled Roger Downs, conservation director at the Sierra Club, who helped rally organizers negotiate the permitting process to secure venues and streets, but declined a formal role in the march itself. "We all shared information, coordinated strategy, and helped shape messaging. Trust was built upon an internal understanding that no one would ultimately sell out—even though the public narratives [between the groups] may have been very different—even contradictory."

In the days preceding the event, organizers gathered at a church in downtown Albany for rehearsals, workshops, the making of signs and props, and counseling on the dos and don'ts of civil disobedience. Efforts were coordinated by dozens of team leaders and helpers representing regional and national grassroots organizations including 360.org, Frack Action, New Yorkers Against Fracking, Catskill Mountainkeeper, Catskill Citizens for Safe Energy, Food and Water Watch, Green Umbrella, and others. Bill McKibben, the founder of 360.org, and Steingraber embodied the social consciousness of the event, and they were joined by actress Debra Winger and Josh Fox as rally headliners.

On the morning of the rally several thousand protesters—by then they were known as "fractivists"—arrived via bus and carpool at the Corning Preserve on the Hudson River. After an impassioned speech from McKibben on fracking as a source of global warming, they began marching to the capitol. From the east-facing windows of the fourteenth floor of the DEC executive offices, Commissioner Joe Martens could hear and see the animated throng crowding the street and spilling onto the sidewalk as it approached behind a mounted police escort. Soon the rollicking mass would envelop the base of the agency's headquarters. Martens was scheduled for a meeting outside the building and to avoid being trapped or inadvertently becoming a featured part of the rally, he hastily exited before the crowd overtook the city block. A few minutes later, he ducked unnoticed into a building on a side street to watch the spectacle unfold.

On cue, the front ranks of the protest, thick with people, banners, and signs, came to a halt in front of the DEC headquarters and for several minutes the call and response of "Don't Frack NY . . . CUOMO!" reverberated through the streets while the rear of the procession accordioned into the block. A contingent broke off and space was cleared for a street theater performance. An 8-foot-tall Andrew Cuomo puppet ambled in front of a drilling derrick while musing over the prospects of gas development. A devil to his left urged him to allow drilling, while tucking bundles of cash into the governor's pockets and, sensing hesitation, threatening the governor's political aspirations. To the governor's right, an angel urged the governor to "listen to the scientist and medical experts . . . you will poison the water forever and people will suffer just for money." Loud cheers and boos filled the streets for fifteen minutes. Then the marchers—on cue and in synch—resumed their shouting and continued to the capitol, where there was more chanting and speechmaking. At another stop, demonstrators linked arms and chanted: "DEC, if you will not protect us, we will protect ourselves! Governor Cuomo, if you sell us out, we will force you out of office! Gas industry, first we will block you, then we will remove you from our state!"

At the capitol, Steingraber and a trailing swarm of picketers flowed through the executive wing—the Hall of Governors—and hand-delivered a special package to Cuomo's office: 3,300 pledges to take nonviolent "direct action" should the governor allow gas companies to drill in the state.

It was a spectacle, even in a city renowned for political circus. The morning after, the political liaisons of the state's established environmental groups received calls from the DEC commissioner's office announcing a meeting in the boardroom on the fourteenth floor of the agency's headquarters. Kate Sinding, an attorney for the NRDC, remembered the call. "Attendance was 'strongly advised,'" she said. "That meant it was mandatory."

Following long-held Albany protocol, the administration sought input from established lobbies on policy issues and this instance was no different. Representatives from the Sierra Club, the Environmental Defense Fund, Environmental Advocates, Riverkeeper, and NRDC were summoned to the table. None of these groups were formally connected with the previous day's march—condoning a pledge of resistance was outside their political comfort zone—but some of their leadership and many of their members supported the event. As far as the Cuomo administration was concerned, this contingent of mainstream environmental NGOs represented a critical mass of political influence that shaped the fracking dilemma, and it was prudent to get their reading on matters in light of the opposition that was taking the political fight to the streets.

DEC commissioner Martens was a well-respected figure within the environmental community owing to his years as head of the Open Space Institute, and he was on a first name basis with just about everybody in the room. In addition to one or two representatives from each of the NGOs, attendees included DEC executive deputy commissioner Marc Gerstman (who had replaced Stu Gruskin in heading up the SGEIS review), Department of Health (DOH) commissioner Nirav Shah, and various bureaucrats and aides from the two government agencies—fourteen people in all.

The mood was tense. There was no formal agenda and Martens began with a request that the meeting remain off the record. From there, the two commissioners spent much of the next hour, unprompted, defending the work that went into the SGEIS, the exhaustive public hearing process that lead to revisions, and the overall soundness of the review, by then in its fifth year.

The meeting room—decorated with pictures of eagle's nests as tribute to the agency's American eagle recovery program—could hold sixteen people, although some of those in attendance recall it feeling cramped as the commissioners waxed extravagantly on their talking points. The early part of the meeting was "dominated by assurances by both agencies that they were confident of the job they had done with the SGEIS and they could properly mitigate all public health impacts associated with fracking" a representative from one of the groups recalled. "We felt unnerved about these claims." Shah reminded the attendees that poverty itself was a health risk, and that the public health equation warranted consideration of the shale gas proposal for the Southern Tier as an antidote against the ills of poverty. With this, one source told me, "there was some eye-rolling, because even then we could see that the Pennsylvania boom was not reaching the rural poor."

Frustrations mounted, but representatives from the environmental NGOs sensed that the commissioners were not acting on their own, and that the meeting was intended as something more than damage control in the wake of the tectonic response to the leak about the phased drilling plan. "There were expectations that the Big Green environmental lobby—the NRDC's of the world—might go along with the plan to roll out fracking on a limited basis," Sinding recalled. "And they learned clearly this was not the case."

"How do you see this unfolding?" Martens finally asked the group near the end of the meeting. There was consensus from the NGO delegation: an independent study of health impacts from fracking would best serve public interest. By this, they meant an Independent Health Assessment, or IHA, as it's formally known—an exhaustive and legally binding administrative tool, like the SGEIS, that requires formal public review. It would likely take years to design and complete.

The request for a health review was not new. Public health advocates began lobbying for this soon after Governor Paterson announced the environmental review in 2008. With the second draft of the SGEIS released in the fall of 2011, the call grew stronger from many qualified groups and individuals. In January 2012 two groups representing county health departments prepared separate reports on the burden natural gas drilling could place on their operations. The analysis included this blunt assessment by the New York State Conference of Environmental Health Directors: The "program proposed by the (DEC) is inadequate to protect public health and to detect and remediate contamination of drinking water aquifers." The New York State Association of County Health Officials urged "an expert, independent, and evidence-based study of potential public health impacts, preventive approaches to mitigate human health risks, and estimated related costs prior to lifting the moratorium on permits."

The authors of the SGEIS were against this approach because, according to an internal report, it would "involve making a large number of assumptions about the many scenario-specific variables that influence the nature and degree of potential human exposure and toxicity." With proper oversight "the Department expects that human chemical exposures during normal HVHF [high-volume hydraulic fracturing] operations will be prevented or reduced below levels of significant health concern."

At the meeting in the commissioners' boardroom, Shah clung to the argument that the SGEIS accounted for all human exposure pathways and, with proper regulation, they would be eliminated or "mitigated." Beyond that, he continued, the state's ability to track health impacts—changes in asthma rates, birth weights, cancer instances, venereal diseases, accidents—in real time through a sophisticated database of hospital records was "unmatched." With this resource, the state would be able to assess health patterns within drilling communities and flag any problems.

Throughout the meeting, there was a tacit dynamic at work that added to the tone of defensiveness. It was later explained by one of the attendees this way: "No one needed to talk about lawsuits. It was generally accepted that any move forward with fracking would bring on litigation from both sides."

The DEC, not the DOH, was the lead agency, and the final decision over shale gas development would fall to Martens, or somebody higher up. Martens and Shah had confidence in the plan for a phased approach to gas development. But there seemed to be reason for more deliberation, Martens suggested at the conclusion of the meeting. The agency would consider a more extended review of the health questions. As the meeting broke up, he added that the agency had been acting in good faith to account for all the variables and a little praise for its work would be appreciated.

Three weeks later on September 20, the DEC issued a statement announcing that Martens had asked the state's health commissioner to assess whether the administration had sufficiently covered the potential for health problems. To help with this, Martens directed Shah "to identify the most qualified outside experts" to advise him in his review. Martens cited two reasons for the reassessment. One was, of course, health. The other was lawsuits: "I want to make sure that we have done the most thorough review possible, especially when it comes to public health concerns. In addition, I want to ensure that the Department has the most legally defensible review so that when the Department issues its final determination on this matter, protracted litigation is avoided, whatever the outcome."

Reactions were mixed. An internal review fell well short of the public IHA that the groups had asked for. But it was still a marked departure from the plan for a fracking roll-out by the end of the year. It would help inoculate the administration from legal claims that the agency had failed to do its due diligence in assessing health impacts. It would also keep the decisionmaking process in the hands of the agency and, by extension, the governor.

Moreover, it would buy the governor more time.

THE COURT'S DECISION

While a decision over shale gas permits by New York's executive branch dragged on into a sixth year with no clear endpoint, the home rule case hastened through the judicial branch in just over three years—swift by precedent-setting legal standards.

In late June 2014, Nicole Dillingham heard from Helen Slottje who had heard from a law clerk that a decision on home rule from New York State's highest court was imminent. Final victory or defeat was at hand for the battle over home rule that had begun with Orville Cole's 2008 prospecting well. Dillingham's training and disposition inclined her to position herself for such uncertain moments, so she set about writing two press releases, one anticipating each outcome, and then held her breath.

The case had made news each time it progressed through the New York court system and the high-profile saga drew an impressive ensemble of legal talent. Deborah Goldberg, the EarthJustice attorney from Manhattan, had joined Helen Slottje as a principal figure representing the towns. Goldberg, barely five feet tall with the carriage of an elite gymnast, had earned the nickname "Lady Chainsaw" for her potent mix of refinement and ruthlessness in the legal arena. Tom West had joined Scott Kurkoski to head the legal team representing the industry and landowners. West was a young grandfather who still enthusiastically partook of his lifelong passions for downhill skiing, waterskiing, and kiteboarding. He possessed charm and a contagious energy, and had a smile that could be seen as winsome or devious, depending on one's opinion of gas lobbyists. Articulate, sharp, and photogenic, the attorneys got plenty of practice arguing the case on radio shows, television programs, and in law clinics when not in the courtroom.

Legal careers and organizational interests staked to the outcome of the case reached far and wide, and many of the latter were expressed in the amicus curiae briefs filed with the court. The term, Latin for "friends of the court," refers to legal opinions filed by stakeholders to help judges weigh public interest in the case. In the home rule case, more than fifty towns and villages filed briefs, along with dozens of organizations representing planners, mayors, businesses, landowners, farmers, sportsmen, environmental causes, and industry. Parties ranged from the American Petroleum Institute to Theodore Gordon Flyfishers.

Rulings from the lower courts came quickly, and were decisive. On February 21, 2012, eight months after West filed the complaint against Dryden in Tompkins County, State Supreme Court Justice Phillip R. Rumsey ruled that the language of the Oil, Gas and Solution Mining Law—including the supersession clause—was too vague to prohibit Dryden from banning oil and gas operations. The state lacks "clear expression of legislative intent" to override local zoning controls with respect to oil and gas production, Rumsey wrote. The ruling also noted that the state law was last amended more than thirty years ago, "long before the potential use of hydrofracking to recover natural gas from the Marcellus Shale in New York could have been anticipated." Just three days later, Acting Supreme Court Justice Donald F. Cerio Jr. upheld the Town of Middlefield's ban. Cerio ruled that the state's jurisdiction applies only to "the method and manner" of oil and gas production but does not "preempt local land-use control."

West and Kurkoski fared no better on appeal, where the two cases were heard as one. In May 2013 a four-judge panel in the Appellate Division unanimously upheld the lower court rulings. That left the lawyers representing Huntington and Anschutz—landowner and industry—with one last recourse. West knew that a hearing before the state's highest court was a long shot. In 2012, the Court of Appeals heard just 64 cases out of nearly 1,000 requests. But West counted on the case's stature and sweep to attract the high court's attention and subsequent events would prove him correct.

An announcement by the Court of Appeals on August 29, 2013 that it would hear the case gave industry one final winner-take-all chance. But it also gave home rule advocates a chance to remove any ambiguity in the lower court rulings, closing the door to challenges in other districts. "Lawyers might be interested in this dragging on. But it gets expensive and it's a long process," Slottje told me. "There's no quicker or easier way than having the case go to the Court of Appeals." West characterized Slottje's remarks as "a rare point of agreement between us. . . . This will be definitive. However the Court of Appeals rules, that's it. "

With much of the heavy lifting done first in the trial courts and then with the appellate briefs, the attorneys made their oral arguments before the high court on June 3, 2014. In characteristic form, the appellate judges seemed short on patience with both sides, often cutting off the attorneys in mid-sentence with their own questions and observations, many rhetorical, some sincere. But several of the seven judges struck Dillingham, who watched the proceedings online, as especially tough on West and Kurkoski. The industry attorneys were attempting to show why the *Frew Run* precedent, which subjected gravel mining to local zoning despite state control over operations, should not apply to shale gas development, in part because shale gas encompassed a more crucial public interest. The judges tended to redirect the arguments to the semantics of the oil and gas law, and the intent of those who legislated it. "It's pretty drastic to change long-time residential zoning restrictions," Judge Victoria Graffeo told West at one point. "One would think the legislature would be a little more explicit [in the oil and gas law] if that's what they intended to do." Chief Judge Jonathan Lippman and Judge Jenny Rivera also marshaled their questions to probe the issue of legislative intent.

A ruling was expected within a month, and with typical punctuality the court handed down its decision on June 30, 2014. By a 5-2 margin, the court ruled in favor of the towns. Graffeo, writing for the majority, validated the argument that banning an industry was altogether different than regulating it. "Plainly, the zoning laws in these cases are directed at regulating land use generally and do not attempt to govern the details, procedures or operations of the oil and gas industries," Graffeo wrote. The court found it significant that the towns "both studied the issue and acted within their home rule powers in determining that gas drilling would permanently alter and adversely affect the deliberately cultivated, small-town character of their communities." The majority ruled there was nothing in the language of the oil and gas law, plainly interpreted, to allow the state to supersede duly enacted zoning laws. The law also lacked the "legislative intent" and "clear expression" necessary to preempt zoning.

Judge Eugene F. Pigott Jr. wrote a dissenting view supporting the complaints filed by Kurkoski and West. The terms of the zoning ordinances, Pigott wrote, "go into great detail concerning the prohibitions against the storage of gas, petroleum exploration and production materials and equipment in the respective towns." In essence, "The zoning ordinances of Dryden and Middlefield do more than just regulate land use, they regulate oil, gas and solution mining industries under the pretext of zoning."

When the decision came, Dillingham was en route from Hawaii, where her husband does business as a dealer of antiques and fine art. Shortly after arriving in Los Angeles on a red-eye flight, she received a message from Goldberg declaring victory. Between flights, Dillingham was on the phone to her colleagues and putting the finishing touches on the winning press release. By the time she landed in Washington, D.C., and embarked for Albany, the staff and volunteers at Otsego 2000 were arranging for a victory celebration.

It was six years almost to the day since XTO Energy had treated Deposit landowners, organized by Dewey Decker and represented by Jackie Root, to a white-linen banquet at the Binghamton Regency Hotel. That celebration, featuring tenderloin tips and chocolate mousse, was one of anticipation— that the industry Decker described as the "giant coming over the mountain" would shower Southern Tier landowners with financial abundance. Now, on an early July day in 2014, the community organizers of the home rule movement gathered to celebrate the barrier they had erected in the giant's path.

The site for the event was Origins Cafe, a vegetarian restaurant in the courtyard of an organic nursery ten minutes south of Cooperstown. It was a mild evening, with the sun still high above the horizon as people began to gather. Women in sundresses were accompanied by men in jeans or khakis and open collar shirts. A few wore sport coats. They strolled on tile walks amid hanging baskets lush with summer flowers and gazed out over acres of green lawn and well-tended gardens. A jazz trio played. Larry Bennett, one of the first to raise the alarm when seismic testing began near his Brewery Ommegang, drew frothy mugs of Belgian-style beer. Food soon arrived: salads topped with edible flowers, lasagna, curries and rice, local cheeses, fruits, and drink choices of smoothies, herbal teas, or chai. As people enjoyed their meals, the jazz trio began packing their instruments and a cover band that played Jefferson Airplane and Buffalo Springfield songs set up for the after-dinner entertainment.

Dillingham led an informal program recognizing certain neighbors, council members, merchants, health care providers, parents, and activists. There were speeches and toasts celebrating grassroots political activism and the community's strength in the face of exploitive outside threats. The speeches were both sincere and humorous. "It was a kumbaya moment if there ever was one," Dillingham said, referencing the spiritual folk song popularized in the 1960s by Pete Seeger and Joan Baez, and a term intimately connected with the civil rights movement, war protests, and various left-leaning causes. The gathering in the Origins garden, she continued, "expressed everything about who we were . . . and why local control was so important."

There was a similar party for community organizers in Dryden, but not everybody within the anti-fracking movement was celebrating the home rule victory. For Walter Hang, the advocate who had built his case on the deficiencies of the SGEIS, the reliance on towns to deliver New York from fracking was a step in the wrong direction. Governor Cuomo, who had successfully managed to avoid the politically loaded issue as the 2014 elections neared, had suggested in 2012 that drilling could begin on a trial basis in the towns along New York's border with Pennsylvania, where gas reserves were thought to be the richest. Now, Hang feared, the courts had handed Cuomo a fresh opportunity to seize on a town-by-town approach to gas development, rather than establishing comprehensive statewide safeguards. Hang was concerned that the logic that would prevent drilling in one town would encourage it in another. "The reason there is no drilling now in New York has nothing to do with home rule, and that's not going to keep it from coming in the future," he said. "It's about tactics and strategy with state government. It's about getting a fatally flawed policy, the SGEIS, off the table and going back to the drawing board."

Whether it happened town-by-town or not at all, Cuomo had the exclusive authority to set shale gas development in motion in New York. If his administration approved the SGEIS, then the high court's home rule decision would become keenly relevant—even in towns *without bans*. There was a simple reason for this: in upholding the prohibitions in the towns of Middlefield and Dryden, the high court set precedent that zoning applied to shale gas development in *any town*. With that, operators were effectively prohibited from drilling in areas not explicitly zoned for the type of industrial activity that comes with unconventional gas drilling and fracking.

That was a problem for drilling supporters in the Town of Sanford, where the world's largest energy company held rights to 50,000 acres. Dewey Decker, the Town of Sanford supervisor and head of the Deposit landowner coalition that had leases with the Exxon Mobil subsidiary, XTO, had been waiting for more than six years for approval from Albany, but he never expected that a fracking ban by towns in another part of the state would affect prospects in his own town.

"It caught everybody by surprise," he told me. "It changes everything."

More than 80 percent of the Town of Sanford, including most of the land leased to XTO, is zoned as agricultural. With the home rule precedent, shale gas development would have to be somehow reconciled with the existing land use plan, which recommends provisions that "preserve agricultural character of the Town by encouraging compatible agricultural uses and limiting inappropriate, non-agricultural uses."

Dewey Decker's problem was also the problem of landowners in other rural communities eager to lease their land to prospectors. More than three-quarters of the state's 1,511 municipalities have zoning or land use plans, and that percentage includes virtually all towns sitting over the part of the Marcellus Shale along the Pennsylvania border—thought to be most viable in New York. In this Southern Tier region, the largest expanses were zoned as agricultural and residential, and none of the zones identified fracking as a permitted use. "You can't just stick an unconventional gas well—and all the industrial activity that goes with it—in an area zoned as agricultural, even if there is no ban," Slottje explained. If that happens "it will be subject to legal challenge as incompatible with the comprehensive plan."

Wholesale zoning changes—across multiple towns and multiple zoning districts—would therefore be necessary to enable shale gas development in any economically feasible way, and that would require lengthy reviews, public hearings, and town board action. In short, New York's fight over fracking, in the absence of a statewide ban or further delays by Cuomo, would return to municipal halls, where it had all begun.

THE EPA RESULTS

In contrast to Orville Cole's short-lived and ultimately fruitless venture in upstate New York, Cabot's operations just across the border were booming. By the end of 2011, as the EPA was embarking on its study in Dimock and the home rule case was wending its way through the courts in New York,

Cabot had drilled more than two hundred wells in Susquehanna County and they were producing tens of billions of cubic feet of natural gas. The number of Marcellus wells throughout Pennsylvania approached six thousand, and eight of the top ten producers were Cabot wells in and around Dimock. These were true blockbusters—the pride of the company and a showcase for investors—and CEO Dan Dinges wanted to drill more.

Cabot's success came despite a prohibition—imposed by John Hanger's DEP and still in effect—from drilling in the 9-square-mile zone around the Carter wells. The prohibited area was a lucrative one—the bull's-eye of the gas bonanza in that neck of the woods. Negotiations with the DEP under Michael Krancer were going much better for Cabot; absolution from water deliveries to litigants was an auspicious turn of events. But the next step—the resumption of drilling in the sweet spot around Carter Road—was complicated by the EPA investigation. The outcome of those water tests could affect public perception of fracking and, more broadly, advance the argument for more federal involvement in Dimock and elsewhere.

Three months after they collected water samples from Craig and Julie Sautner's basement, a team of federal officials returned to their home on Carter Road to report the results. It was a damp, overcast morning in late April 2012 and Richard Rupert, EPA's site coordinator; Trish Taylor, the community involvement coordinator; and Karl Markiewicz, a toxicologist from the ATSDR, removed their shoes at the door. They were followed by Lisa Barr and Vera Scroggins, videographers and anti-fracking activists, who joined the party settling in at the kitchen table. Taylor announced that the EPA would also keep its own tape of the meeting and the first matter of discussion concerned who would film from where, followed by a brief arranging of cameras, plugs, and seating. Taylor, distributing handouts, said that she hoped the filming by outside parties would not change the dynamics of the conversation between the Sautners and the EPA. Craig Sautner said he was glad to have the records "for protection" against misrepresentation.

Trish handed Craig and Julie a copy of a letter with a packet of data. The letter, she explained, was from an agency toxicologist who reviewed the water results to flag potential health issues. Taylor summed up the toxicologist's assessment in a single sentence: "She didn't see anything."

There was a brief silence as Craig and Julie scanned the letter. Then Craig spoke: "So this is telling me that this is okay to drink?" he asked about his water. Taylor said yes, and began explaining the process of the review. Craig broke in, pointing to a line in the letter. "It says—'for consumption' . . . So in other words, I'm allowed to drink this water? And it's not going to cause me any harm?" Taylor said that was correct, and she returned to her explanation of the protocol. "I beg to differ with you," Craig interrupted again, his voice rising. "I just can't believe it . . ."

Julie Sautner asked if the health review took into consideration the history of problems at the household. "It was Cabot's very own test that sent you guys in here to begin with," she reminded them. Rupert, listening at the head of the table with his legs crossed and fingers laced over a knee, leaned forward and rested his forearms on the table. He picked up a pen and turned it in his fingers. "My understanding of the basis on which we came in actually had to do with some error in the data," he said. "I've looked at all your historical data all the way back to 2008. And other than some bacteria that seemed to be somewhat persistent early on we didn't see anything in there that raised any alarms in our eyes. I've seen your water. I've tasted your water. It didn't taste unusual. Actually, compared to the water at the hotel, it didn't have all the chlorine."

"You're serious?" Craig asked.

"I'm absolutely serious," Rupert responded.

Craig, his voice rising in pitch and intensity, recounted the history of problems beginning with the discoloration and odor in the water pumped from their well after Cabot drilled nearby, the report from the DEP linking the problem to Cabot's operation, the ineffective filtration system installed by Cabot, the stains on the laundry, and rashes his kids developed after showering. The conversation turned to the paperwork in front of them—pages filled with columns and rows of compounds, values, ranges, and detection limits. For ten more minutes, the conversation went back and forth about thresholds, qualifiers, estimated values, and scientific uncertainties. Tempers flared. "It's a natural system," Rupert said. "This is nature. Okay? You are not going to pull water from the ground and not find iron and all these different minerals. It's flowing in the rocks."

Craig jumped up. "You're saying my water's safe. Okay. I'm going to get some. I want you guys to drink it." Rupert flatly declined. "Why not?" Craig yelled. "You think I can drink it. Why can't you drink it?" He returned to the table and slammed an assortment of plastic bottles of water on the table one by one. "Which one do you want to drink? Come on! Take your choice!"

"I'm not drinking your water," Rupert said again. "It has bacteria problems." Craig explained that his septic system was on the opposite side of the house from his well. "What the hell is wrong with you people?" he shouted.

Rupert exchanged glances with Taylor. "I think we're done here," he said. After several more verbal offensives from Craig and now Julie, the EPA team began collecting folders, donning coats, and working their way toward the door. Their exit was delayed by the task of putting their shoes on and a latch on the storm door that seemed to be jammed. Craig paced furiously about the kitchen and the adjoining living room shouting that he could not believe what was happening. "You people have changed!" The EPA team exited through the back door to a cacophony of dog barks. A moment later, Craig met them in the driveway with a jug of water and a lighter.

"I know you have methane in your water" Rupert said after he climbed into the driver's seat of a small sedan.

"And that's okay?" Craig asked.

"A fire and explosion danger . . ." Rupert paused. "I don't think levels are high enough for that." His colleagues climbed into the car and buckled up as Craig again challenged them to drink the water. Rupert put the car in gear and backed out of the driveway.

Barr and Scroggins posted footage of the encounter online, where it was immediately seized by partisans to support opposing arguments: There was nothing wrong with Dimock water versus the EPA was covering for industry. The investigation had promised a reliable reference point falling somewhere within the limits of science and outside the reach of propaganda, and now the results were in. It was about science. But it was also about policy, and therefore it was about politics.

President Obama, whose endorsement for natural gas development was spelled out in the report "Blueprint for a Secure Energy Future" in 2011, was preparing a platform for the 2012 campaign that included natural gas development as "a priority" in a broader energy plan to stimulate middle-class opportunities, economic recovery, and national independence. Inconveniently, Dimock was one of three EPA investigations into fracking's impact on water and all of them were providing grist for some of the president's most powerful critics.

The EPA investigation of the Wind River Reservation in Pavillion, Wyoming—begun before Obama was elected—had produced damning results. After nearly three years of study, the agency's draft report released in late 2011 confirmed its preliminary analysis: the presence of dozens of hazardous chemicals and carcinogens in the aquifer, including glycol ethers, benzene, and "an assortment" of other organic components "is explained as the result of direct mixing of hydraulic fracturing fluids with ground water in the Pavillion gas field." This was the first high-profile scientific evaluation to explicitly link fracking operations to water contamination. That it came from the federal government was significant and the agency was submitting its findings for peer review, which would strengthen the report's standing. The results provoked immediate condemnation from congressional officials representing gas interests. Senator James Inhofe, a ranking member of the Senate Environment and Public Works Committee, took the lead by stating that the Pavillion study was "part of President Obama's war on fossil fuels and his determination to shut down natural gas production."

A similar battle was about to come to a head in Parker County, Texas, where the EPA linked contamination of an aquifer to nearby gas production by Range Resources. In 2010, the EPA issued an emergency order under the Safe Drinking Water Act for the company to provide water to affected homes. The EPA report, notably, ran counter to findings by the Texas Railroad Commission.

The president's staff was keenly aware that the public perception of the EPA investigations could provide material on the campaign trail for those eager to paint him as a proponent of government overreach at the expense of jobs and energy independence. The regional director of EPA's office in Texas, Al Armendariz, had provided the president's adversaries with deadly ammunition in this war of perceptions. During the Dimock investigation a story broke that Armendariz had explained to a community group in 2010 that the EPA would handle polluters by making examples out of them. To illustrate his point he made an analogy to the Roman armies of antiquity. "They'd go into a little Turkish town somewhere and they'd find the first five guys they saw and they'd crucify them," Armendariz said on tape. "Then that little town was really easy to manage for the next few years." Obama readily accepted Armendariz's resignation on April 26, 2012, days after Inhofe introduced the administrator's remarks to the country in a speech on the Senate floor. As the previously obscure video went viral, Speaker of the House John Boehner tweeted "Obama admin admits 'crucify' strategy for energy job creators." Congressman John Fleming, a Republican from Louisiana, characterized the comment as "enviro-fascism at its worst."

Perhaps the most sustained and influential criticism of the EPA's forays into natural gas issues, however, came from Darrell Issa, a ranking House Republican and chairman of the Committee on Oversight and Government Reform. With the hearty support of John Cornyn of Texas and several other

Republican senators, Issa had begun a probe in the summer of 2011 into the Obama administration's handling of shale gas development. The effort culminated with hearings scheduled in the spring of 2012, just as the EPA investigation unfolded in Dimock. The committee summoned industry parties who portrayed the EPA as secretive and subversive. Michael Krancer was a star figure. "My prime advice [for the EPA] would be to back off," Krancer told the committee. "Trust the states . . . the states in which this is happening know exactly what they are doing."

The president left these public political thrashings mostly unanswered, but the action of his administration spoke volumes. The EPA withdrew its emergency order against Range Resources and, in March 2012, abruptly closed its investigation in Parker County. Four months later, the agency would announce an end to its involvement in Dimock. Lisa Jackson, under attack and with no public support from the president on the fracking investigations, would resign her post as EPA chief at the end of 2012. In late June 2013, the EPA would also drop its work in Pavillion and abandon plans for a peer review of its findings.

I spoke with officials at various levels within the agency. On and off the record, the message was the same: fracking was politically charged. Decisions about local investigations were made at the highest levels and rank-and-file operations, including interaction with the media, were micromanaged from above. "When we first started, there was more open exchange of thoughts and ideas," said Greg Oberley, a team coordinator and author of the draft report for the Pavillion investigation. "I know it sounds bureaucratic, but we're pretty much limited to the official statement and press releases." He added, "The decision [to drop the Pavillion investigation] was made above me."

Ken Ely's son Scott, who had worked for Cabot, also had a lot riding on the EPA probe into Dimock's water quality. It was both a question of health and a matter of reputation. After Scott reported to the Pennsylvania DEP in late 2009 that he had personal knowledge of and was witness to various spills and cover-ups, the agency referred the matter to Cabot and the company's environmental contractor, URS. Cabot released the results of that investigation at the end of 2011: URS found no evidence of spills or impacts that would require cleanup at any of the thirteen work sites in question. "All of Scott Ely's allegations are now confirmed by the study to have lacked substance," Cabot announced in a press release. A follow-up by the DEP later found the URS report contained "unclear, misleading or only partial" information at one well site, and failed to acknowledge evidence of spills and seeps from at least eight of the sites. But the DEP took no formal action on the matter.

Like the Sautners, Ely was expecting retribution when the EPA team visited his home in the spring of 2012 to report the outcome of his water tests. The party gathered at the kitchen table, by a broad window facing the forested slope of the mountain where Scott and his late father had hunted. Ely lit a cigarette as Rupert opened his file and handed him pages of raw data. Some values were highlighted in yellow and Rupert explained these showed elevated levels of some "naturally occurring" compounds. Among them were chromium and lithium. Ely knew lithium and chromium are metals typically found in high concentrations with other compounds in geologic layers well below freshwater zones. Although lithium is among hundreds of elements in freshwater with no regulated limit, the EPA's "trigger level" to flag potential health problems is 31 ug/l. The concentration recorded in Ely's water was seven times higher than that. Chromium was below the regulated limit of 100 ug/l, but three times higher than the EPA's trigger level of 10.8 ug/l. Other compounds that exceeded trigger levels included sodium and arsenic.

Rupert began explaining the results in the context of natural systems, although Ely needed no lecture about how water wells worked. He understood that the ground contained things that were safe at some levels and unsafe at others. The EPA tests offered no more than a snapshot of an ever-changing system subject to influences from nature, from man, and from man's interaction with nature. While working for Cabot, Ely had seen thousands of pounds of chemicals mixed, injected, and occasionally spilled above ground. He also understood the mechanics below ground. Drill bits perforated the aquifer and churned through miles of highly pressurized natural systems, providing conduits for elements to move from one zone to another. Cement casings and other controls worked, except when they didn't.

"Is my water safe or not?" Ely finally asked. Rupert responded that it was safe with treatment. That task was being handled by Cabot and the state. The water never needed treatment before Cabot began drilling, Ely replied. Rupert could not address that. Determining the source of pollution was "outside the scope of the investigation," he said.

Through the spring and early summer of 2012, the EPA delivered reports to the other sixty-four households that had been tested and, on July 25, the agency announced the end of its investigation in Dimock. That same day the EPA issued a one-page press release which contained this sentence: "An evaluation of the particular circumstances at each home did not indicate levels of contaminants that would give EPA reason to take further action." That message, coupled with a rush of press releases from industry interests declaring victory, was widely interpreted in mainstream news coverage to mean that the federal agency had found Dimock water to be safe. A *Wall Street Journal* report captured the essence of the public message: "The Environmental Protection Agency has given the all-clear to drinking water in Dimock, Pa., a tiny town that received national attention after people there voiced concern about contamination from hydraulic fracturing." This and similar reports by the Associated Press, *USA Today,* and other mainstream outlets did not pick up on the agency's finding of "hazardous substances . . . at levels that could present a health concern" in well water sampled at five homes. The problem wells represented 8 percent of the total number tested. The compounds were fairly consistent—methane, barium, arsenic, sodium, lithium, and chromium. The contaminants happen to be both naturally occurring and a common result of drilling. The EPA justification for abandoning the investigation was summed up in the very last sentence of its press release: "In all cases the residents have now or will have their own treatment systems that can reduce concentrations of those hazardous substances to acceptable levels at the tap."

QUIET ENDING

The contentious public fight between Cabot and the Carter Road 15 came to a quiet ending on July 25, 2012. Hours after the issuance of the EPA press release, Dan Dinges told investors in a conference call that the company had reached a verbal agreement to settle the claims of Dimock plaintiffs. The agreement would save the company significant legal fees, Dinges said, while the payouts "were not a material item with respect to Cabot's financial statements." After years of warring in the public eye, Cabot had nothing else to say on the matter, nor did the plaintiffs' counsel.

By the time of the settlement, the number of plaintiffs had increased from the original Carter Road 15 to 36 households. Of these, 32 agreed to Cabot's terms. In conversations that followed, I sensed that many were conflicted, battle weary, and resigned to accept the deal on the advice of their attorney. The new offer, with few exceptions, was no better and in some instances was worse than the old one brokered through the Pennsylvania DEP that had stipulated payments of twice the assessed value of the affected properties. Now, one third of the settlement went to legal fees and an undetermined amount to expenses.

At the beginning of the fight, Victoria Switzer had shared her wariness of involving lawyers in what she saw as a civic campaign: "Settlements are made behind closed doors and everybody walks away with money, but there is no shakedown. Nothing changes, and the industry does not get the spanking it deserves." Now, three years later, I took down the last words she shared publically about her fight with Cabot before she signed the nondisclosure agreement that was a condition of the settlement: "You can be forced into this game forever," she told me. "It sucks the life out of you. I know the outcome for those living over this gas, and it doesn't matter if you cast your lot with or against the industry: the house always wins."

The settlement was also signed by the Fiorentinos, the Carters, the Far-nellis, and the Sautners. The terms were never officially announced. Little was heard from the Carter Road 15 on the subject of Cabot and three weeks would pass before reporters, acting on tips, teased the basics of the settlement from the transcript of Cabot's second-quarter earnings call with investors and heavily redacted court documents.

A full year would pass before another relevant document turned up. On the first anniversary of the settlement, an unpublished EPA assessment of the Dimock investigation was leaked to the Washington bureau of the *Los Angeles Times*. It was in the form of a PowerPoint presentation featuring data from eleven gas wells in Dimock, with focus on the Baker, Gesford, and Costello wells along Carter Road. Laden with technical information, the conclusion noted this finding: "Methane and other gases released during drilling (including air from the drilling) apparently cause significant damage to the water quality." The report noted methane "at significantly higher concentrations in the aquifers after gas drilling and perhaps as a result of fracking and other gas well work." Staff writer Neela Banerjee reported that the unpublished presentation was a warning by staff members to agency higher-ups that "several wells had been contaminated with methane and substances such as manganese and arsenic, most likely because of local natural gas production." EPA spokeswoman Alisha Johnson confirmed the report was authentic work by an unnamed staff member. She dismissed it as a "preliminary evaluation that requires additional assessment."

The story of the leaked document made national news, but it gained little traction. The presidential election was over. Although the turf war between local and state control over fracking in New York was still pending in courts at the time, the battle over fracking between state and federal governments had been decided in Pennsylvania, Texas, and Wyoming. By the time the

EPA report made it to the media, the Sautners had moved to Ithaca, and the new owner of 1101 Carter Road was Cabot Oil and Gas.

Before the Sautners vacated their home, they were advised by their attorney to remove the signs in their front yard. On a damp September morning, Craig walked through the gardens of precise stonework and magnificent rhododendron bushes and considered the home where, five years earlier, he had envisioned a bright future with his family. He stopped and pulled up a placard that said "You Can't Drink Gas" and tossed it in the back of his pickup. He retrieved a dozen or so others from around the house: "Water & Fracking Don't Mix" . . . "Our Future Is Our Water" . . . "No Fracking Way" . . . "Make Cabot Pay." Later that day, he delivered the signs to a neighbor down the road—Ray Kemble—who with Scott Ely was among the few plaintiffs who refused the Cabot settlement. Kemble wanted the signs to add to the collection on his own property.

When the moving van pulled out of the Sautner driveway, all visible signs of complaints from the previous owners were gone. Not long after, a white company pickup truck arrived at the residence. A man stepped out, walked to the end of the driveway, and hammered a single sign into the ground by the road. It read "Dimock Proud! Where the water IS clean and the people are friendly."

The house remained vacant for more than a year; Cabot appeared in no hurry to sell.

At face value, the Sautner's house at 1101 Carter Road looked like a good buy. The three-bedroom ranch, sitting tidy and plumb under a canopy of maple trees had, in real estate parlance, curb appeal. Though potential buyers inclined to inquire about water and septic would find matters complicated, that, in the end, would be academic because the property was never listed.

One day in early September 2013, I received a tip that a demolition excavator and several dumpsters had been delivered on flatbed trucks to the property. I called George Stark, Cabot's spokesman, who told me the land was more marketable without the house. He would not elaborate and was under no pressure to explain. Months earlier, this would have been a big story. Now, after the EPA withdrawal and Cabot's settlement with the plaintiffs, the regional media spotlight had shifted from the industry's fight with the national government in Dimock, Pennsylvania, to its fight with local governments in upstate New York.

The only person on the scene when the demolition began was Vera Scroggins. She stood solemnly on the edge of the road, her video camera trained on the house. The excavator's boom rose and swung toward the side of the garage and, with much splintering and cracking, the machine began biting into the asphalt shingles, roof boards, and rafters. Within an hour, the two-car garage was mostly gone and the excavator continued tearing apart the house and packing wads of siding, insulation, wiring, and splintered timber into dumpsters. When I arrived on the second day of the demolition, Vera was gone and the machine was chewing through the living room and kitchen where the Sautners had confronted the EPA crew, and then the bedrooms and bathrooms beyond. Cross sections of the home's interior become briefly visible through clouds of dust and heaps of wreckage before disappearing altogether.

I returned to Dimock a year later, in the fall of 2014, after the home rule question was decided in New York but before Governor Cuomo issued a definitive policy on fracking. By then, the federal government's intervention in Pennsylvania was old news and unlikely to be remembered as a historic event. At 1101 Carter Road, I found no evidence that a house had ever existed. The basement had been filled and seeded, and a seasonal lawn display with an antique tractor and a few scarecrows blended innocuously with the leaf-littered landscape. But the story of 1101 Carter Road did not end here, and finding the rest required a search through real property records at the Montrose courthouse. There, the deed showed that Cabot had sold the property, for which it paid $165,000, for $4,000. Its new owners, along with all future owners, were bound by restrictions known as a land covenant—a legal instrument sometimes employed by regulators or developers to reduce exposure risks and related liability on chronically polluted parcels. The deed for 1101 Carter Road forbids a "residence or dwelling for human habitation" on the land. The timeframe for this and other restrictions is "forever," to be observed by future generations as "covenants running with the land."

With this stipulation, Cabot sold the property to the Mayes, who live next door. The Mayes, too, had complained to Cabot and the Pennsylvania DEP that their water turned bad soon after drilling began in 2008. Debbie Maye had found it unfit for her family to drink, and she was sure it was making her animals sick. But unlike the Sautners, who became public faces of the anti-fracking campaign, the Mayes quietly settled their grievances without litigation sometime before 2011. Later, they became supporters of shale gas development. With the Sautner house gone, the Mayes were able to expand their yard and ensure a pleasant view.

On a subsequent visit, I learned that some stories in Dimock ended more happily then others. Don and Shirley Lockhart sold their lunch counter for $475,000 to a limited partnership called Rocks Investment. Don, now secure for retirement but not ready to sever ties with his livelihood, still came to the store to greet and serve customers for several hours every morning. Norma Fiorentino moved to a new home in Elk Lake Township near her sister and members of her extended family. The Carters moved to a bigger house on the opposite side of Carter Road. Although Ron had spent a chunk of settlement money on a reverse-osmosis filtration system, Jeannie still could not bring herself to drink the water.

Victoria and Jimmy Switzer remained in their home and likewise did not drink the water. With her fight against Cabot settled, Victoria delved into electoral politics and the upcoming gubernatorial race. Victoria and Jimmy had recently hosted an event with John Hanger, who was seeking the Democratic Party nomination for governor with a platform that included stronger oversight and enforcement for shale gas operations.

Before I left the Switzers we walked out the French doors of the great room to a bluestone patio and Zen garden they had recently completed as a finishing touch to their home. The stone, from Ken Ely's quarry, was a commemoration of sorts to her late neighbor, and our conversation turned briefly and abstractly to the early days of the drilling boom and Victoria's efforts to organize the community. Switzer, perhaps mindful of the non-disclosure deal she had made with Cabot, tended to redirect my queries about the past to the present and future. "Frankly, at this point, I don't think standing in the middle of a shale gas field with a 'no fracking' sign is going to do anything," she told me.

This position, along with her settlement and support of a candidate who was not categorically opposed to fracking, had drawn pointed criticism from some members of the anti-fracking camp who branded her a sellout. By now, she was somewhat hardened to public criticism. She was resuming the role of consensus builder she had embarked on before lawyers got involved. Then, as now, she simply wanted to live peacefully in the community that had been the setting for her dream home and she did not relish fighting with neighbors. The fact that the drilling remained on hold in the 9-square mile area around her home, at least for now, contributed significantly to her peace of mind.

Scott Ely refused to settle. Going against the advice of his lawyers, he sought new counsel, with limited success. When I last saw him in late 2014, his complaints were languishing in various courts and he was representing himself. Not only had Cabot effectively countered his legal moves, the company had bought the neighboring home and 12-acre property of his cousin, Mike Ely, for $140,000. That purchase came in November 2013, months after the Pennsylvania DEP documented explosive levels of methane in water wells—including Mike's—in an area around the Costello 1 gas well, near the intersection of Carter Road and State Route 3023. State regulators had determined the Costello gas well was "unviable" and would have to be plugged. Soon after, work crews dismantled and removed the doublewide modular home where Mike had lived, leaving the lot empty, while they attempted to save their investment in the Costello gas well. "That's their answer," Scott said. "Keep buying up polluted properties. Then they won't have any problems to interfere with their drilling."

Leaving Scott's house, I drove across the small bridge over Burdick Creek, to where Carter Road intersected with the state road, and pulled into the driveway of the Civil War–era ancestral home occupied by Bill and Sheila Ely. Bill answered the door and came out on the porch to talk. Like Victoria, he was conscious of what he could and could not say, having accepted the settlement offer, but summed it up for me this way: "I'm not leaving!" From his side yard, he could see crews working on the towering service rig at the Costello well next door. He could only guess its status and even the public record was unclear. Pennsylvania DEP spokeswoman Colleen Connolly reported that the well remained "essentially unviable," but she was "not aware of the gas well having been officially plugged." Bill, like his late brother Ken, was not an especially loquacious source, so it was hard to gauge his comfort level with this uncertainty. Nevertheless I returned to my car and headed back to New York with the firm impression that at least two Ely families would be staying on their land along Carter Road; and while others had settled, one would fight on.

Pennsylvania's modern-day gas boom—exploitive, risky, powerful, and productive—embodies the state's approach to mineral extraction from the time of the anthracite coalmines of the twentieth century and the Drake oil fields of the nineteenth century. Citizens and prospectors had to live with the consequences, for better and for worse, while the economic ramifications rippled throughout the nation. As in Pennsylvania, the shale gas conflict in New York reflected local and regional values about what land is for and how it should be used. Stakeholders would also have to live with the outcome, but it would be a far different one.

By fall 2014, the policy document that would make or break shale gas development in New York, in its third draft, had grown from 500 to more than 5,000 pages to take into account more than 160,000 comments from the public. To gas supporters, the SGEIS embodied an exhaustive accounting of all considerations. To critics, it was an industry-wrought claptrap of uninformed conclusions and weak rationalizations.

For much of the remainder of his abbreviated tenure as health commissioner, Nirav Shah remained confident that the state could monitor health impacts while beginning a phased approach to issuing well permits in Southern Tier counties. This "learn as you go" approach, which he ardently defended in the meeting called by Martens the day after the Albany march, also passed muster with the review by outside health officials ordered by Martens at the end of 2012.

The three experts hired to review the plan added their own recommendations: the creation of a health advisory panel to inform regulators, encouragement of public participation in the permitting process, and transparency in the collection and monitoring of data, among other suggestions. But none of the reviewers advised against the rollout. Lynn Goldman, dean of the Milken Institute School of Public Health at George Washington University, agreed with "the notion of a phased approach to HVHF [high-volume hydraulic fracturing] gas-development that would allow public health problems to be identified earlier, and reduce problems resulting from overly rapid growth." She found "many of the proposed mitigation measures are a model for other states." Goldman also stressed her approval of a prohibition on fracking in the watersheds that feed Syracuse and New York City.

John Adgate, chair of the Department of Environmental and Occupational Health at the Colorado School of Public Health, called the proposal "a viable approach to addressing the main public health issues associated with shale gas development." Adgate added that tracking respiratory, asthma, and neurological symptoms "are the place to begin evaluation" because of the prevalence of these conditions and their threat to "sensitive populations." Richard R. Jackson, chair of Environmental Health Sciences at the University of California, Los Angeles, struck a less encouraging but still supportive note. Jackson advised the administration to leave open the door for an HIA (health impact assessment) if it decided to begin issuing permits. "What is needed is an aggressive 'learn as you go' HIA during a carefully phased rollout," Jackson wrote in his evaluation. He felt it necessary to state that his analysis reflected "not an endorsement" of high-volume hydraulic fracturing but his belief that the review represents "substantial due diligence" on the state's part to protect public health.

"Due diligence" was just what Martens had asked for. The support and recommendations from the panel was a critical step, and it had been completed on schedule.

But then came another unexpected development. On February 12, 2014, the administration released a letter from Shah notifying Martens that the review of the SGEIS was "ongoing" and the health department was still considering impacts on water, air, and communities. Shah, in an apparent change from his earlier position, stressed that "the time to ensure the impacts on public health are properly considered is before a state permits drilling. Other states began serious health reviews only after proceeding with widespread HVHF. In my view, that is not the right approach . . . if we are serious that public health is the paramount question in making the HVHF decision." Shah referenced several major studies designed to examine correlations between fracking and public health, including the Congressionally directed study by the EPA, championed by Congressman Maurice Hinchey, on fracking's impact on water, and a study by Geisinger Health System using hospital records to evaluate health data in drilling communities. Shah concluded, cryptically, that after taking these studies into consideration, the health department review would be completed "within a few weeks." However, the Geisinger and EPA studies were both years away from producing results.

Two weeks passed, and then two months, and then nearly a year went by. There were no public traces of the review. Throughout this period, members of the administration reported they were "considering all available information" but refused to elaborate on the status of the project. Despite calls from both the industry and fractivists to make the review process transparent, Adgate, Goldman, and Jackson were bound by nondisclosure agreements that forbid them from discussing or releasing their work.

Facing criticism for these indefinite delays and lack of transparency, Shah announced his resignation from the health department in April 2014 to take an administrative job with Kaiser Foundation Health Plan in southern California. The responsibility for the review that would determine the outcome of New York's fracking fight now fell to Howard Zucker, a deputy commissioner promoted to acting commissioner. Zucker, in his mid-fifties, was something of a prodigy in the health field. He became one of the country's youngest doctors when, at twenty-two, he received his MD from George Washington University. He also holds law degrees from Fordham and Columbia universities. Zucker, clinically trained as a pediatric heart specialist, is familiar with the craft of public policy in political settings, having served as medical liaison to the World Health Organization, the White House, and the United Nations.

Soon after Zucker took over, Cuomo confirmed that there would be no decision on the outcome of the review until after the gubernatorial election of 2014. Cuomo's Republican challenger was Rob Astorino, Westchester County Executive and a fracking proponent. Astorino, overmatched by Cuomo in funding, organization, name, and experience, was never considered a serious threat to unseat the governor and, predictably, he lost in a lopsided contest that had little to do with fracking. Far less predictable, however, were events within Cuomo's own party preceding the election. Zephyr Teachout, a Fordham University law professor and political unknown, challenged Cuomo in the primary on an anti-fracking platform that gained the enthusiastic backing of the Sierra Club along with a motivated network of other environmental lobbies and grassroots anti-fracking groups upstate.

Fracking protesters, meanwhile, continued to dog Cuomo, and they provided the Cuomo family ample material for discussion at family functions. One of the governor's daughters told her father that if they arrived at a public event with no fracking protesters, it must mean they were in the wrong place. On the morning of the primary, Cuomo climbed out of a black SUV to cast his ballot at his polling place—Presbyterian Church of Mount Kisco—seven hours ahead of his announced schedule. Yet it was not early enough to avoid a throng of chanting fractivists. "They're everywhere," he remarked later. "They are literally the most prevalent protest group in the state by far."

The presence of the protesters at the poll was no surprise. But the results of the primary were shocking. Underfunded, unknown, and untested, Teachout managed to take 182,000 votes—34 percent of the total—away from a popular incumbent governor with a $35 million war chest. Teachout won thirty upstate counties, with significant support in the Hudson Valley, the Capital District, the Adirondacks, the Catskills, the Finger Lakes and a good portion of Central New York. In Albany County, Teachout took 10,000 votes to Cuomo's 6,000, and she ran close in many others. Her achievement—especially against a governor with presidential aspirations—was unprecedented. The primary result was aptly described by one reporter as leaving "a sizable gash in his [Cuomo's] left flank." It was the most potent message yet that the anti-fracking movement had for the governor, and a demonstration of its ability to quickly and effectively organize, not just in the street, but at the polls.

Soon after the primary Cuomo announced that his staff would complete the fracking review by the end of the year. Anticipation of an imminent decision from Cuomo had by this time become a repetitive and fatiguing story line but, given the stakes, it always warranted serious consideration. By mid-December, with the holidays approaching, many, including Gannett's Albany reporter Jon Campbell, did not rule out another delay. But if news was going to happen, the year-end cabinet meeting would be the likely place.

On the morning of December 17, Campbell rose from bed and punched out the first news nugget of the day to his Twitter followers: "Gov. Cuomo holding cabinet meeting today. I'm not saying there's a #fracking decision today, but all key players will be there." A few hours later, Campbell and about twenty other reporters took their assigned spots in the plush and spacious mahogany-paneled "Red Room" of the Capitol. The press section faced Cuomo, flanked by several dozen members of his cabinet seated at long tables arranged in a horseshoe shape.

Over the years, I have kept in touch with Campbell, a colleague at Gannett. We share some of the same sources and sometimes compare notes. A frequent topic of conversation in Albany involves the effectiveness with which Cuomo, compared to his predecessors, is able to control information—a skill predicated on exceptional loyalty from his administration encouraged by a sense of both fear and respect for the boss. The lid on the fracking story over the years was a salient example. Campbell later reflected, it was "crazy" that the news about to happen was not leaked or even announced as an agenda item.

After introducing new cabinet members, Cuomo summarized the agenda, including various year-end budget issues and preparations for the State of the State address. "And then," he mentioned offhandedly, "we will have an update on fracking." About an hour later, with the other agenda items covered, Cuomo gave a five-minute preamble to the matter that everybody had waited for. "The fracking issue concerns everybody" he warned members of his cabinet. "You will all be asked about fracking. I've been asked about fracking 14,600 times at various decibel levels. . . . You talk to the anti-frackers and in thirty seconds they are yelling. Pro-frackers, same thing." He characterized the issue as "the most emotion-charged issue that I've ever experienced."

The governor then conspicuously washed his hands of the decision that was about to come down. "My position is simple," he said. "This is a highly technical question. . . . Obviously everybody is for creating jobs. Obviously everybody is against creating a danger to public health or the environment.

So the question becomes, which is it? . . . My answer is—I don't know. And it's not what I do. I'm a lawyer. I am not a scientist. I'm not an environmental expert. I'm not a health expert. I'm not a doctor. So let's bring the emotions down and ask the qualified experts what their opinion is. And all things being equal, I will be bound by what the experts say." Cuomo introduced Commissioner Martens and Acting Commissioner Zucker and explained their respective roles. "I will defer to these gentlemen." He put his finger to his chin and looked across the table at Martens and Zucker, who were seated next to each other. "Who goes first?" the governor asked.

Martens raised his hand and cleared his throat. He began by recounting the development of the SGEIS drafts in 2009 and 2011, and the corresponding public hearings that produced first tens of thousands and then more than 160,000 comments on the plan outlining New York's approach to permit high-volume hydraulic fracturing. "Unlike any other activity regulated by DEC, the potential adverse environmental impacts by HVHF are wide-ranging and widespread," he said. He spent the next several minutes itemizing the potential for air pollution, water pollution, and land degradation; the problems with waste management and risks from spills; and impacts on communities, including infrastructure damage, noise, light, demands on emergency services and health care providers, and the industrialization of landscapes. Costs related to managing these problems could "easily overwhelm local, county, and state governments."

As risks became known, drafts of the SGEIS incorporated restrictions in certain areas, including key watersheds feeding New York City and Syracuse, primary aquifers, certain public lands, and flood plains. By the time the second draft of the SGEIS was released in 2011, more than half of Marcellus land, about 6 million acres, had been designated as off limits for drilling as a result of policy decisions Martens attributed to "evolving science."

Then, in June of 2014, came what Martens characterized as a "legal game-changer"—the Court of Appeals decision to uphold the drilling bans in Dryden and Middlefield. This plus the other restrictions would take at least 63 percent of the Marcellus Shale region out of play in New York. The home rule decision, he noted, went well beyond towns considering bans. The fallout would reach every town in the state with stakes in shale gas development. Approximately 253 towns within the Marcellus acreage would have to amend local zoning ordinances and land use plans to allow or ban fracking— a lengthy and potentially costly legal process that would require public hearings town-by-town. Another 145 towns with no zoning "would still have to decide whether to allow HFVH virtually everywhere, or to adopt zoning to prescribe where drilling could occur." The home rule decision would cause "a patchwork of local land-use rules which industry claims would utterly frustrate the rational development of the shale resource," Martens continued. "Clearly the court's decision shifted the battleground to town boards."

Martens turned the presentation over to Zucker. The health commissioner began with a summary of the agency's work—more than 4,500 hours spent interviewing experts and stakeholders, and reviewing "anecdotal reports, scientific papers, institutional reports, government documents, industry analysis, and environmental assessments." He gestured to stacks of bound documents several feet high on a table behind him. Through this effort, the staff learned primarily that the body of science assessing health impacts from shale gas development was thin. Zucker noted the absence of longitudinal studies—the kind of conclusive scientific research that tracks variables within large populations over long periods—the gold standard in the health profession.

In short, there wasn't much to go on. But existing studies raised many "red flags," Zucker continued. He held up, one by one, copies of relevant reports: A study by Duke University researchers showing a correlation between methane migration and shale gas development. The draft of the EPA study in Pavillion, Wyoming, showing water pollution associated with fracking. A study by National Institution for Occupational Safety and Health documenting worker exposure to particulates from silica sand and diesel exhaust, both carcinogens, at fracking sites. A study by Geisinger Health Systems showing increases in traffic fatalities near drilling sites. A study published in *Environmental Health Perspectives* that tracked 125,000 births over a thirteen-year period in Colorado and found higher rates of heart and nervous system defects in babies born to mothers who live near gas wells. The studies were not strong enough to prove or disprove health claims, Zucker said. "But the cumulative concerns, based on the information I've read—a mere fraction of which I've cited this morning—gives me reason to pause." He compared the status of the science assessing risks of fracking to that of secondhand smoke decades ago—once thought to be harmless and later found to be deadly.

Reporters in the room sensed where Zucker was heading, but few could believe it. After expectations of another delay, a new review, or possibly a phased approach to permitting, it now appeared that Martens and Zucker were setting the stage for an altogether different and more final decision.

Campbell grabbed his phone and thumbed another dispatch to his followers: "So the NY health and enviro commissioners have laid out the negatives of #nyfracking and cast doubt on the positives. Read between the lines."

Zucker moved on to his conclusion: "Would I live in a community with HVHF based on the facts I have now? Would I let my child play in the school field nearby? Or let my family drink the water from the tap or grow their vegetables from the soil? After looking at the plethora of reports you see behind me and others I have in my office, the answer is no." Leaning forward in his seat, he delivered his final words on the matter: "I consider the people of the state of New York as my patients. We cannot afford to make a mistake. The potential risks are too great. In fact, they are not even fully known. . . . I cannot support high-volume hydraulic fracturing in the great State of New York."

Martens, in a terse concluding statement, made it official: "Considering the research, public comments, relevant studies, Dr. Zucker's report, and the enormous record DEC has amassed on this issue. I will direct my staff to complete the SGEIS early next year. After it is finalized, I will issue a legally binding findings statement prohibiting HVHF in New York State at this time."

Six years, four months, and twenty-four days after Governor David Paterson ordered a review that resulted in a de facto moratorium on shale gas amid grand visions and grave doubts, the prospects for the Marcellus Shale in New York were officially a bust.

All eyes turned to Cuomo. "Let's talk about this a little bit," the governor said. "You're right," he continued, addressing Zucker. "Sometimes things can be very complicated and sometimes they can be very simple. I thought that was very powerful and poignant when you said you wouldn't let your children live there, and if you don't believe your children should live there, then I agree your duty is to suggest that no child lives there."

Cuomo then picked up Martens' point that the economics grew more tenuous as more of the Marcellus became off limits due to restrictions, and future prospects in every town became a wild card due to zoning uncertainties yet to play out. "Did you find any towns ready to move forward?" Cuomo asked. Martens replied that, due to the home rule precedent, battle over zoning would become "a very severe local issue" as fracking supporters and opponents took the fight in front of town boards.

In response to a question after the meeting, Martens, who once publicly said he thought shale gas could proceed safely, explained his change of position: "I said way back in the beginning that with all the proper mitigations in place that I thought it could be done safely. This is where it's beneficial to take a long-term view and watch what has been going on around the country. And that's exactly what we've done."

New York became the second state in the nation, after Vermont, to enact a fracking ban, and the first state to do so that sat over viable shale reserves. There was much news, analysis, and reaction following the decision. Fractivists continued their spirited rallies in Albany, but they were now to thank Cuomo for the ban, and to urge the state to pursue alternative energy sources. The governor continued to distance himself from both the praise and the criticism that came with the decision that will shape his political legacy—a decision ostensibly made by his staff.

Lost in the excitement over the decision, for the most part, was the reason fracking had never started in New York to begin with. The ban was an endpoint to a seven-year journey. It grew from opinions expressed in town halls along the Pennsylvania border as events unfolded in Susquehanna County—fortuitous or ill-fated depending on one's perspective—that gave New Yorkers a chance to see what unconventional shale gas development looked like apart from industry annual reports and commercials. Events that gave rise to the SGEIS, which became the vehicle that provided time and space for the anti-fracking movement to flourish, were set in play long before "frack" had been recast as a bad word by Josh Fox's *Gasland*, or before the fractivist movement achieved a critical mass. It started when landmen went door to door offering money for leases, and residents started attending town meetings to educate themselves. Whether they were politically active or not, many of these residents became angry and motivated when—due to ignorance or by intention—regulators they entrusted to oversee the industry failed to give straight answers, and instead presented the image of a field of wildflowers growing over a reclaimed well site while offering assurances that there would be "no problems."

No matter how much science was brought to bear on the question of fracking in New York, in the end it came down to a political decision. Science, of course, has no will or conscience. It's a tool to inform policymakers who gauge its usefulness, completeness, and sources, and weigh it against myriad factors.

Throughout his presentation, Martens cited the economics of the shale play as a consideration. Those economics are driven in turn by variables of global supply and demand, and the cost and availability of competing energy sources ranging from coal and fuel oil to wind and solar. Some factors, such as access to petroleum reserves under a given town, are local. Some, like New York State's protracted controversy and eventual ban on shale gas drilling, are regional. Others, like the EPA investigation in Dimock and its resulting scientific and political fallout, are national. And still others, like Russia's 2014 annexation of the Crimea and threats to Ukrainian sovereignty, tie incentives for domestic gas production to unpredictable global events. All are connected to and affect the value of natural gas. While the battle in New York was over extraction, the larger battle remains one of consumption. The more we consume, the more incentive we provide operators and investors to chase rewards and to marginalize risks, and the more leverage we give politicians to approve policy that encourages production.

The history of mineral extraction is famous for boom-bust stories, and the fate of the Marcellus Shale from 2008 through 2014 in New York State is another to add to the list. By the time the ban was announced, the supply of natural gas due to production in Pennsylvania and elsewhere had outpaced demand, and the price was a fraction of what it was when prospecting began. Landmen were no longer at the doors of Southern Tier homeowners offering to make them rich. Industry lawyer Tom West had told me that, given the dismal return on investment to develop the Marcellus north of the Pennsylvania border, the industry "had written off New York," and all indications were that he was correct.

When five hundred farmers in the Town of Sanford leased their land at a premium to XTO Energy in June of 2008, nobody thought that seven years would pass with no signs of drilling. Regardless, few could argue that those farmers were worse off for the deal. But a $110 million windfall for a group of landowners just prior to a drilling moratorium—and then a ban—did not represent the standard course of affairs, and the land coalition movement that it inspired did not work out for everybody.

Jim Worden, the patient farmer and leader of the Windsor landowner coalition, was within a pen stroke of cashing in on the same kind of wealth that had enriched Dewey Decker. Worden kept a copy of a 2009 lease offer on his desk that he and others turned down on Chris Denton's counsel that it failed to offer acceptable safeguards to his property. By the standards of a small upstate family farm, it would have brought a career's worth of income to Worden.

Worden seldom looked at the lease, but he couldn't throw it away. Simply knowing it was there, poking from the cubbyhole of the drop-front oak desk where his father had paid bills, planned his will, bought cows, and executed other family business, gave rise to powerful emotions. Engaged from dusk to dawn with his farm and family, Worden is a man disinclined to muse over what could be or might have been. On several occasions over the years Worden had told me "the gas will keep under the ground." But when I caught up with him at the end of 2014, he shared doubts, if not regret. He had spent considerable time and energy organizing the coalition that could have been spent on family and work. The clinic in Boston that treated his daughter's mitochondrial illness was closing and now he was trying to manage resources for treatment in Pittsburgh. "There have been extreme highs and extreme lows," he said. "We thought that money would have helped us, and it would have. It would be nice to not be in debt. But looking back on it, my life would have been better if I never heard of natural gas under my land."

Dan Fitzsimmons was less philosophical and far less patient. In addition to his work organizing landowners in the towns of Conklin and Binghamton, Fitzsimmons represented the Join Landowners Coalition of New York on statewide policy and legislative matters. He blamed farmers' economic struggles on the anti-fracking movement and he resented those responsible for it. Where they saw danger, he saw opportunity. Where they saw righteous resistance, he saw conspiracy. On certain nights, from his farmhouse on a high hill in Conklin, Fitzsimmons could see the glow on the southern horizon from flaring wells just across the state border in Susquehanna County. To him, they were plain evidence that fears about them were exaggerated or altogether fabricated. "They're right there!" he said. "They are not hurting anybody!"

Fitzsimmons knew that nothing is done politically that cannot be undone politically, and he was sure political circumstances with the Marcellus ban in New York would change. "We will not leave those assets in the ground," he told me when we last talked in early 2015. "They will get drilled someday." After all these years, Fitzsimmons was still fully expecting roughnecks and rigs to cross the Pennsylvania border and begin drilling on his land—and he would not tire of the fight to bring them.

This book is based on knowledge I gained since 1999 while covering natural gas exploration and construction of the Millennium Pipeline as an environmental reporter for the *Binghamton Press & Sun-Bulletin*; and on my reporting after leaving the paper in June 2010 to work fulltime on the book. I have also drawn on the extensive work of others, including newspaper, magazine, television, and independent media reports noted throughout the text. I owe a special debt of gratitude to Laura Legere of the *Scranton Times Tribune* and Abrahm Lustgarten of *ProPublica*, both of whom delivered aggressive and thorough coverage of the gas rush and related issues that complemented my own reporting. I also relied on work from an array of independent videographers, photographers, bloggers, and coalition organizers who documented meetings, forums, press events, and debates. Vera Scoggins and Dan Fitzsimmons are two of many examples of volunteers whose dedication to their respective visions represents the grassroots influence that characterizes this story.

Much of my education on shale gas development I credit to anti-drilling advocates, industry proponents, and scholars on both sides of the issue who helped me with their views on the record, and by providing background information on legal, technical, and policy matters. These people include Chris Tucker of Energy in Depth, Tony Ingraffea of Cornell University, Terry Engelder of Penn State, Walter Hang of Toxics Targeting, Deborah

Goldberg of Earthjustice, Lindsay Wickham of the New York Farm Bureau, Brady Russell of Clean Water Action, and Chris Denton of the Denton Law Firm. I also owe thanks to Ken Komoroski of Cabot Oil and Gas, Dennis Holbrook of Norse Energy, and Brian Grove of Chesapeake Energy for arranging tours of drilling rigs and explaining operations and the workings of a gas play.

I am grateful to Bruce Selleck, Professor of Geology at Colgate University, for looking over the manuscript and providing essential advice on technical (and sometimes on non-technical) matters.

I also owe special thanks to Walter Hang and Terry Engelder for sharing files with me that chronicle important aspects of the story explained and referenced in chapter 4. Terry provided records ranging from call transcripts to investors to e-mails to his colleagues, as well as personal notes he logged as he calculated the potential of the Marcellus in the early days of the rush. Walter provided records he uncovered in newspaper archives and through searches of files in New York State health and environmental agencies.

In addition to these files, expository aspects of the book are based on publicly available policy documents, studies, reports, and lawsuits. These are noted throughout the text, but it is worth mentioning that I obtained information documenting the explosion of Norma Fiorentino's well from my review of records and internal emails at the Department of Environmental Protection's Northcentral Regional Office in Meadville, as well as from accounts of people who witnessed the damage.

The book features narrative scenes based on my firsthand observations and on multiple interviews with Ron and Jeannie Carter, Pat Farnelli, Jackie Root, Victoria and Jim Switzer, Julie and Craig Sautner, Ken, Scott, and Bill Ely, Dan Fitzsimmons, Chris Denton, Ashur Terwilliger, Dewey Decker, Lindsay Wickham, Scott Kurkoski, Don and Shirley Lockhart, Walter Hang, Stuart Gruskin, Tony Ingraffea, Terry Engelder, Dewey Decker, and many others, including state and local elected officials and industry representatives. Dialogue in narrative scenes where I was not present, and where there was no available transcript or recording, were based largely on people's recounting of it, and supported in some cases by information of record. Victoria Switzer's speech to New York State Senator Antoine Thompson in chapter 6, for example, was based on her prior letter to Pennsylvania State Senator Gene Yaw, and on my interviews with Victoria and others prior to and after the meeting with Thompson. Details from other scenes derived from third-party sources include James Underwood's visit to the Carters in chapter 1,

Ken Ely's first encounter with Victoria Switzer in chapter 1, Victoria and Ken's tour of Ken's property in chapter 5, meetings Victoria organized with Tina Pickett and with Cabot representatives in chapter 5, and Victoria's bus tour with Cornell University officials in chapter 6. The demonstration outside of the Lewis property in chapter 6 is based on interviews with people at the scene, photographs and accounts circulated on the Susquehanna County Gas Forum, and from footage and reporting by WBRE-TV. Ken Ely's observations on his land in chapter 3 and chapter 5 are based on my interviews with Ken Ely, Laura Legere, Victoria Switzer, Ken Ely's family members, and my personal understanding of the features of the land based on maps and visits to his property. The standoff between Ken Ely and Cabot workers in chapter 5 is based on court records and interviews with family members. The depiction of the initial farm bureau meeting on natural gas in chapter 2 is derived from interviews with Jackie Root, Ashur Terwilliger, Chris Denton, and Lindsay Wickham, and from my own observations from later meetings.

The people who inspired the writing of this book include my former colleagues at the *Binghamton Press & Sun-Bulletin*. Metro Editor Ed Christine, Assistant Managing Editor Al Vieira, Executive Editor Calvin Stovall, and Online Editor Jeff Platsky, were among the first in the mainstream news world to recognize the importance of Marcellus developments and commit time and resources to pursue the story. I also thank my successor at Gannett, reporter Jon Campbell, for his solid daily coverage of shale gas policy development in New York State.

The idea of the book originated with Michael McGandy, acquisitions editor at Cornell University Press, and it was Michael's careful nurturing that brought the project through its various stages. The book is largely a product of his initial suggestions, with improvements by Karen Laun, senior production editor, and others at Cornell University Press. Finally, I thank Bill Keegan, who created graphics and maps for the book, and the host of volunteers who contributed to the effort of refining the manuscript. These include wordsmiths within my own family—Julie Boyd, Tracy Wood, and Jim Wilber—who volunteered their time to read, reread, and suggest improvements while encouraging me through the labors of writing.

FIGURES AND MAPS

CHAPTER 1. AN AGENT OF DREAMS

11 *The Marcellus* A report by the USGS in 2002 estimated mean undiscovered gas reserves in the Marcellus at 1.9 trillion cubic feet; Robert C. Milici, "Assessment of Undiscovered Oil and Gas Resources of the Appalachian Basin Province," Fact Sheet 009-03, United States Geological Survey, 2002, table 1, Appalachian Basin Province Assessment Results, http://pubs.usgs.gov/fs/fs-009-03/FS-009-03-508.pdf. At the beginning of 2008, Terry Engelder, Pennsylvania State University geosciences professor, estimated the Marcellus held between 168 and 516 trillion cubic feet of gas; Terry Engelder and Gary Lash, "Unconventional Natural Gas Reservoir Could Boost U.S. Supply," *Penn State Live,* January 17, 2008, http://live.psu.edu/story/28116 (accessed November 6, 2011).

12 *The best known of these* United States Energy Information Administration Independent Statistics and Analysis provides comprehensive data on the Oriskany, Trenton Black River, and other domestic oil and gas reserves (www.eia.doe.gov/). See also Fenton H. Finn, "Geology and Occurrence of Natural Gas in Oriskany Sandstone in Pennsylvania and New York," *American Association of Petroleum Geologists Bulletin* 33(3) (1949): 303–35. See also New York State Energy Research and Development Authority, *New York's Natural Gas and Oil Resource Endowment: Past, Present, and Potential* (Albany, N.Y.: NYSERDA, 2006).

12 *The discovery* As discussed in the prologue, hydraulic fracturing, also known as "fracking," had been used to produce gas from vertical wells since the first part of the twentieth century. It did not become feasible to produce shale gas, however, until advancements in horizontal drilling, chemistry, mechanical engineering, and production techniques in the late twentieth and early twenty-first centuries. Hydraulic fracturing is examined in more detail in chapter 2.

12   *Primeval forests* James Elliott Defebaugh, *History of the Lumber Industry of America,* vol. 2 (Chicago: American Lumberman, 1906), 560–61, 564–74.

13   *Just prior to World War I* Thomas Dublin and Walter Licht, *The Face of Decline: The Pennsylvania Anthracite Region in the Twentieth Century* (Ithaca: Cornell University Press, 2005).

13   *Between 1869 and 1999* Tabulated from Bureaus of Mining and Reclamation, District Mining Operations and Deep Mine Safety, *Pennsylvania's Annual Report on Mining Activities* (Harrisburg: Department of Environmental Protection, Commonwealth of Pennsylvania, 1999), table 1, Anthracite Coal Statistical Summaries, 1870–1999.

13   *Waste from mine shafts* About 4,785 miles of streams with low pH in the mid-Atlantic region have been impacted by the extraction of resources, primarily coal. West Virginia and Pennsylvania each have about 2,200 stream-miles impacted. U.S. Environmental Protection Agency, *Mining Operations as Nonpoint Source Pollution* (Washington, D.C.: EPA, 2009).

13   *The earth beneath entire* According to a survey by the Office of Surface Mining Abandoned Mine Land Inventory System, there were forty uncontrolled mine fires known in Pennsylvania in 2004. In addition to other toxicants, the Centralia fire emits "small but detectable" amounts of mercury. Bureau of Air Quality, *Centralia Mine Fire Mercury Study Final Report* (Harrisburg: Commonwealth of Pennsylvania Department of Environmental Protection, 2008). See also Kevin Krajick, "Fire in the Hole," *Smithsonian Magazine,* May 2005, http://www.smithsonianmag.com/travel/firehole. html.

13   *Another disaster* Dublin and Licht, *Face of Decline,* 110–12. See also Robert Wolensky, Kenneth Wolensky, and Nicole Wolensky, *The Knox Mine Disaster, January 22, 1959: The Final Years of the Northern Anthracite Industry and the Effort to Rebuild the Regional Economy* (Harrisburg: Pennsylvania Historical and Museum Commission, 1999).

14   *Coupled with advances* For an analysis of how the whaling industry was displaced by advances in petroleum, see Amory Lovins, *Winning the Oil Endgame: Innovations for Profits, Jobs, and Security* (Boulder: Rock Mountain Institute, 2004), 4–6.

14   *The worst was* Accounts of this event are found in newspaper reports of the time. A compilation of reports and pictorial history is presented in *The Valley That Changed the World* (WQED and the Oil Region Alliance, Pittsburgh, 2009), a documentary of the early oil industry funded by the Oil Region Alliance of Business, Industry, and Tourism; the Pennsylvania Department of Environmental Conservation; and the National Parks Service.

15   *The family made enough* The USDA National Agricultural Library lists a chronological bibliography of books, book chapters, and reports addressing the evolution of farm and land management practices. According to a USDA summary, modern organic and sustainable agriculture grew out of "attitudinal and scientific changes" following a slow, steady shift to chemical fertilizers and technologically intensive farming in the first half of the century. "During this period Americans were confronted with evidence of deteriorated rangelands, soils and forests. The first critics of the new 'industrial' agriculture emerged and a heightened conservation ethic began to take root." National Agricultural Library, *Tracing the Evolution of Sustainable Agriculture* (Washington,

D.C.: U.S. Department of Agriculture, 2008), www.nal.usda.gov/afsic/pubs/tracing/ TESA1900.shtml (accessed June 8, 2011).

15 *The Susquehanna River watershed* The Susquehanna basin drains 27,510 square miles, covering half the land in Pennsylvania as well as portions of New York and Maryland, and provides half of the Chesapeake Bay freshwater flows. It also feeds 4,582 "direct-water-dependent businesses" in the basin, ranging from textile mills to wood products manufacturing. Collectively, these businesses employed 237,500 people with an annual payroll of $6.8 billion, according to a 2006 report. The largest was food manufacturing, with 521 establishments employing 42,169 people and having an annual payroll of $1.4 billion. That was followed by the fabricated metal products industry, with 948 establishments employing 32,969 workers and having an annual payroll of $1 billion. Susquehanna River Basin Commission, "Economic Value of Water Resources: Direct Water Dependent Business in the Susquehanna Basin" (Harrisburg: Susquehanna River Basin Commission, 2006), www.srbc.net/pubinfo/docs/FactSheetEconValue1106.pdf (accessed June 8, 2011). For more about the history and significance of the Susquehanna River, see Jack Brubaker, *Down the Susquehanna to the Chesapeake* (University Park: Pennsylvania State University Press, 2002).

15 *Across the United States* National Commission on Small Farms, *A Time to Act: A Report of the USDA National Commission on Small Farms* (Washington, D.C.: U.S. Department of Agriculture, 1998). See also Stewart Smith, "Farming—It's Declining in the U.S.," *Choices* 7(1) (1992): 8–11.

16 *By the early 2000s* Census data available in 2010 listed the average per capita income in Dimock in 2000 as $15,216, compared to $21,587 for the United States.

17 *Many of them worked for national* This information is in part based on an interview with Robin Forte, executive vice president of the American Association of Professional Landmen, July 16, 2010. Forte estimated that between 2,000 and 3,000 landmen were working in the Marcellus Shale play in New York and Pennsylvania. Many had moved from Texas and other places with a rich legacy of petroleum development to pursue opportunities with the Marcellus prospecting rush in the Northeast.

18 *They were privy to data* Tom Wilber, "Landowners Cry Foul over Seismic Searches: Trespassing Claims Stir Debates," *Binghamton Press & Sun-Bulletin*, September 7, 2008.

22 *The problems were not isolated* In July 2010, I checked to see if anything had come of the investigation into other companies or of the code of conduct the attorney general's office had urged the industry to adopt. (At that time, Cuomo was in the middle of his gubernatorial bid.) A spokesman for the attorney general's office told me officials were still seeking "reform for the entire scope of business practices" used by landmen.

22 *In Pennsylvania, contract disputes* The attorney general's handling of lease disputes is one example of contrasting political cultures and regulatory approaches addressing the drilling industry in New York and Pennsylvania. In New York, where mineral exploitation was never a significant part of the economy, Governor David Paterson signed a bill in summer 2008 prohibiting Marcellus permits until the state conducted a review of the environmental impact of the industry and updated its policy accordingly—a process that ended up stalling the progression of Marcellus development at the border. By contrast, Governor Edward Rendell of Pennsylvania immediately embraced the

industry. The industry also received outspoken support from Tom Ridge, who was hired as spokesman for the Marcellus Shale Coalition after he served as Pennsylvania governor and secretary of U.S. Homeland Security, and from Tom Corbett, who in 2011 succeeded Rendell as Pennsylvania's governor. Chapters 2 and 4 address differing social and political settings influencing Marcellus development across state lines.

22   *Residents from fifteen households* Norma J. Fiorentino, et al. v. Cabot Oil & Gas Corp., et al., No. 09-2284, M.D. Pa.

22   *Industry representatives did not deny* Tom Wilber, "Early Signees Feel Cheated by Landmen," *Binghamton Press & Sun-Bulletin*, September 16, 2008.

23   *As with any industry* Ibid.

24   *In late 2006, the Carters* Policies differ in New York and Pennsylvania regarding how gas is assigned to a rightful owner. In Pennsylvania, the "rule of capture" applies. This rule favors a driller's claim to the gas, even if it flows from an adjoining property. Further complicating the picture is the phenomenon of methane migration. This is the inadvertent, hazardous, and sometimes untraceable flow of gas to places where it isn't wanted, including basements and water supplies.

24   *Beyond that, prospective returns* The deduction by gas companies of postproduction costs prior to calculating landowner royalties is the subject of lawsuits. A landmark case is *Kilmer v. Elexco Land Services, Inc.,* No. 2008-57 (Susquehanna Ct. Com. Pl. Mar. 3, 2009). In *Kilmer*, the Pennsylvania Supreme Court affirmed the trial court's ruling in favor of the company. For analysis, see Benjamin F. Hantz, "Royalty on Sweet or Sour Gas? The Supreme Court of Pennsylvania's Interpretation of the Guaranteed Minimum Royalty Act ..." *Duquesne Business Law Journal*, vol. 13(2), www.duquesneblj.com/volume13-2/ (accessed November 6, 2011). Case law related to royalty disputes is catalogued by Penn State Law, Natural Gas Exploration Agricultural Law Resource and Reference Center, http://law.psu.edu/academics/research_centers/agricultural_law_center/resource_areas/natural_gas_exploration (accessed June 8, 2011).

29   *With the Delaware River watershed* The Delaware, running 330 miles from the confluence of its east and west branches at Hancock, New York, to the mouth of the Delaware Bay, is the longest undammed river east of the Mississippi. The basin encompasses 13,539 square miles, draining waters from Pennsylvania, New Jersey, New York, and Delaware. It supports a concentrated population base. Nearly 15 million people rely on its waters, although the watershed drains only four-tenths of 1 percent of the total continental U.S. land area. The population figure includes about 7 million people in New York City and northern New Jersey who live outside the basin. New York City gets roughly half its water from three large reservoirs located on tributaries to the Delaware in the Catskills. See annual reports for 2008, 2009, and 2010 from the Delaware River Basin Commission, www.state.nj.us/drbc/public.htm.

CHAPTER 2. COMING TOGETHER

35   *Because the gas of the Barnett proved* Useful summaries of geological and economic aspects of Barnett development include Kent A. Bowker, *Barnett Shale Gas Production, Fort Worth Basin: Issues and Discussion, AAPG Bulletin* 91(4) (April 2007); Jeff Hayden and Dave Pursell, *The Barnett Shale: Visitors Guide to the Hottest Play in the U.S.* (Houston: Pickering Energy Partners, 2005).

37   *A client who worked* The Jimerson well was part of the Wilson Hollow field developed in the Town of Hornby by Pennsylvania General Energy.

37   *Led by the western New York Trenton Black River wells* New York State Energy Research and Development Authority, *New York's Natural Gas and Oil Resource Endowment: Past, Present, and Potential* (Albany: NYSERDA, 2006).

40   *"Compulsory integration"* There are several options for landowners faced with compulsory integration. They can buy into ownership of the well, for example, taking on costs and liabilities in return for a greater share of the revenues, or they can form a limited partnership company and lease the land to themselves, thereby entitling them to a greater share of the royalties once the operator has recouped the expenses of drilling. Information can be found on the DEC's website, www.dec.ny.gov/energy/1590.html (accessed July 20, 2011).

41   *It provided gas* Eileen Lash and Gary Lash, "Kicking Down the Well," SUNY Fredonia Shale Research Institute, http://www.fredonia.edu/shaleinstitute/history.asp (accessed June 9, 2011).

42   *Yet gas production* In 2007, prior to the Marcellus boom, New York ranked twenty-first out of the fifty states, with just under 55,000 million cubic feet of production. In contrast, Pennsylvania ranked fifteenth with 182,000 million cubic feet, and Texas was first with 6,091,000 million cubic feet. "Top Natural Gas Producing States, 2007," U.S. Energy Information Administration, www.eia.gov/neic/experts/natgastop10.htm.

42   *These developments were* "New York's Natural Gas and Oil Resource Endowment, Past, Present and Potential," New York State Energy Resource Development Authority, www.dec.ny.gov/docs/materials_minerals_pdf/nyserda1.pdf (accessed July 22, 2011).

42   *But it was in 1998* For elaboration on the expansion of the pipelines associated with Marcellus development, see Office of Oil and Gas, "Expansion of the U.S. Natural Gas Pipeline Network: Additions in 2008 and Projects through 2011," Energy Information Administration, September 2009, 9–10, ftp://ftp.eia.doe.gov/pub/oil_gas/natural_gas/feature_articles/2009/pipelinenetwork/pipelinenetwork.pdf (accessed June 9, 2011).

43   *Its path eastward* For more on the Oriskany's geology, history of exploitation, and capacity for storage of both CO_2 and natural gas, see Jamie Skeen, "Basin Analysis and Aqueous Chemistry of Fluids in the Oriskany Sandstone, Appalachian Basin, USA," Eberly College of Arts and Sciences, West Virginia University, 2010, www.wvcarb.org/oriskany/skeen_jamie_thesis.pdf (accessed July 21, 2011).

44   *"You have whole communities* Associated Press video report from Ted Shaffrey, "PA Farmers Hope for Natural Gas Windfall," July 14, 2008, www.youtube.com/watch?v=6-vcRZv9sTo (accessed July 23, 2011).

44   *Governor Ed Rendell likened* See Andrew Maykuth, "Rendell Warns Natural Gas Industry That Resistance to Tax Will Backfire," *Philadelphia Inquirer*, March 30, 2010; Cheryl K. Chumley, "Rendell Speaks Out against Proposed Drilling Moratorium," *Environment and Climate News,* November, 2010, www.heartland.org/policybot/results/28534/Rendell_Speaks_Out_Against_Proposed_Drilling_Moratorium_Proposes_New_Taxes.html (accessed June 9, 2011).

46   *By the time Marcellus prospecting* For a comprehensive list of operations, see the searchable database for oil and gas operations compiled by the DEC, www.dec.ny.gov/cfmx/extapps/GasOil/search/wells/.

46   *It was no accident* Tom Wilber, "New Riches Found North of Broome," *Binghamton Press & Sun-Bulletin*, October 12, 2008.

46   *Schlumberger, a $27 billion well development firm* Tom Wilber, "Natural Gas Promise Attracts a Giant Suitor," *Binghamton Press & Sun-Bulletin*, August 29, 2009.

47   *A line of thinking developed* Matthew Wald, "The Energy Challenge: Utilities Turn from Coal to Gas, Risking Price Increase," *New York Times*, Business sec., February 5, 2008.

47   *Energy producers, motivated by high prices* Kenneth S. Deffeyes, *Hubbert's Peak: The Impending World Oil Shortage,* new ed. (Princeton: Princeton University Press, 2008).

54   *The next meeting was held* Tom Wilber, "Lease Deals Please Landowners," *Binghamton Press & Sun-Bulletin*, May 2008.

54   *I wrote about Kathi Albrecht* Tom Wilber, "Farm Family Embraces Gas 'Miracle,'" *Binghamton Press & Sun-Bulletin*, May 25, 2008.

55   *"It's going to be all around us,"* Quoted in Tom Wilber, "Gas Rush Spurs 200 to Sign Land Deals," *Binghamton Press & Sun-Bulletin*, August 2, 2008.

56   *It gave the company access* Richard Nyahay, James Leone, Langhorne Smith, John Martin, and Daniel Jarvie, "Update on the Regional Assessment of Gas Potential in the Devonian Marcellus and Ordovician Utica Shales in New York," Paper presented at the American Association of Petroleum Geologists (AAPG) Eastern Section Meeting, September 16–18, 2007, Lexington, Kentucky.

58   *Since declaring parts of the region a preserve* New York State Department of Environmental Conservation, "Catskill Park State Land Master Plan," Albany, 2008.

58   *The impoundment on the unspoiled west branch* New York City water is impounded in three systems in the watershed, composed of nineteen reservoirs and three controlled lakes with a total storage capacity of approximately 580 billion gallons. The Delaware Aqueduct was completed in 1944, Rondout Reservoir in 1950, Neversink Reservoir in 1954, Pepacton Reservoir in 1955, and Cannonsville Reservoir in 1964. For an overview of the system, see New York City Department of Environmental Protection, "History of New York City's Water Supply System," www.nyc.gov/html/dep/html/drinking_water/history.shtml (accessed June 10, 2011).

58   *Tompkins County, encompassing the southern part of Cayuga Lake* Median household incomes are Tompkins County, $48,547; Broome County, $42,619; Delaware Country, $39,821; Chemung County, $41,909; Sullivan County, $43,467; and Pennsylvania $41,514. U.S. Census Bureau, "State and County Quick Facts," http://quickfacts.census.gov/qfd/states/36000.html (accessed June 10, 2011).

59   *With a collective body of students* Women's suffrage and its ties to abolition and temperance movements hold a celebrated place in the region's history. In 1848, Elizabeth Cady Stanton presented the Declaration of Sentiments at a convention in Seneca Falls, which many credit as the genesis of the suffragette movement. See Sally McMillen, *Seneca Falls and the Origins of the Women's Rights Movements* (New York: Oxford University Press, 2009

59   *Citizens of Ithaca elected a socialist mayor* Benjamin Nichols, a member of the Democratic Socialists of America, served as the mayor of Ithaca from 1989 to 1995. Regarding the 2000 presidential election, results recorded by the New York State Board of Elections show that the Green Party captured 11 percent of the vote in Tompkins

County. Kevin Harlin, "Tompkins Greens Express No Regrets," *Ithaca Journal*, November 9, 2000.

59   *The Marcellus fairway covers* Although the distribution of shale gas formations tends to be broader and more uniform than conventional plays, finding and developing their sweet spots require capital-intensive engineering and operations. As with all natural gas development, shale gas exploration is largely influenced by the price of gas and economic factors. See Kent Perry and John Lee, "Unconventional Gas Reservoirs—Tight Gas, Coal Seams, and Shales," NPC Oil & Gas Study, February 21, 2007, www.npc.org/Study_Topic_Papers/29-TTG-Unconventional-Gas.pdf. See also "U.S. Shale Gas: An Unconventional Resource. Unconventional Challenges," Halliburton, July, 2008, www.halliburton.com/public/solutions/contents/shale/related_docs/H063771.pdf.

60   *Expectations for economic returns* Bernard L. Weinstein and Terry L. Clower, "Potential Economic and Fiscal Impacts from Natural Gas Production in Broome County, New York," report prepared for the Broome County Legislature, July 2009, www.gobroomecounty.com/files/countyexec/Marcellus-Broome%20County-Preliminary%20Report%20for%20distribution%207-27-09.pdf (accessed June 10, 2011).

60   *Tens of thousands more wells would be developed* Hazen and Sawyer Environmental Engineers, "Final Impact Assessment Report: Impact Assessment of Natural Gas Production in the New York City Watershed," NYC-DEP, December, 2009, 23, fig. 3.2. Information is derived from well completion, permitting, and rig activity data from regulatory and industry sources.

60   *For all this to happen* The Oil, Gas and Solution Mining Law, Article 23 of the New York Environmental Conservation Law, establishes standards for well spacing that vary according to the target formation and depth, with a flexibility of plus or minus 10 percent to account for site-specific circumstances that may require the movement of operations on the surface. Although the act also establishes setback distances for each unit to lessen the likelihood that wells drain oil and gas from under adjacent spacing units, spacing is a hypothetical exercise that does not calculate the actual area from which gas is extracted. A table of unit sizes and setbacks for each formation is available at the DEC website, www.dec.ny.gov/energy/1583.html (accessed July 20, 2011). For a comprehensive analysis of the bill that amends the act for the Marcellus Shale play, see John F. Spinello and B. David Naidu, "Significant Developments in Oil and Gas Drilling in New York State," Oil and Gas Alert, K&L Gates, September 4, 2008, www.klgates.com/files/Publication/bc835239-8c65-4500-9854-20710843cdda/Presentation/PublicationAttachment/9ad05cba-5283-4b44-ad99-36e726d42dfc/OGA_090408.pdf (accessed July 15, 2011).

62   *The amendment, S-8169-A* The motives behind the spacing bill would become part of the larger controversy about shale gas development in New York State. In an editorial that appeared in various newspapers in 2008, DEC Commissioner Peter Grannis stated the new law "has been widely misreported as allowing a new type of drilling, or somehow making it easier to get the environmental permits necessary for drilling. In fact . . . It authorizes nothing new or in any way reduces the environmental review needed before a drilling permit is issued." "Commissioner's Editorial on Marcellus Shale,"

August 11, 2008, DEC website, http://www.dec.ny.gov/energy/46570.html (accessed June 10, 2011). Drilling opponents argued that, by streamlining the process, the bill eliminated a level of public review and input that would have been required to grant variances for horizontal wells under the old system.

62 *In 1984, the U.S. Environmental Protection Agency (EPA)* U.S. Environmental Protection Agency, Region 2, Water, *Federal Register* Notice, Monday, January 14, 1985, 50(9), 2025.

63 *Before it can begin producing* Draft Supplemental Generic Environmental Impact Statement on the Oil, Gas and Solution Mining Regulatory Program, NYDEC, September, 2009, sec. 5.7, www.dec.ny.gov/energy/58440.html (accessed July 24, 2011).

63 *Some of the water* This is considered a "consumptive water" use because the water is removed from the ecosystem.

63 *At 4 million gallons* Calculation are based on four hundred wells per year, each requiring 4 million gallons, and on water production data provided by "Water Supply and Sewage Disposal Systems in the Southern Tier East Region," Southern Tier East Regional Development Planning Board, 2009, 72, table 18, www.steny.org/usr/Publication/FINAL%20WS%20REPORT%20WITH%20TOC.pdf (accessed November 6, 2011). Estimations of water requirements for well development in the Delaware River watershed are between 1 billion and 2 billion gallons per year; Hazen and Sawyer Environmental Engineers, "Final Impact Assessment Report," 33.

63 *Although the exact chemical recipes* For a list of known additives in hydraulic fracturing solutions, see Draft Supplemental Generic Environmental Impact Statement on the Oil, Gas and Solution Mining Regulatory Program, sec. 5.4.

63 *Sewage treatment plants* Tom Wilber, "Treatment Plants' Capabilities Lacking," *Binghamton Press & Sun-Bulletin*, July 27, 2008.

64 *Landowners, following the Deposit model* In June 2008, wellhead prices spiked at $10.8 per million cubic feet; "Natural Gas Year in Review, 2008," U.S. Energy Information Administration, April 2009, ftp://ftp.eia.doe.gov/pub/oil_gas/natural_gas/feature_articles/2009/ngyir2008/ngyir2008.html (accessed July 24, 2011).

65 *At these hearings* Both Dahl's and Collart's presentations followed themes presented on the agency website. The relevant Web page also includes the photo described here and states, "As a result of New York's rigorous regulatory process, the types of problems reported to have occurred in states without such strong environmental laws and rigorous regulations haven't happened here." New York DEC website, www.dec.ny.gov/energy/46288.html (accessed June 10, 2011).

65 *A person in back* At the time of this writing, federal regulations governing the injection of substances into the ground under the Clean Drinking Water Act do not apply to drilling. The recipes for concentration and specific chemical compounds of fracking solutions are seen as proprietary by companies, which have successfully resisted lobbies for federal mandates that would require public disclosure.

66 *Drawing from these streams* In May 2008, drill operators working for Range Resources and Chief Oil & Gas illegally diverted water from rural streams to large-scale drilling operations in Lycoming County, west of Scranton, according to a report from the Pennsylvania Department of Environmental Protection (DEP). The DEP partially shut down the operation. Tom Wilber, "Drilling Carries a Hefty Environmental Price," *Binghamton Press & Sun Bulletin*, June 8, 2008.

66   *After the meeting* Regarding events in Chenango Town Hall Meeting see Tom Wilber, "Gas Drilling Questions Unanswered," *Binghamton Press & Sun-Bulletin*, July 17, 2008.

67   *The high school auditorium* Tom Wilber, "Drilling Regulations May Be Needed," *Binghamton Press & Sun-Bulletin*, July 18, 2008.

68   *He was interrupted* Martha Goodsell, "Gas Drilling: NY State DEC Meets with Concerned Landowners," *Broader View Weekly*, July 25, 2008.

CHAPTER 3. GAS RUSH

74   *Then he handed her* A version of this map and list of wells is included in the Pennsylvania DEP Consent Order and Agreement with Cabot Oil and Gas, November 4, 2009; see map 3, which is based, in part, on the map from the consent order. By that time, sixty-three wells had been drilled at twenty-nine well pads.

74   *Even with these limits* Cabot press release, "Cabot Operations Update for February 13, 2008." Reuters, U.S. News & Markets, www.reuters.com/article/2008/02/14/idUS10059+14-Feb-2008+PRN20080214 (accessed June 10, 2011).

74   *In summer 2008, Dan Dinges,* "Cabot Q2 2008 Earnings Call Transcript," July 25, 2008, Seeking Alpha, http://seekingalpha.com/article/87129-cabot-oil-amp-gas-corp-q2-2008-earnings-call-transcript (accessed June 10, 2011).

75   *In January 2008, Terry Engelder* Penn State issued Engelder's findings in a press release in January 2008. In summer 2009, Engelder revised the figure to 489 trillion cubic feet after a statistical analysis of the initial production data. Terry Engelder and Gary Lash, "Unconventional Natural Gas Reservoir Could Boost U.S. Supply," *Penn State Live,* January 17, 2008, http://live.psu.edu/story/28116 (accessed July 24, 2011). See also, Terry Engelder, "Marcellus 2008: Report Card on the Breakout Year for Gas Production in the Appalachian Basin," *Fort Worth Basin Oil & Gas Magazine,* August 2009, 19–22, http://fwbog.com/index.php?page=article&article=144 (accessed July 24, 2011).

76   *The company responded* Cabot press release, "Cabot Operations Update, Establishes 2009 Plans," October 29, 2008, Reuters, U.S. News & Markets, www.reuters.com/article/2008/10/29/idUS309527+29-Oct-2008+PRN20081029 (accessed June 10, 2011).

76   *In 2009, the company added* "Cabot Oil & Gas Corporation Q4 2009 Earnings Call Transcript," February 22, 2010, Seeking Alpha, http://seekingalpha.com/article/189950-cabot-oil-amp-gas-corporation-q4-2009-earnings-call-transcript (accessed July 24, 2011).

77   *From 2005 to 2010* Christopher Helman, "Range Resources Is King Of The Marcellus Shale," *Forbes,* August 9, 2010, www.forbes.com/forbes/2010/0809/companies-energy-range-resources-bp-gas-blowout-beneficiary.html (accessed October 14, 2011).

80   *In his year-end conference call* "Cabot Oil & Gas Corporation Q4 2009 Earnings Call Transcript," February 22, 2010.

80   *An acceleration in Marcellus permit approvals* Figures are from the Pennsylvania Department of Environmental Protection Bureau of Oil and Gas Management, "2010 Year End Workload Report," January 25, 2011, www.dep.state.pa.us/dep/deputate/minres/oilgas/2010%20Year%20End%20Report%20as%20of%2012-31-2010.pdf (accessed June 10, 2011).

80   *Craig Lobins,* Pennsylvania Department of Environmental Protection, "Minutes of the Oil and Gas Technical Advisory Board Meeting," October 30, 2008, www.dep.state. pa.us/dep/subject/advcoun/oil_gas/Minutes%20103008.pdf (accessed June 10, 2011).

81   *Testifying before the Pennsylvania House* See J. Scott Roberts, testimony before the Pennsylvania House Appropriations Committee Subcommittee on Fiscal Policy, April 3, 2009, Philadelphia.

81   *These and other revelations* See *Damascus Citizens for Sustainability v. Commonwealth of Pennsylvania,* Environmental Hearing Board, docket 2010-102M, March 24, 2011. See also Michael Rubinkam, "Marcellus Shale Links," Associated Press, April 13, 2011.

81   *In West Virginia* "W. Va. regulators Scramble to Keep Up with Drilling," Associated Press, June 21, 2010, http://abclocal.go.com/kfsn/story?section=news/technology&id=7511352 (accessed January 6, 2012).

82   *With permits outpacing* Laura Legere, "Gas Well Scrutiny by DEP Knocked," *Scranton Times Tribune,* December 15, 2008.

82   *EPA officials, while lacking the jurisdiction* U.S. Environmental Protection Agency, "EPA Announces 'Eyes on Drilling' Tipline," press release, January 27, 2010, http:// yosemite.epa.gov/opa/admpress.nsf/0/E4BFD48B693BCF90852576B800512FF2.

82   *A single well requires* New York Department of Environmental Conservation, Draft Supplemental Generic Environmental Impact Statement on the Oil, Gas and Solution Mining Regulatory Program, September 2009, www.dec.ny.gov/energy/58440.html (accessed June 10, 2011).

83   *In early July 2008, a truck knocked over* Relying on reports from Cabot, the Pennsylvania DEP originally estimated 800 gallons had spilled. Officials said later they were unsure about the amount. Josh Mrozinski, "Diesel Fuel Spills at Dimock Township Natural Gas Drilling Site," *Times Tribune,* June, 11, 2008. See also Tom Wilber, "Drilling Carries a Hefty Environmental Price," *Binghamton Press & Sun-Bulletin,* June 8, 2008.

83   *Roberts, deputy secretary for mineral resources management* J. Scott Roberts, testimony before the Pennsylvania House Republican Policy Committee, May 20, 2010, www. pagoppolicy.com/Display/SiteFiles/112/Hearings/5_20_10/5_20_10_Roberts_ Testimony.pdf (accessed June 10, 2011).

85   *They soon learned their water* Conditions the Carters and Sautners described are indicative of an infestation of bacteria that thrive in water with high iron levels. This condition, which can be a flag for other problems, is sometimes but not always associated with drilling contamination.

85   *A letter they sent* Letter to Scott Perry of the Pennsylvania DEP from Ronald and Anne Teel, May 5, 2010, on file at the DEP Northwest Regional office in Meadville

86   *One of the treatment destinations* Tom Wilber, "Treatment Plants' Capabilities Lacking," *Binghamton Press & Sun Bulletin,* July 27, 2008.

87   *"You hear that?"* Laura Legere's interview with Ken Ely, December 15, 2008.

88   *A drill that jammed* Problems at Gesford 3 are documented in internal Pennsylvania DEP correspondence from early 2009, including a January 23 memo from Craig Lobins, "Re: Cabot update." Komoroski later explained the problems at Gesford 3 during a presentation to the League of Women Voters, "LWV Gas Forum, Water Quality," Dimock, Pa., March 12, 2010.

90   *A report by the Pittsburgh Geological Society* Pittsburgh Geological Society, "Natural Gas Migration Problems in Western Pennsylvania," n.d., www.pittsburghgeologicalsociety. org/naturalgas.pdf (accessed June 10, 2011).

90   *The DEP files also contained* These and other summaries of accidents listed here are from a draft report, Pennsylvania Department of Environmental Protection Bureau of Oil and Gas Management, "Stray Natural Gas Migration Associated with Oil and Gas Wells," October 28, 2009, www.dep.state.pa.us/dep/subject/advcoun/oil_gas/2009/ Stray%20Gas%20Migration%20Cases.pdf (accessed June 10, 2011). In response to my request under the Pennsylvania Right to Know Law for more information about the Armstrong County case, DEP Open Records Officer Dawn Schaef replied that the agency had no further records. However, Schaef added that program staff recalled a fatal explosion in Burrell Township in 2003 caused by a leak from the deteriorating casing of an old gas well.

91   *The study of methane migration* Tests can determine whether the gas is biogenic or thermogenic and sometimes even distinguish generally among gases from different reservoirs.

CHAPTER 4. FIGURES, FACTS, AND INFORMATION

93   *This report piqued* Results were reported in a press release issued by Range Resources on December 10, 2007, announcing a conference call scheduled for the same day.

94   *His calculation of the Marcellus* The Associated Press, UPI, and the *Wall Street Journal* were among the first to feature stories of Engelder's work in early 2008. The *Time Magazine* story, "This Rock Could Power the World: Why Shale Gas Could Solve the Energy Crises," ran in the April 11, 2011, issue.

95   *The heart of this information* Since Sheldon, others have contributed to this field. Among those noteworthy to Engelder was J. M. Parker, author of "Regional Systematic Jointing in Slightly Deformed Sedimentary Rocks," *Geological Society of America Bulletin* 53(3) (1942): 381–408. For a more complete list of references, see Terry Engelder and Gary Lash, "Systematic Joints in Devonian Black Shale: A Target for Horizontal Drilling in the Appalachian Basin," paper presented at the meeting of the American Association of Petroleum Geologists, 2008.

95   *Understanding the pattern* Orientating a well bore perpendicular to the fracture lines allows the well to interface a maximum number of fractures over a given distance. Engelder called this "playing the fractures." He explained this concept in a December 12, 2007, email to companies testing the Marcellus: "Word on the street is that Range's success is based on playing J1 fractures which constitute the ENE joint set whose orientation is controlled by the Appalachian-wide stress field of the late Paleozoic (cf., Engelder, 2004; Engelder and Whitaker, 2006; Engelder and Lash, 2008)."

96   *I think it's fair* Transcript of Jefferies conference call, December 14, 2007.

97   *After analyzing production data in 2009* Terry Engelder, "Marcellus 2008: Report Card on the Breakout Year for Gas Production in the Appalachian Basin," *Fort Worth Basin Oil & Gas Magazine* (August 2009), http://fwbog.com/upload/file/ EngelderlayoutLowRes.pdf (accessed June 13, 2011).

97   *"It's almost divine intervention* Quoted in Clifford Krauss, "Drilling Boom Revives Hopes for Natural Gas," *New York Times*, Business sec., August 24, 2008, www.

nytimes.com/2008/08/25/business/25gas.html?scp=1&sq=clifford%20krauss%20 drilling%20boom%20august%2024&st=cse (accessed June 13, 2011).

99 *Pennsylvania was the only significant gas-producing state* New York also lacked a severance tax, but it was not counted as a significant gas-producing state prior to Marcellus development.

99 *"The report makes plain* "PSU Report Credits Hydraulic Fracturing, Marcellus Shale with 30K PA Jobs in '08, Openly Questions Sen. Casey's Anti-Frac Bill," *Energy in Depth*, July 28, 2009, www.energyindepth.org/2009/07/psu-report-credits-hydraulic-fracturing-marcellus-shale-with-30k-pa-jobs-in-08-openly-questions-sen-caseys-anti-frac-bill/ (accessed June 13, 2011).

100 *The Pennsylvania Department of Labor and Industry* Pennsylvania Department of Labor and Industry, "Marcellus Shale Industry Snapshot," table 2, Center for Workforce Information and Analysis. April, 2010, http://tricountywib.org/TCWIB/images/ PDFs/industryclusters/Marcellus%20Shale%20PA%20Statewide.pdf (accessed July 23, 2011).

100 *A report by the Marcellus Shale Education and Training Center* James R. Ladlee, "Marcellus Shale: Direct Workforce Needs Assessment (Preliminary Findings)," Marcellus Shale Education and Training Center, Penn State Cooperative Extension, June 30, 2010, www.marcellus.psu.edu/resources/PDFs/workforce.pdf (accessed June 13, 2011).

100 *Most of the direct gains* Marcellus Shale Committee Public Meeting, Penn College, Williamsport, Pennsylvania, September 23, 2009, video, http://wn.com/ (accessed May 15, 2011).

101 *A study commissioned* Bernard L. Weinstein and Terry L. Clower, "Potential Economic and Fiscal Impacts from Natural Gas Production in Broome County, New York," presented to the Broome County Legislature, July 2009, www.gobroomecounty.com/ files/countyexec/Marcellus-Broome%20County-Preliminary%20Report%20for%20 distribution%207-27-09.pdf (accessed June 13, 2011).

101 *Tourism jobs, by comparison* U.S. Census Bureau, "County Business Patterns," 2009, database, www.census.gov/econ/cbp/.

101 *Barth, who holds a doctorate* Jannette M. Barth, "Unanswered Questions about the Economic Impact of Gas Drilling in the Marcellus Shale: Don't Jump to Conclusions," J. M. Barth and Associates, Croton on Hudson, N.Y., March 2010.

101 *Christian Harris, a senior economist* Tom Wilber, "Job Projections Vary Widely," *Binghamton Press & Sun-Bulletin*, March 27, 2010.

102 *Local workers lacked* Ibid.

102 *Larry Milliken, director of energy programs* Quoted in Elizabeth Skrapits, "So Far, Gas Jobs Mainly in Related Fields," *Scranton Times-Tribune*, November 7, 2010.

102 *Tracy Brundage, a director* Quoted in ibid.

102 *Marcellus development would produce* Pennsylvania Department of Labor and Industry, "Marcellus Shale Fast Facts," June 2011, www.paworkstats.state.pa.us/gsipub/index. asp?docid=775 (accessed July 23, 2011). See also Stephen Herzenberg, "Drilling Deeper into Job Claims: The Actual Contribution of Marcellus Shale to Pennsylvania Job Growth," Keystone Research Center, June 20, 2011, http://keystoneresearch.org/ sites/keystoneresearch.org/files/Drilling-Deeper-into-Jobs-Claims-6-20-2011_0.pdf (accessed July 23, 2011).

103 *State Representative David Levdansky* Bill Toland, "Unions Want Their Piece of Shale Business," *Pittsburgh Post-Gazette*, August 29, 2010.

103 *Whether by design* Local union leaders in both New York and Pennsylvania—mindful of alienating those who held the keys to job prospects—walked the line between support and criticism of the industry. In June 2010, Alex Parillo, a representative from Local 785 Laborers International Union of North America, spoke on behalf of organized labor across the Southern Tier, advocating Marcellus development in New York. In April 2011, the New York State Laborers Union launched a public campaign against the Central New York Oil and Gas project, criticizing the company for awarding a construction contract to a West Virginia contractor instead of a local one for work to build a gas compression station in the town of Owego.

103 *Labor officials complained* See written testimony from the Pennsylvania House Labor Relations Committee hearing, September 16, 2010. See also Laura Legere, "Unions Say They Are Left Out of Marcellus Shale Jobs," *Daily Review*, September 17, 2010.

108 *With control over* Tom Wilber, "Three Power Brokers Take Control of Unleased Land," *Press & Sun-Bulletin*, November 21, 2009, www.pressconnects.com/article/20091121/NEWS01/911210352/Marcellus-Shale-Three-power-brokers-take-control-unleased-land (accessed January 9, 2012).

109 *It was a permit application to the New York DEC* Fortuna Energy received a permit to inject drilling waste into a former Trenton well on Rumsey Hill Road in Van Etten in spring 2009. In February 2010, Chesapeake Energy sought permits to inject 181,000 gallons of waste per day into a defunct well in Steuben County. Fortuna and Chesapeake stopped pursuing their respective plans in the face of opposition from residents in both localities. For a summary of the controversies, see Tom Wilber, "Chesapeake Waste Disposal Plan Puts Small Steuben Town at Center of Drilling Dispute," *Binghamton Press & Sun-Bulletin*, February 14, 2010, www.pressconnects.com/article/20100214/NEWS01/2140335/Chesapeake-waste-disposal-plan-puts-small-Steuben-town-center-drilling-dispute (accessed July 23, 2011). For more details on the Fortuna plan, see Sue Smith-Heavenrich, "Fortuna Explains Injection Testing of Mallula Well in Van Etten," *Broader View Weekly*, March 18, 2009, www.tiogagaslease.org/images/BVW_03_18_09.pdf (accessed November 7, 2011).

110 *He was on a first-name basis* Before replacing Grannis, Martens was president of the Open Space Institute, a land preservation agency. Enck was a staff member of both NYPIRG and, later, Environmental Advocates. Gruskin was a student board member of NYPIRG. Before being appointed as commissioner by Spitzer, Grannis had built a pro-environmental record as an assemblyman. He championed many initiatives supported by NYPIRG, including the passage of State Environmental Quality Review Act (SEQRA), the original "Bottle Bill," and the cleanup and revitalization of the brownfields in the state.

112 *"For major spills* Memorandum from Robert Leary (Spills Unit, Buffalo) to Joe Yarosz (Mineral Resources, Olean), "Spill Unit/Mineral Resources Cleanup Responsibility," DEC, June 16, 1987.

112 *Another memorandum, also written* Memorandum from C. Bruce McGranahan (Mineral Staff), New York DEC, December 3, 1985.

113 *On January 1, 2005* New York Department of Environmental Conservation, Spills database, Spill no. 0485474.

113 *I asked officials* Tom Wilber, "State Files Show 270 Drilling Accidents in Past 30 Years," *Binghamton Press & Sun-Bulletin,* November 8, 2009.

114 *The New York DEC also sent* Letter from Peter Grannis to William R. Parment and other lawmakers, December 30, 2009.

115 *"A representative I spoke with* Memorandum from William T. Boria, Chautauqua County Department of Health, "Re: Impact of Gas Drilling on Water Wells," July 30, 2004.

115 *Soon after the operation began* Memorandum from Thomas E. Hull, Allegany County Department of Health, to Michael J. McCormick, legislator, Allegany County Legislative District 4, June 16, 2009.

115 *In spring 1984, the* Post Journal John Moore, "Levant, Warren Families Tell Their Stories," *Jamestown Post Journal,* March 24, 1984.

116 *Angry residents turned to the town board* John Moore, "Levant Family Evacuates Gas-Plagued Home," *Jamestown Post Journal,* May 9, 1984.

116 *The problem dragged on* John Moore, "State Environmental Theory Challenged," *Jamestown Post Journal,* January 8, 1985.

116 *Five years after* Brian Bashinski, "Levant Gas Leak Cause Couldn't Be Determined," *Jamestown Post Journal,* June 1, 1989.

118 *It is true that hydraulic fracturing* Commercial wells were first stimulated with hydraulic fracturing in 1948, with improvements and refinements made over the next forty years. For an industry chronology of the development of hydraulic fracturing and related regulatory policy, see "History of Hydraulic Fracturing," Energy in Depth, www.energyindepth.org/in-depth/frac-in-depth/history-of-hf/ (accessed 7/21/2011).

118 *In 1992 when Cheney* "Contingency Operations: Army Should Do More to Control Contract Cost in the Balkans," Report to Subcommittee on Readiness and Management Support, Committee on Armed Services, U.S. Senate; U.S. General Accounting Office (GAO), September 2000.

119 *The two-paragraph clause* The Halliburton Loophole consists of the following amendments to federal water pollution laws in the Energy Policy Act of 2005:

SEC. 322. HYDRAULIC FRACTURING: Paragraph (1) of section 1421(d) of the Safe Drinking Water Act (42 U.S.C. 300h(d)) is amended to read as follows: UNDERGROUND INJECTION.—The term "underground injection"—(A) means the subsurface emplacement of fluids by well injection; and (B) excludes— (i) the underground injection of natural gas for purposes of storage; and (ii) the underground injection of fluids or propping agents (other than diesel fuels) pursuant to hydraulic fracturing operations related to oil, gas, or geothermal production activities.

SEC. 323. OIL AND GAS EXPLORATION AND PRODUCTION DEFINED. Section 502 of the Federal Water Pollution Control Act (33 U.S.C. 1362) is amended by adding at the end the following: OIL AND GAS EXPLORATION AND PRODUCTION.—The term "oil and gas exploration, production, processing, or treatment operations or transmission facilities" means all field activities or operations associated with exploration, production, processing, or treatment operations, or transmission facilities, including activities necessary to prepare a site for drilling and for the movement and placement of drilling equipment, whether or not such field activities or operations may be considered to be construction activities.

119 *Working from the premise* U.S. Environmental Protection Agency, "Evaluation of Impacts to Underground Sources of Drinking Water by Hydraulic Fracturing of Coalbed Methane Reservoirs," Washington, D.C., June 2004.

119 *Evoking protection as a whistleblower* Various outlets reported on Wilson's letter and his subsequent public appearances criticizing the EPA. See Todd Hartman, "He's Either Loved or Reviled: EPA Whistle-Blower Stands Up to Agency," *Rocky Mountain News*, May 31, 2005, http://m.rockymountainnews.com/news/2005/may/31/hes-either-loved-or-reviled/ (accessed July 16, 2011). DeGette joined with Maurice Hinchey and U.S. Congressman Jared Polis (D-Colo) to sponsor the Fracturing Responsibility and Awareness of Chemicals Act in June 2009.

119 *"It wasn't meant to be a bill of health* See the transcript of Abrahm Lustgarten's interview with Benjamin Grumbles, "Former Bush EPA Official: 'Fracking' Exemption Went Too Far," *ProPublica*, March 9, 2011, www.propublica.org/article/former-bush-epa-official-says-fracking-exemption-went-too-far.

120 *According to an ensuing congressional investigation* The results of the investigation are detailed in a letter from Henry A. Waxman (D-Calif.), Edward J. Markey (D-Mass.), and Diana DeGette to EPA Administrator Lisa Jackson, January 31, 2011, http://democrats.energycommerce.house.gov/index.php?q=news/waxman-markey-and-degette-investigation-finds-continued-use-of-diesel-in-hydraulic-fracturing-f (accessed June 23, 2011).

120 *On December 15, 2007* Ohio Department of Natural Resources, "Report on the Investigation of the Natural Gas Invasion of Aquifers in Bainbridge Township of Geauga County, Ohio," Sept. 1, 2008, www.dnr.state.oh.us/Portals/11/bainbridge/report.pdf (accessed June 21, 2011).

120 *Not long after the Ohio incident* The Agency for Toxic Substances and Disease Registry (ATSDR) issued a bulletin that advised, "Inorganic constituents in the water including sodium, magnesium, iron, selenium, sulfate, and nitrates could cause health effects" and "petroleum hydrocarbons, which are not usually found in drinking water, were found in many wells." U.S. Environmental Protection Agency (Region 8), "Pavillion, Wyoming, Groundwater Investigation: January 2010 Sampling Results and Site Update," August 2010, www.epa.gov/region8/superfund/wy/pavillion/PavillionWyomingFactSheet.pdf (accessed June 13, 2011).

121 *Monitoring wells installed by the EPA* Although only two of the residential wells had pollution levels that exceeded EPA regulatory limits, data collected from three monitoring wells showed that the aquifer from which residents drew water was connected to a "highly contaminated" shallow aquifer that tested positive for levels of benzene and total petroleum hydrocarbons "well above health-based levels of concern," according to the ATSDR analysis. Samples collected from the monitoring wells showed maximum values for total petroleum hydrocarbons such as diesel-range organics and gasoline-range organics of 62,100 micrograms per liter (μg/L) and 2,720 μg/L, respectively. Total extractable hydrocarbons and total purgeable hydrocarbons were measured in all three wells with maximum values of 42,000 μg/L and 3,790 μg/L. The source of the pollution was undetermined, but federal investigators were working with

the hypothesis that it involved natural gas extraction. In the area immediately around the polluted aquifer, there are 211 active gas wells, 30 plugged and abandoned wells, 20 wells identified as "shut-in," and 37 pits that had once held drilling fluids. According to an evaluation by the ATSDR, "The United States Geological Survey (USGS) has reported that contaminants associated with oil and gas production have the potential to affect the water resources of the area. The potential contaminants associated with oil and gas production include petroleum hydrocarbons, brines, and trace metals, and in some cases naturally occurring radioactive material. Sources of these contaminants include overflowing, failing, or unlined pits, leaking tanks, leaking well heads, and interaction between the groundwater and petroleum or brine zones inside well bores." For an assessment of the analytical results and recommendations, see David Dorian and Dana Robison, "Evaluation of Contaminants in Private Residential Well Water, Pavillion, Wyoming," ATSDR, August 31, 2010, www.atsdr.cdc.gov/hac/PHA/ Pavillion/Pavillion_HC_Well_Water_08312010.pdf (accessed June 30, 2011).

121 *The report recommended* Tom Wilber, "EPA's Study of Gas Drilling in Wyoming Could Impact Local Operations," *Binghamton Press & Sun-Bulletin,* September 13, 2009.

121 *Representatives of Encana Oil and Gas* "Encana: Drilling Did Not Taint Water in Pavillion," *Wyoming Energy News,* September 18, 2010, http://wyomingenergynews. com/2010/09/encana-drilling-did-not-taint-water-in-pavillion/ (accessed June 24, 2011).

121 *Cathy Behr was treating* The *Durango Herald* reported that state officials were unaware of the accident because it happened on tribal land south of Bayfield, over which they had no jurisdiction. The spill involved a contractor working for British Petroleum. Joe Hanel, "Drilling on Tribal Property Not Covered by State Regulations," *Durango Herald,* August 1, 2008.

122 *Behr did recover* Joe Hanel, "Secrets Surround Gas-Field Chemicals: Exposure Sickens Nurse; State Agency Pushing for Transparency," *Durango Herald,* July 20, 2008. See also Susan Green, "Oil Secret Has Nasty Side Effect," *Denver Post,* July 24, 2008; Abrahm Lustgarten, "Buried Secrets: Is Natural Gas Drilling Endangering U.S. Water Supplies?" *ProPublica,* November 13. 2008, www.propublica.org/article/buried-secrets-is-natural-gas-drilling-endangering-us-water-supplies-1113 (accessed June 13, 2011).

122 *In August,* Newsweek *reported* Jim Moscou, "A Toxic Spew?" *Newsweek,* August 20, 2008.

123 *When well service companies* My own analysis of the MSD sheets listing fracking compounds filed for conventional wells in New York found forty-eight substances generally listed as hazards, many of which are flammable, explosive, or caustic. These sheets tend to list the products by their trade names (e.g., Flomax 50, SandWedge WF, and BioClear) and mostly lacked the specific chemical blueprints of these products. Tom Wilber, "Safety, Disposal of Fracking Fluid Raise Concerns, Gas-Drilling Companies Keep Chemical Formulas a Secret," *Binghamton Press & Sun-Bulletin,* August 17, 2009.

124 *Workers at a steel mill* Joaquin Sapien, "What Can Be Done with Wastewater?: Rapid Expansion of Gas Drilling Has Led to Problems with Disposal, Contamination," *ProPublica,* October 4, 2009.

124 *By October 2008, TDS levels* Pennsylvania DEP test results showed "unusually high" TDS levels of up to 852 milligrams per liter (mg/L) along approximately 70

stream-miles on the Monongahela River, beginning at the West Virginia border and continuing to the confluence with the Youghigheny River. State and federal standards limit TDS to 500 mg/L. Pennsylvania Department of Environmental Protection, "DEP Investigates Source of Elevated Total Dissolved Solids in Monongahela River," press release, October 22, 2008, www.portal.state.pa.us/portal/server.pt/community/ newsroom/14287?id=2024&typeid=1 (accessed June 24, 2011). See also PADEP Bureau of Water Standards and Facility Regulation, "Coordinating National Pollutant Discharge Elimination System Permitting in the Monongahela River Watershed," May 1, 2010, www.elibrary.dep.state.pa.us/dsweb/Get/Document-79820/362-2100-001.pdf (accessed June 24, 2011).

125   *Subsequent investigations* Louis Reynolds, "Update on Dunkard Creek," EPA Region 3, Office of Monitoring and Assessment, November 23, 2009.

125   *Illegal dumping of drilling wastewater* In March 2011, the Pennsylvania state attorney general accused Allan Shipman and his company, Allan's Waste Water Service, of illegally dumping millions of gallons of drilling wastewater, from 2003 to 2009, at Morris Run and in remote tributaries, including those feeding Dunkard Creek, sometimes under cover of darkness or during heavy rains. The attorney general's office filed ninety-eight criminal counts against Shipman and seventy-seven counts against his company, located in Greene County. Shipman's lawyer said he was innocent and that the charges would be disproved. Don Hopey, "DEP Reviewing Permit for Hauler Charged with Illegal Dumping," *Pittsburgh Post-Gazette*, March 19, 2011, www.post-gazette.com/pg/11078/1133161-113.stm#ixzz1Q8NQpVm4 (accessed June 23, 2011). See also "Pa Waste Hauler Charged for Illegal Dumping," Associated Press, March 18, 2011, www.whsv.com/home/headlines/PA_Waste_Hauler_Charged_for_Illegal_Dumping_118243504.html (accessed June 23, 2011).

125   *In December 2010* Pennsylvania Department of Environmental Protection, "Biennial DEP Report Shows 80 Percent of Streams, Rivers Attaining Use Designation—Challenges Remain; Report to EPA Also Recommends Streams, Rivers for 'Impaired' Status," press release, December 23, 2010, www.portal.state.pa.us/portal/server.pt/ community/newsroom/14287?id=15710&typeid=1 (accessed May 12, 2011).

125   *The Monongahela River* Pennsylvania Department of Environmental Protection, "Permitting Strategy for High Total Dissolved Solids (TDS) Wastewater Discharges PA DEP," April 11, 2009. See also, Pennsylvania Department of Environmental Protection, "Comment and Response Document for Wastewater Treatment Requirements, 25 Pa. *Code* Chapter 95," April 28, 2010, 19.

126   *Industry proponents* Joaquin Sapien, "What Can Be Done with Wastewater?" *ProPublica*, Sunday, October 4, 2009, www.post-gazette.com/pg/09277/1002919-113. stm#ixzz1Q8xK2ie0 (accessed May 31, 2011).

126   *Drilling activity accounted* "Evaluations of High TDS Concentrations in the Monongahela River," Tetra Tech, January 2009, http://marcelluscoalition.org/ wp-content/uploads/2010/06/Tetra_Tech_TDS_Report.pdf (accessed June 23, 2011).

126   *The new rules limited* Title 25—Pennsylvania Waste Water Treatment Requirements, Pa.; Code Ch. 95, Environmental Quality Board, *Pennsylvania Bulletin*, August 21, 2010, www.pabulletin.com/secure/data/vol40/40-34/1572.html.

127   *In August 2009, Tapo Energy* West Virginia Department of Environmental Protection, "Buckeye Creek Discharge Final Report," Office of Oil and Gas, www.dep.wv.gov/

oil-and-gas/Documents/Buckeye%20Creek%20Discharge%20Final%20Report.pdf (accessed June 24, 2011).

127 *In October of that year* Pennsylvania Department of Environmental Protection, "DEP Penalizes Range Resources $141,175 for Spill in High Quality Waterway," press release, May 14, 2010, www.portal.state.pa.us/portal/server.pt/community/newsroom/14287?id=11412&typeid=1 (accessed June 24, 2011).

127 *That event was followed* Pennsylvania Department of Environmental Protection, "DEP Fines Atlas Resources for Drilling Wastewater Spill in Washington County," press release, August 17, 2010, www.portal.state.pa.us/portal/server.pt/community/newsroo m/14287?id=13595&typeid=1 (accessed June 24, 2011).

127 *Investigators were faced* Pennsylvania Department of Environmental Protection, "Fracking Fluid Spill at XTO Energy Marcellus Well; Spill Impacted Spring, Unnamed Tributary to Sugar Run," press release, November 22, 2010.

127 *In the first six months* See John Hanger, testimony before the Pennsylvania Senate Environmental Resources and Energy Committee, Wednesday, June 16, 2010, http://cbf.typepad.com/files/hanger.pdf (accessed July 24, 2011).

127 *An investigation by the Associated Press* David B. Caruso, "Pennsylvania Allows Gas-Well Wastewater to Flow into Rivers," Associated Press, January 3, 2011, www.pennlive.com/midstate/index.ssf/2011/01/pennsylvania_allows_gas-well_w.html (accessed June 24, 2011).

128 *According to policymakers* Pennsylvania Department of Environmental Protection, "Comment and Response Document," April 28, 2010, p. 10.

CHAPTER 5. ACCIDENTAL ACTIVISTS

132 *A fountain of pressurized gas* Flaring burns off impure gas surging from a new well and moderates pressure. This is necessary before a well is hooked up to pipelines, when the pressurized gas coming from the hole is too contaminated with flowback to be commercially usable.

133 *He told her "enforcement actions"* Consent orders are a common tool used in many states by regulators and polluters to settle pollution problems without getting tied up in lengthy and expensive court battles.

133 *O'Donnell was one* Laura Legere, "Investigation Reveals Little State Oversight, Looming Problems with Gas Drilling," *Times Tribune,* June 20, 2010.

133 *Many of these reports* A review of records by DEP officials on February 18, 2009, found Cabot had failed to submit required well records for twenty-two wells, including Gesford 3, the malfunctioning well near Pat Farnelli's home later blamed for methane migration. Pennsylvania Department of Environmental Protection, Consent Order and Agreement with Cabot, November 4, 2009.

135 *At the time, she was* Tina Pickett, along with state Representative Dave Reed, introduced House Bill 1050, "Energize Pa." on March 24, 2009. Pickett and Reed are co-chairs of the Pennsylvania House Republican Energy Task Force.

135 *Debbie and Tim Maye* Jon Hurdle, "U.S. Energy Future Hits Snag in Rural Pennsylvania," Reuters, News & Markets, March 13, 2009, www.reuters.com/article/2009/03/13/us-energy-gas-town-idUSTRE52C07920090313 (accessed June 13, 2011).

138 *The court granted* See *Cabot Oil & Gas Corporation v. Kenneth R. Ely,* U.S. District Court for the Middle District of Pennsylvania, March 25, 2009.

141 *After the tour* I found the letter dated January 15, 2010, from Stalnaker to Means "re: Scott Ely Allegations" in a review of files at the DEP Northcentral Regional Office in Harrisburg on December 9, 2010.

141 *It had hired URS* The URS investigation, released in December 2012, found traces of substances that could indicate past spills from natural gas operations—including surfactants, chlorides, and compounds associated with diesel fuel—at six of eleven well sites identified by Ely. Those substances, which have no state mandated limits, "were not commonly present" in soil and water at the sites and posed no health threats, the report concluded. Cabot officials said the report showed no condition that required cleanup, and that it confirmed that Ely's allegations "lacked substance." The DEP was reviewing the report and had not made an assessment at the time of this writing. The URS report is available at www.cabotog.com/pdfs/FINAL%20Wellsite%20Evaluation%20Report%20 -%202011-12-09.pdf (accessed January 8, 2012).

141 *As events unfolded* Problems at Gesford 3 are outlined in records and emails on file in the DEP Meadville office and the defective wells are outlined in the DEP Consent Order and Agreement with Cabot

142 *In March 2010, Komoroski* League of Women Voters Gas Forum, Water Quality, Dimock, Penn., March 12, 2010. The meeting was taped by Vera Scroggins and a copy was provided to me courtesy of Julanne Skinner, Susquehanna County League of Women Voters.

142 *In addition to the complaints* Tom Wilber, "Gas Drilling Workers Suspended Following Drug Tests," *Binghamton Press & Sun-Bulletin*, December 11, 2009.

144 *Fiala was so confident* Broome County was not part of the deal under negotiation with Hess. It later received a bid from Inflection Energy.

145 *The deal was collectively worth* Tom Wilber, "Landowners Latest Gas Deal Worth Millions," *Binghamton Press & Sun-Bulletin*, June 20, 2009. See also Tom Wilber, "Kirkwood Hess Deal Nears," *Binghamton Press & Sun-Bulletin*, July 18, 2009.

146 *Rally-goers, perhaps 1,500–2,000 of them* Tom Wilber, "2,000 Attend Gas Drilling Rally," *Binghamton Press & Sun-Bulletin*, August 24, 2009. The number of rally-goers I reported was challenged by environmental groups. My estimates were based on the number of vehicles (close to nine hundred) driven to the event, and my observation that many of them carried more than one passenger, and on a rough count of the clusters of groups in front of the stage.

149 *Chris Denton had a message* Tom Wilber, "Landowners: New $5,500-per-Acre Gas Deal 'Exactly What We Needed,'" *Binghamton Press & Sun-Bulletin*, September 14, 2009.

149 *Most of the population* Rural communities made up of small populations that control large tracts of land are strategic hot spots for energy companies looking to establish landholds. The first big deal in New York (between XTO and the Deposit coalition) was anchored by tracts controlled by relatively few landowners in the town of Sanford, geographically one of the biggest towns in Broome County but populated by the fewest people. Conversely, more densely populated urban populations presented technical problems pertaining to the acquisition of leases, logistical problems due to infrastructure and density, and strategic problems in community relations stemming from the potential for social conflicts and organized resistance. Nonetheless, shale gas development has extended into urban areas, including Fort Worth, Texas. For more on this and an

analysis of "social stratification and inequality," and conflicts involving "haves versus have-nots," and "rural versus urban residents," see Kathy Brasier, "Community Impacts of Marcellus Shale Development: A Research Update," Penn State University, Department of Agricultural Economics & Rural Sociology, September 2010.

151 *"If they're so sure* Tom Wilber, "EPA's Study of Gas Drilling in Wyoming Could Impact Local Operations," *Binghamton Press & Sun-Bulletin*, September 13, 2009.

151 *With this goal, the Oil and Gas Accountability Project* Oil and Gas Accountability Project, "Colorado Oil and Gas Industry Spills: A Review of COGCC Data (January 2003–March 2008)," Earthworks, http://cogcc.state.co.us/RuleMaking/PartyStatus/FinalPrehearingStmts/OGAPExh1.pdf (accessed June 13, 2011). Data filed with New Mexico Conservation Division were analyzed by the Oil and Gas Accountability Project and presented in a report by Earthworks, "Groundwater Contamination," www.earthworksaction.org/NM_GW_Contamination.cfm (accessed June 13, 2011).

152 *Stories of sudden wealth and bitter disillusionment* Gannett's Central New York Newspaper Group ran daily coverage of Marcellus development in the Twin Tiers for papers in Ithaca, Elmira, and Binghamton, which published more than four hundred Marcellus-related articles in 2008 and 2009. The Times-Shamrock Communications, with its flagship *Times-Tribune* in Scranton, covered regional developments equally aggressively during this period, as did local radio and television news programs. In late 2009 through 2011, Dimock began appearing in news features and exposés about shale gas development appearing periodically in larger outlets, including the *New York Times*, the *Philadelphia Inquirer*, *Vanity Fair*, Bloomberg, and *Playboy*. Crews from CNN, CBS *60 Minutes*, *NBC Nightly News,* and some outlets from Canada came to Dimock to interview the Carters and other residents for shows that aired in 2011.

155 *Victoria's efforts* The scene is based on interviews with Heitsman, Fiorentino, Jennifer Carney, and the Carters.

157 *Brad Gill, executive director of IOGA* IOGA NY press release, September 18, 2009, http://readme.readmedia.com/IOGA-of-NY-Press-Statement-on-Spill-at-Gas-Drilling-Site-in-Dimock-PA/954211 (accessed October 17, 2011).

159 *Komoroski was quoted as saying* This specific quotation is from Abrahm Lustgarten's report, "DEP Issues Citation to Pennsylvania Driller as a Third Spill Occurs," *ProPublica*, September 23, 2009, www.propublica.org/article/dep-issues-citation-to-pennsylvania-driller-as-a-third-spill-occurs-923 (accessed June 13, 2011). Komoroski was also quoted by other outlets.

159 *In this regard* The insufficiency of the MSDS and the need for more tests are outlined in PADEP internal memos and correspondence after the spill, including a narrative report filed in the North Central Regional Office by Eric Rooney, September 18, 2009.

160 *The conversation then turned* Chapter 5, Sec. 501, Act 2 gives a party involved in a cleanup protection from liability related to the pollution if they meet the standards set forth in the program.

160 *But "inconsistent responses and record keeping* Laura Legere, "State Lacks Consistent Record Keeping for Natural Gas Drilling Contamination," *Scranton Times Tribune*, June 21, 2010.

161 *Internal DEP emails* George Basler, "Pa. Orders Shutdown of Cabot Drilling," *Binghamton Press & Sun-Bulletin*, September 25, 2009.

162 *In the report* Oil and Gas Technical Advisory Board, "Proposal to Modify 25 Pa Code Chapter 78 to Address Stray Gas Migrations," Pennsylvania DEP Bureau of Oil and Gas Management, September 17, 2009.

163 *The industry reaction* Reported by Laura Legere, "Gas Drilling with Catastrophic Results," *Citizens Voice*, October 26, 2009.

163 *In November 2009, the agency issued* Pennsylvania Department of Environmental Protection, Consent Order and Agreement with Cabot.

CHAPTER 6. THE DIVISION

165 *The Pennsylvania DEP had logged* A precise number of spills is hard to determine, due to the inconsistencies and disorganization in DEP records and uncertainty over unreported spills. The approximation here is derived from my review of records at the Harrisburg office, information in the DEP Consent Order with Cabot on November 4, 2009, and from other media reports. Laura Legere, "Investigation Reveals Little State Oversight, Looming Problems with Gas Drilling," *Scranton Times Tribune*, June 20, 2010, http://thetimes-tribune.com/news/troubled-promise-little-oversight-looming-problems-for-pa-gas-industry-1.855759#axzz1QyX9WNLy (accessed July 2, 2011).

166 *The cameras rolled* The case would later be turned over to Napoli Bern Ripka Shkolnik and Associates, the New York City firm that settled the case of respiratory ailments related to the World Trade Center disaster. The Napoli firm hired Lewis to continue working on the case.

166 *As a result of Cabot's negligence* *Norma J. Fiorentino et al. v. Cabot Oil & Gas Corp. et al.*, No. 09-2284, M.D. Pa.

166 *The resulting article* Laura Legere, "Nearly a Year after a Water Well Explosion, Dimock Twp. Residents Thirst for Gas-Well Fix," *Scranton Time Tribune*, October 26, 2009, http://thetimes-tribune.com/news/nearly-a-year-after-a-water-well-explosion-dimock-twp-residents-thirst-for-gas-well-fix-1.365743#axzz1PYaWiy23 (accessed June 17, 2009).

166 *The article recounted the explosion* In subsequent coverage by another newspaper, Switzer credited Legere's reporting with pressuring Cabot into making water deliveries and disputed a statement by Komoroski that the company had actually agreed to the water deliveries in the September 16, 2009, meeting that Switzer organized at the VFW Post. Robert Baker, "News Yields Clean Water," *Wyoming County Press Examiner*, October 27, 2009, http://wcexaminer.com/index.php/archives/news/7597 (accessed June 17, 2009).

167 *In response to the DEP consent order* Robert W. Watson, "Report of Cabot Oil & Gas Corporation's Utilization of Effective Techniques for Protecting Fresh Water Zones/Horizons during Natural Gas Drilling: Completion and Plugging Activities," Cabot Oil & Gas, June 23, 2010, www.cabotog.com/pdfs/Dr_Bob_Watson_WhitePaper_101010.pdf (accessed June 28, 2011).

168 *A few weeks after the press conference* Jad Mouawad and Clifford Krauss, "Dark Side of a Natural Gas Boom," *New York Times*, December 7, 2009, www.nytimes.com/2009/12/08/business/energy-environment/08fracking.html (accessed July 2, 2011).

169 *In March 2010, the fledgling Obama administration* The bill, sponsored by Senators Lindsey Graham (R-S.C.), Joseph Lieberman (Indep. D-Conn.), and John Kerry

(D-Mass.) promised to transform the U.S. approach to energy and climate change through the cap and trade system. The bill languished and then died amid resistance from industry and amid the pressing issues of health-care reform and economic revival. Ryan Lizza, "As the World Burns," *New Yorker*, October 11, 2010, www.newyorker. com/reporting/2010/10/11/101011fa_fact_lizza?currentPage=10 (accessed July 2, 2011).

170 *In addition to the Marcellus* Petroleum development is measured in "proved reserves." U.S. natural gas proved reserves increased by 11 percent in 2009, to 284 trillion cubic feet, their highest level since 1971. The trend was expected to continue with the acceleration of shale gas development in subsequent years. U.S. Energy Information Administration, "Summary: U.S. Crude Oil, Natural Gas, and Natural Gas Liquids Proved Reserves 2009," November 2010, www.eia.gov/pub/oil_gas/natural_gas/data_ publications/crude_oil_natural_gas_reserves/current/pdf/arrsummary.pdf (accessed June 29, 2011).

170 *The Exxon and Shell purchases* Tom Wilber, "Analysts: Marcellus to Produce Fuel of Choice: Exxon Deal Seen as More Proof of Ambitious Future Prospects," *Binghamton Press & Sun-Bulletin*, December 19, 2009, www.pressconnects.com/article/20091219/ NEWS01/912190347/Analysts-Marcellus-produce-fuel-choice (accessed July 2, 2011).

170 *"I've read the newspapers* Robert L. Baker, "Residents: Concern for Yaw's Inattention to Water Problems," *Wyoming County Press Examiner*, January 13, 2010.

171 *At the Towanda site* Information here is from Senator Thompson's office and based on my own experience on one of these tours while reporting for Gannett. Tom Wilber, "Chesapeake Tries to Dispel Marcellus Shale Drilling Fears," *Binghamton Press & Sun-Bulletin*, April 29, 2010, www.pressconnects.com/article/20100429/ NEWS01/4290415/Chesapeake-tries-dispel-Marcellus-Shale-drilling-fears (accessed June 18, 2011).

172 *Thompson learned that the Carter Road residents* The account of Thompson's visit draws on information provided to me by people who attended the meeting. It was also documented by Michael Lebron of the New Yorkers for Sustainable Energy Solutions Statewide on an email posted on various list serves, including www.dangerdrilling. com/?p=319 (accessed July 2, 2011).

177 *After the release of the report* Tom Wilber, "Marcellus Question: Who Will Pay to Monitor Gas Drilling?" *Binghamton Press & Sun-Bulletin*, October 10, 2009.

177 *For its part, the NYC-DEP itemized* See written comments on the SGEIS, submitted in a letter from Lawitts to the NYC-DEC Bureau of Oil & Gas Regulation on December 22, 2009.

177 *City of Syracuse officials* See written comments on the SGEIS from Syracuse Water Commissioner Michael Ryan, submitted to the DEC on December 30, 2009.

178 *Officials from the New York State Department of Health* Written comments from the New York State Department of Health Bureau of Environmental Radiation Protections submitted to the New York DEC regarding the SGEIS, July 21, 2009.

178 *The New York Farm Bureau* "Delaware County Farm Bureau Quarterly Connection," June 2010, www.nyfb.org/img/county_docs/newsletter_oyblox6s61.pdf.

180 *Between July 1, 2009, and June 30, 2010* The Marcellus production figures became public in 2010 after Pennsylvania lawmakers passed legislation introduced by Senator Eugene Yaw. Act 15 of 2010 requires Marcellus operators to submit production figures

every six months for publication on the DEP website. Prior to the legislation, oil and gas companies had to submit production statistics annually to the DEP, but those figures, by law, could not be made public for five years. Act 15 was designed to provide landowners receiving royalties faster and more complete information regarding well production. Production reports and figures are available at the DEP website, www. paoilandgasreporting.state.pa.us/publicreports/Modules/Welcome/Welcome.aspx.

180 *Even with gas prices suppressed* In February 2011, after leaving the DEP, John Hanger projected that Pennsylvania would hit the 1 trillion cubic feet mark in 2012, based on production figures with the DEP. See "Marcellus Data off the Charts," http://johnhanger. blogspot.com/2011/02/marcellus-production-data-off-charts.html (accessed October 25, 2011). Storage values in the lower forty-eight states between July 2009 and July 2011 were typically at or near historical maximums for a five-year period between 2006 and 2010. At the end of October 2010, working natural gas inventories marked an all-time high of 3,847 billion cubic feet. See "Working Gas in Underground Storage Compared to Five-year Range," Weekly Underground Natural Gas Storage Report; U.S. Energy Information Administration Independent Statistical Analysis, August 11, 2011, http:// ir.eia.gov/ngs/ngs.html (accessed August 12, 2011).For a summary of the contribution of shale gas development to a global natural gas glut see "An Unconventional Glut," *Economist*, March 11th 2010, www.economist.com/node/15661889 (accessed October 25, 2011).

180 *Although shale gas wells had proven prolific* The rate at which a given well produces gas over time—both actual and projected—is called a decline curve. Interpretations and analyses of decline-curve projections for shale vary greatly. For examples, see Arthur E. Berman and Lynn Pittinger, "Realities of Shale Play Reserves: Examples from the Fayetteville Shale," Petroleum Truth Report, September 17, 2009, http:// petroleumtruthreport.blogspot.com/2009/09/realities-of-shale-play-reserves.html. See also, Terry Engelder, "Marcellus 2008: Report Card on the Breakout Year for Gas Production in the Appalachian Basin," Basin Oil & Gas, January 2011, http://fwbog. com/index.php?page=article&article=144.

180 *Critics pointed to evidence* For an analysis of the problems associated with projecting decline curves, see Greg McFarland, "Shale Economics, Watch the Curve," Oil & Gas Evaluation Report, February 25, 2010, www.oilandgasevaluationreport.com/2010/03/ articles/oil-patch-economics/shale-economics-watch-the-curve/ (accessed July 24, 2010). See also Mike Markes, "Does I.P. Mean Investor Problems?" Oil & Gas Evaluation Report, February 25, 2010, www.oilandgasevaluationreport.com/2010/02/ articles/asset-valuation/does-ip-mean-investor-problems/ (accessed July 24, 2010).

181 *That possibility of inflated shale gas economics* Ian Urbina, "Insiders Sound an Alarm Amid a Natural Gas Rush," *New York Times,* June 26, 2011.

182 *In March 2010, a pit containing 400,000 gallons* Summer Wallace-Minger, "Atlas Energy Gas Well Fire Leaves $375,000 in Damages," *Weirton Daily Times,* April 2, 2010, www.weirtondailytimes.com/page/content.detail/id/537136.html?nav=5006 (accessed July 24, 2011). See also Iris Marie Bloom, "Fire at Pennsylvania Fracking Site," *Philadelphia Weekly Press,* April 7, 2010, http://weeklypress.com/fire-at-pennsylvania-fracking-site-p1872-1.htm (accessed July 24, 2011).

183 *The torrent continued unchecked* EOG Resources paid the Pennsylvania Fish and Boat Commission (PF&BC) $208,625 in a settlement for the three spills of substances

"deleterious, destructive or poisonous to fish." See "Texas Company Pays $208,625 in Settlements for Polluting Creeks in Clearfield County," PF&BC press release. April 29, 2011, www.fish.state.pa.us/newsreleases/2011press/EOG_settle.htm. See also Laura Legere, "Hazards Posed by Natural Gas Drilling Not Always Underground," *Scranton Times Tribune*, June 21, 2010, http://thetimes-tribune.com/news/hazards-posed-by-natural-gas-drilling-not-always-underground-1.857452#axzz1T2gfNE2j (accessed July 24, 2011).

183 *Five days after the blowout* Vicki Smith, "West Virginia Gas Well Blast Injures 7," Associated Press, June 7, 2010, http://thetimes-tribune.com/news/west-virginia-gas-well-blast-injures-7-flames-now-40-feet-1.834576#axzz1T2gfNE2j (accessed July 24, 2011).

183 *An entire field of engineering* For an analysis of the chemistry of drilling mud, see Jerry M. Neff, "Composition, Environmental Fates, and Biological Effects of Water Based Drilling Muds and Cuttings Discharged to the Marine Environment," Petroleum Environmental Research Forum and American Petroleum Institute, January 2005, www.perf.org/pdf/APIPERFreport.pdf (accessed October 24, 2011).

183 *In the first decade of the development of the Barnett Shale* Data is from Texas Railroad Commission, Data and Statistics, Blowouts and Well Control Problems, updated September 11, 2011, www.rrc.state.tx.us/data/drilling/blowouts/allblowouts11-15.php (accessed October 20, 2011).

185 *"Any decision Cornell makes* Cornell University, "Ad Hoc Advisory Committee Report on Leasing of Land for Exploration and Drilling for Natural Gas in the Marcellus Shale," Ithaca, N.Y., May 28, 2010.

186 *The Kennedy meeting represented* Accounts of the Kennedy meeting are from interviews with people who attended and media reports. Laura Legere, "Robert Kennedy Jr., Environmentalists Hear of Gas Woes in Dimock," *Citizens Voice*, June 4, 2010, http://citizensvoice.com/news/robert-kennedy-jr-environmentalists-hear-of-gas-woes-in-dimock-1.830053#ixzz1Rjf2a4yh.

187 *The bill also drew* The bill passed the New York Assembly and was vetoed by Paterson on the grounds that it would ban fracking of conventional wells and hurt the industry that traditionally operated in New York. Paterson, instead, placed an executive moratorium to ban high-volume fracking of vertical wells until June 31, 2011.

194 *But the company had addressed problems* The DEP's Consent Order with Cabot in November 2009 required Cabot to provide the agency with a list of all the households where it was providing water. The thirty-six wells noted here are based on a list on file in the DEP's Northwest Regional Office.

196 *Hanger responded* Email exchanges on September 29, 2010, reflect concern by DEP officials regarding a complaint by the Sautners that Cabot representatives were coming on landowners' property with armed guards. Amy Barrette of K&L Gates, a law firm representing Cabot, informed Scott Perry of the DEP that Cabot had hired an off-duty Montrose police officer for security detail, and that company representatives would not be going back to the Sautners after being ordered off the property.

197 *"Everything is back to normal,"* Reported by Trish Hartman, "Cabot Responds to DEP Accusations," WNEP TV, October 20, 2010, www.wnep.com/news/countybycounty/wnep-susq-cabot-responds-to-dep,0,850221.story (accessed October 20, 2011).

199 *The report concluded that Cabot* Robert Watson, "Report of Cabot Oil & Gas Corporation's Utilization of Effective Techniques for Protecting Fresh Water Zones/ Horizons during Natural Gas Well Drilling: Completion and Plugging Activities," Cabot Oil and Gas, www.cabotog.com/pdfs/WatsonRpt.pdf.

203 *Days after the election* Rove was the keynote speaker on the opening day of a two-day shale gas conference sponsored by Hart Energy Publishing LLP. Andrew Maykuth, "In Pa., Rove tells Marcellus Shale Drillers: Expect 'Sensible Regulation,'" *Philadelphia Inquirer,* November 4, 2010, http://articles.philly.com/2010-11-04/ business/24953218_1_marcellus-shale-obama-climate-change-legislation (accessed July 29, 2011).

CHAPTER 7. SUPERIOR FORCES

206 *With Dimock as a high-profile example* In May 2011 a NY1/YNN-Marist College poll showed 41 percent of adults statewide opposed fracking, 38 percent supported it, and 21 percent were undecided. A Siena College poll released in July 2011 found 45 percent of voters supported the DEC's recommendations to allow permits for drilling operations under specific conditions, with 43 percent opposed. See Jon Campbell, "Poll: New Yorkers Evenly Split on Hydrofracking," *Binghamton Press & Sun-Bulletin,* May 18, 2011; and Jon Campbell, "Poll: New Yorkers Divided Over Hydrofracking Report," *Binghamton Press & Sun-Bulletin,* July 14, 2011.

207 *According to insiders, Larry Schwartz* The account of the conflict between Schwartz and Grannis is from my own interviews and from multiple newspaper reports documenting the Grannis firing and events leading up to it. See Brian Nearing, "DEC Commissioner Fired After Layoff Memo Leak," *Albany Times Union,* October 21, 2010, www.timesunion.com/local/article/DEC-commissioner-fired-after-layoff-memo-leak-717567.php (accessed September 3, 2011).

209 *The policy proposal* Letter from Gill to DEC Commissioner Joe Martens, September 2, 2011

210 *"I'm looking to see* Scott Detrow, "One Month Later, Corbett Stays Vague on Marcellus Report's Recommendations," StateImpact Pennsylvania, August 23, 2011, http:// stateimpact.npr.org/pennsylvania/2011/08/23/one-month-later-corbett-stays-vague-on-reports-recommendations/ (accessed July 9, 2012).

210 *At one of its first meetings* See Jon Campbell, "DEC: 226 New Workers Needed for Fracking Enforcement," *Binghamton Press & Sun-Bulletin,* September 13, 2011, www. pressconnects.com/article/20110913/NEWS01/109130379/DEC-226-new-workers-needed-fracking-enforcement (accessed September 18, 2011).

211 *Another twenty towns* Home rule cases began making their way to the courts in the fall of 2011. In September, landowners filed suit against the town of Middlefield, in Otsego County, arguing the town had no authority to ban drilling. At the same time, a pro-drilling coalition in Dryden announced it would file a similar law suit against the town. See Campbell, "Second Suit Contests Drilling Ban," *Binghamton Press & Sun-Bulletin,* September 16, 2011, www.pressconnects.com/article/20110916/ NEWS01/109160358/2nd-suit-contests-drilling-ban-energy-group-criticizes-Dryden-s-action?odyssey=tab%7Ctopnews%7Ctext%7CFRONTPAGE (accessed

September 18, 2011). Meanwhile, a similar ban in Morgantown, West Virginia, was struck down by a West Virginia state judge in a ruling that reinforced the state's longstanding and comprehensive role in regulating the oil and gas industry. See *Northeast Natural Energy, et al. v. The City of Morgantown, West Virginia*, No. 11-C-411, Circuit Court of Monongalia County, Division 1, West Virginia, www.frackinginsider.com/Tucker_Marcellus_Order.pdf (accessed September 18, 2011). A similar ban was rescinded in the city council of Wellsburg, West Virginia, in the face of industry opposition.

212   *CNG cost over 40 percent less* See "Clean Cities Alternative Fuel Price Report," U.S. Department of Energy, July 2011, www.afdc.energy.gov/afdc/pdfs/afpr_jul_11.pdf (accessed September 8, 2011).

212   *In December 2010, there were* See National Gas Vehicle Statistics, International Association for Natural Gas Vehicles, April 11, 2011, www.iangv.org/tools-resources/statistics.html (accessed September 8, 2011). There were 256 million registered vehicles in the United States as of 2008. See Table 1-11, Number of U.S. Aircraft, Vehicles, Vessels, and Other Conveyances, Research and Innovative Technology Administration, Bureau of Transportation Statistics, www.bts.gov/publications/national_transportation_statistics/html/table_01_11.html (accessed September 8, 2011).

212   *There were fewer than 1,000* CNG automobile success stories were rare, the noteworthy exception being the Honda Civic GX, which got a foothold in the market of conscientious motorists when it was introduced in 1998. That was before the development of hybrids and electric cars, which, lacking the fueling restrictions of CNG vehicles, quickly gained market share among buyers attracted to low-emissions vehicles. As of 2010, more than 2 million hybrids were sold by the summer of 2011 in the United States, making the country the largest market for hybrids. See Christie Schweinsberg, "U.S. Hybrid Sales Hit 2 Million Mark," WardsAuto.com, June 7, 2011, http://wardsauto.com/ar/hybrid_sales_million_110607/index.html (accessed Sept. 8, 2011).

213   *The announcement, in late July* Chesapeake Energy Corporation Q2 2011 Earnings Call Transcript, Morningstar, www.morningstar.com/earnings/28691279-chesapeake-energy-corp-chk-q2-2011.aspx?pindex=3 (accessed October, 31, 2011). See also Matthew Steffy, "Shale Called Rich in Oil," *Tribune Chronicle*, July 29, 2011, www.tribtoday.com/page/content.detail/id/559801/Shale-called-rich-in-oil.html (accessed October 31, 2011).

214   *The assessment came from* The updated USGS estimate of recoverable Marcellus gas was the agency's first since the shale gas boom began. Although far less than the estimates of Engelder and the EIA, it was far more than its original estimate of 2 trillion cubic feet. See "USGS Releases New Assessment of Gas Resources in the Marcellus Shale, Appalachian Basin," August 23, 2011, USGS, www.usgs.gov/newsroom/article.asp?ID=2893&from=rss_home (accessed January 9, 2012). Also, "World Shale Gas Resources: An Initial Assessment of 14 Regions Outside the United States," U.S. EIA, April 5, 2011, www.eia.gov/analysis/studies/worldshalegas/ (accessed October 23, 2011)

214   *A week prior to the USGS announcement* Schneiderman, an outspoken critic of fracking, was also suing the federal government for failing to review the impact of fracking on communities and water quality in the Delaware River watershed.

214   *The Securities and Exchange Commission*   The investigation by the SEC and the New York attorney general's office into the integrity of shale gas financial claims was reported in several newspapers and outlets. See Jim Efstathiou Jr. and Katarzyna Klimasinska, "U.S. to Slash Marcellus Shale Gas Estimate 80 Percent," *Bloomberg*, August 23, 2011, www.bloomberg.com/news/2011-08-23/u-s-to-slash-marcellus-shale-gas-estimate-80-. html (accessed September 18, 2011). Also G. Michael O'Leary, Gislar Donnenberg, and Lisa Montgomery Shelton, "Oil and Gas Companies Should Expect Increased SEC Scrutiny of Operations and Reserves," *National Law Review*, September 6, 2011, www.natlawreview.com/article/oil-and-gas-companies-should-expect-increased-sec-scrutiny-operations-and-reserves (accessed September 18, 2011); and Ian Urbina, "Regulators Seek Records on Claims for Gas Wells," *New York Times*, July 29, 2011, www.nytimes.com/2011/07/30/us/30gas.html (accessed September 18, 2011).

214   *So good, in fact*   By late 2011, supplies had risen to a point where companies were seeking to export Marcellus gas. On September 1, 2011, Dominion Resources filed for permission from the Department of Energy to export 1 billion cubic feet per day through the company's terminal in Maryland to any country with which the United States does business. The terminal, Dominion Cove Point on the Chesapeake Bay, is well-situated to export gas from the Marcellus Shale and Utica Shale formations. See Matthew Kemeny, "Virginia Firm Wants to Export Marcellus Shale's Gas," *Patriot News*, October 9, 2011, www.pennlive.com/midstate/index.ssf/2011/10/virginia_firm_wants_to_export.html (accessed October 11, 2011).

215   *House Resolution 1380, commonly known*   Officially known as *New Alternative Transportation to Give Americans Solutions Act of 2011*, the bill would give a $100,000 tax break to every service that provides natural gas fueling, and $64,000 for every big rig that converted from diesel to burn natural gas, and $11,500 for every car. The text of the bill and various summaries are available at www.opencongress.org/bill/112-h1380/show (accessed September 6, 2011).

215   *That summer, 108 Democrats*   The White House plan to encourage global switching from oil to natural gas would be executed through the Global Shale Gas Initiative and the APEC Unconventional Gas Census. The president's energy plan also proposed more on- and offshore drilling, an array of alternative energy programs, and the advancement of alternative fuel vehicles. See Blueprint for a Secure Energy Future, March 2011, www.whitehouse.gov/sites/default/files/blueprint_secure_energy_future.pdf (accessed August 14, 2011).

216   *Using natural gas as both a fuel*   See Dow Chemical Company's Statement for the Record, presented at the U.S. Senate's Committee on Energy and Natural Resources hearing on the Future of Natural Gas on July 19, 2011, http://energy.senate.gov/public/_files/BiltzTestimony071911.pdf (accessed September 18, 2011).

217   *Methane is a potent greenhouse gas*   The study compared estimated emissions for shale gas, conventional gas, coal (surface-mined and deep-mined), and diesel oil, taking into account direct emissions of CO_2 during combustion, indirect emissions of CO_2 necessary to develop and use the energy source and methane emissions, which were converted to an equivalent value of CO_2 for global warming potential. The study is the first peer-reviewed paper on methane emissions from shale gas, and one of the few exploring the greenhouse gas footprints of conventional gas drilling. See R. W. Howarth, R. Santoro,

and A. Ingraffea. 2011. "Methane and the Greenhouse Gas Footprint of Natural Gas from Shale Formations." Climatic Change Letters,

217 *Engelder, who had seen* In the same piece, Engelder criticized Cabot for denying culpability of methane migration in Dimock. See Terry Engelder, "Gas Drilling Yields a Gusher of Hogwash," *Philadelphia Inquirer*, April 28, 2010, http://articles.philly.com/2010-04-28/news/25213127_1_pockets-of-methane-gas-cabot-oil-natural-gas (accessed October, 8, 2011).

218 *Unlike most of his presentations* The event was organized by Responsible Drilling Alliance, a group of citizen watchdogs and industry critics.

218 *For the purposes of the discussion* Ingraffea used numbers cited in a report by Nels Johnson for collection of environmental groups. See "Pennsylvania Energy Impacts Assessment," Nature Conservancy, November 15, 2010, www.nature.org/media/pa/tnc_energy_analysis.pdf (accessed October 24, 2011).

220 *During the first nine months* Numbers are from the Pennsylvania DEP's records of Oil and Gas Inspections, Violations and Enforcements available at www.dep.state.pa.us/dep/deputate/minres/oilgas/OGInspectionsViolations/OGInspviol.htm (accessed October 27, 2011). See also Laura Legere, "DEP Inspections Show More Shale Well Cement Problems," *Scranton Time Tribune:* September 18, 2011, http://thetimes-tribune.com/news/dep-inspections-show-more-shale-well-cement-problems-1.1205108#ixzz1ZjFLq7Gk (accessed October 3, 2011).

221 *On September 7, 2011* Ridge's contract with the Marcellus Shale Coalition was reported by multiple news organizations. For summaries, see Anya Litvak, "Tom Ridge, Marcellus Shale Coalition Split," *Pittsburgh Business Times*, August 2, 2011, www.bizjournals.com/pittsburgh/blog/energy/2011/08/tom-ridge-marcellus-shale-coalition-part.html (accessed October 27, 2011); also, Tom Barnes, "Ridge's Firms to Get $900,000 for Pushing Marcellus Drilling," *Pittsburgh Post-Gazette*, August 2, 2010, www.post-gazette.com/pg/10214/1077108-100.stm (accessed October 24, 2011).

221 *As McClendon began his* This account is based on reporting from several news outlets. See Michael Rubinkam, "Chesapeake CEO takes on Anti-drilling 'Extremists,'" Associated Press: September 7, 2011, www.forbes.com/feeds/ap/2011/09/07/general-energy-us-gas-drilling-conference_8663638.html (accessed October 3, 2011). See also David Falchek and Laura Legere, "Chesapeake CEO Skewers Protesters Picketing Event," *Scranton Times Tribune,* September 8, 2011, http://m.thetimes-tribune.com/news/chesapeake-ceo-skewers-protesters-picketing-event-1.1199484 (accessed October 31, 2011). See also Scott Detrow, "Outside Marcellus Shale Coalition's Conference, Chants of 'Shut Them Down,'" *Stateimpact*, National Public Radio, September 7, 2011, http://stateimpact.npr.org/pennsylvania/2011/09/07/outside-marcellus-shale-coalitions-conference-chants-of-shut-them-down/ (accessed October 31, 2011). See also Sandy Bauers, "Protesters Rally at Gas-Drilling Conference in Center City," *Philadelphia Inquirer,* September 8, 2011, http://stateimpact.npr.org/pennsylvania/2011/09/07/outside-marcellus-shale-coalitions-conference-chants-of-shut-them-down/ (accessed October 31, 2011).

CHAPTER 8. TURF WARS

226 *It was January 24, 2012* The EPA's study ordered by Congress to evaluate the relationship between fracking and groundwater was funded and executed separately

from the testing at the Sautner's house. But while the two studies were distinct, their objectives overlapped and the testing at the home of the Sautners and other homes in Dimock was intended to inform the broader Congressionally-ordered study.

227 *Things came to a head one day in September 2010* Email exchanges on September 29, 2010, reflect concern by DEP officials regarding a complaint filed by the Sautners that Cabot representatives were coming onto landowners' property with armed guards. Amy Barrette of K&L Gates, a law firm representing Cabot, informed Scott Perry of the DEP that Cabot had hired an off-duty Montrose police officer for security detail and that company representatives would not be going back to the Sautners after being ordered off the property.

227 *The Sautner's approach* In a rebuttal to Josh Fox's film, *Gasland*, Phelim McAleer, a filmmaker and fracking supporter from Ireland, depicted the Sautners as self-serving and exploitive phonies in his 2013 film *Frack Nation*.

228 *As a precursor to the activity at 1101 Carter Road* Since 2008, at least, staff at the EPA Region 3 office were routinely sharing links to news articles chronicling events in Dimock, as well as notes from residents and politicians, in email correspondence and briefings. Thousands of pages of EPA correspondence and files relating to the Dimock investigation obtained through the Freedom of Information Act are available at: ftp://ftp.epa.gov/epar3gis/FOIA_EMAIL_RELEASE.

228 *A day later, after the media jumped on the story* Internal correspondence suggests spin control was a concern throughout the investigation. Dennis Carney, a director of the Hazardous Site Cleanup Division, wrote EPA staff in Dimock on December 17, 2012: "The action plan proposes that folks work on messaging first thing Monday including what we tell the residents about why EPA involvement now. I'm confident you all would and will handle that well but know there is some concern in the regional office about what we say."

229 *He is on record as a fossil fuel proponent* Krancer told a reporter: "There is no uniformity within the scientific community on how much the warming is occurring. And there's no agreement about how much is attributable to the human part of it and how much is attributable to other factors." Marie Cusick, "DEP Secretary Michael Krancer Clarifies Views on Climate Change," *StateImpact Pennsylvania*, February 21, 2013, http://stateimpact.npr.org/pennsylvania/2013/02/21/dep-secretary-michael-krancer-clarifies-views-on-climate-change/.

229 *"The real issue here is not safety"* Krancer's comments originally appeared in a letter to the editor in the *Chambersburg Public Opinion* and were distributed through major news outlets, including the Associated Press and network affiliates in November 2011. See www.cleveland.com/business/index.ssf/2011/11/lawyers_demand_water_for_taint.html and http://files.dep.state.pa.us/OilGas/OilGasLandingPageFiles/DimockChambersburgLetter.pdf.

230 *During his tenure with the DEP* This and other letters from Krancer to both Garvin and Jackson are summarized by Scott Detrow, "Mike Krancer and the EPA: It's Complicated," *StateImpact Pennsylvania*, May 5, 2012, http://stateimpact.npr.org/pennsylvania/2012/05/15/mike-krancer-and-the-epa-its-complicated/.

230 *In what he characterized as "an urgent matter to the agency"* Email from Jon Capacasa to Robert Helverson and Lora Werner regarding "Formal Request for ATSDR Health Consultation—Dimock PA Drinking Water," December 7, 2011.

230 *With this toxicological evaluation* Details for justification for the Dimock investigation under the Superfund law are detailed in a January 19, 2012, Action Memorandum from EPA onsite coordinator Richard Fetzer to Dennis P. Carney, a director of the Hazardous Site Cleanup Division.

231 *Emblematic of corporate and civic environmental recklessness* Associated Press, "Love Canal to Fracking: Niagara Falls Mulls Waste," *Wall Street Journal,* October 20, 2011, http://online.wsj.com/article/AP6a04a59c72344063a3453584f13b4841. html; "Exemption of Oil and Gas Exploration and Production Wastes from Federal Hazardous Waste Regulations," U.S. Environmental Protection Agency, www.epa.gov/osw/nonhaz/industrial/special/oil/oil-gas.pdf.

234 *She also chairs the board of directors* Joe Capozzi, "Jane Forbes Clark, a Powerhouse behind the Baseball Hall of Fame," *Palm Beach Post*, July 20, 2010, www.palmbeachpost.com/news/sports/baseball/jane-forbes-clark-a-powerhouse-behind-the-baseball/nL8cq/.

234 *He represented Gastem USA* Gastem was one of several companies that drilled twenty-four wells, both vertical and horizontal, in Quebec province to test the Utica Shale between 2006 and 2009. Positive results were reported, although none of the wells were producing at the end of 2009. Susan R. Eaton, "Shale Play Extends to Canada," *AAPG Explorer,* January 2010, pp. 10–24. Quebec placed a moratorium on fracking in April 2012 pending further study of impacts.

234 *About 34,000 acres of Otsego County was under lease* According to the company's web site, Gastem held "80 percent interest on approximately 34,400 acres, mostly in Otsego County, with a scattering of leases in Herkimer and Schoharie counties." See Joe Mahoney, "Gastem Scales Back Drilling Plans," *Oneonta Daily Star,* November 5, 2010, www.thedailystar.com/news/local_news/gastem-scales-back-drilling-plans/article_518b479b-38cf-5882-b77b-b69307cea3e1.html.

235 *He was explaining to the trustees* Events and dialogue at this meeting are based on reports in the *Cooperstown Crier*, the *Oneonta Daily Star*, village meeting minutes, and interviews with attendees.

238 *Just to the left of Northup's view* In *The Deerslayer,* James Fenimore Cooper portrays Council Rock as a meeting place for tribal leaders "to make their treaties and bury their hatchets."

242 *The site was named Ross 1* Descriptions and slides of the operation can be found in a presentation by Nyahay at Drilling and Fracking Lecture Series, University of Buffalo, April 28, 2011, www.glyfac.buffalo.edu/mib/course/marcellus/drillingfracingrn.pdf and www.youtube.com/watch?v=M91qvUqn4FY.

242 *At the time the Sierra Club* Between 2007 and 2010 the Sierra Club accepted over $25 million in donations from the gas industry, mostly from Aubrey McClendon, CEO of Chesapeake Energy. Bryan Walsh, "How the Sierra Club Took Millions from the Natural Gas Industry" *Time,* February 2, 2012, http://science.time.com/2012/02/02/exclusive-how-the-sierra-club-took-millions-from-the-natural-gas-industry-and-why-they-stopped/.

243 *As he shadowed Cole* In December 2014, the Federal Energy Regulatory Commission approved construction of the Constitution Pipeline to increase gas supplies to New York and New England markets. The pipeline route passed within miles of the Gastem project in Otsego County.

243   *Gastem announced to investors* "Multiple Pays Yield Gas in New York," *Oil & Gas Journal*, January 27, 2010, www.ogj.com/articles/2010/01/multiple-pays-yield.html; Tom Grace, "Gastem, USA: Ross Well Results Promising," *Oneonta Daily Star*, February 1, 2010, www.thedailystar.com/archives/article_eca4ff52-9459-57d8-94be-560a83b1d5b4.html; "Ross No.1 Initial Results," Gastem press release, January 27, 2010, www.bloomberg.com/apps/news?pid=newsarchive&sid=alk1ixH3XiqM.

244   *Cole had obtained a permit* A year after Gastem used a slickwater frack on the Utica Shale, the company used nitrogen foam to frack the much shallower Marcellus formation. "Gastem: Initial Marcellus Tests Successful," *Natural Gas Intelligence*, December 20, 2010, www.naturalgasintel.com/articles/1000-gastem-initial-marcellus-tests-successful.

248   *In 1987, the Town of Carroll* *Frew Run Gravel v. Carroll*, 71 N.Y.2d 126 (1987), www.leagle.com/decision/198719771NY2d126_1188.

249   *With Middlfield and Dryden as the test cases* Fractracker, an organization that charts matters related to shale gas development, tallied 80 New York towns with bans, 100 with moratoria, and 86 considering bans or moratoria as of mid-October, 2014, http://www.fractracker.org/map/us/new-york/moratoria/.

249   *The industry, having already spent millions* The history of home rule relating to petroleum development varies from state to state and the law was evolving at the time of publication of this book. Denton became the first Texas town to pass a ban in November 2014, which the industry representatives said they would challenge. In December 2013, Pennsylvania's high court ruled that legislation to restrict local governing of shale gas matters, passed under the Corbett administration in 2012, was unconstitutional.

250   *It was Cole's particular misfortune* Tom Grace, "Local Town Looks to Decide Gas Drilling Fate," *Oneonta Daily Star*, March 12, 2011, www.thedailystar.com/news/local_news/local-town-looks-to-decide-gas-drilling-fate/article_44f002ed-5b04-5544-a0ae-827e479598dc.html.

250   *Huntington runs Cooperstown Holstein* Jennifer Huntington's speech is a summation of her views expressed in newspaper accounts of presentations and interviews she gave over a period in 2011. Peter Applebome, "Drilling Debate in Cooperstown Is Personal," *New York Times*, October 29, 2011, www.nytimes.com/2011/10/30/nyregion/in-cooperstowns-fight-over-gas-drilling-civility-is-fading.html?pagewanted=all&_r=0; Jim Austin, "Middlefield Officials Aim to 'Zone Out' Gas Drilling," *Cooperstown Crier*, February 10, 2011, www.coopercrier.com/news/local_news/article_5f24ce73-69a5-5a0f-990b-fd74446a0c20.html.

251   *That number was indisputably large* Grace, "Local Town Looks to Decide Gas Drilling Fate."

252   *On June 14, after exhaustive research* The towns of Otsego and Springfield passed similar bans days prior to the Middlefield resolution.

252   *Stakes in the home rule case rose unexpectedly* Danny Hakim, "Cuomo Proposal Would Restrict Gas Drilling to a Struggling Area," *New York Times*, June 13, 2012, www.nytimes.com/2012/06/14/nyregion/hydrofracking-under-cuomo-plan-would-be-restricted-to-a-few-counties.html?_r=0.

253   *The leak—confirmed by documents* The plan for a roll-out was referenced in correspondence with independent scholars hired to review it. For a summary of the correspondence, see: Jon Campbell, "State Mulled Fracking Rollout, Experts

Supported Implementation," *Binghamton Press & Sun-Bulletin*, January 10, 2015, www.democratandchronicle.com/story/news/local/2015/01/10/new-york-hydrofracking-phase/21530069/. The correspondence can be downloaded at the Toxics Targeting website: www.toxicstargeting.com/MarcellusShale/documents/2015-1-07/letters-to-DOH-regarding-fracking-public-health-review.

253 *The governor himself declined to elaborate* Jon Campbell, "Cuomo: Hydrofracking Review "Will Be Done Shortly," *Politics on the Hudson,* June 22, 2012, http://polhudson.lohudblogs.com/2012/06/22/cuomo-hydrofracking-review-"will-be-done-shortly/.

254 *During a policy summit on economic development* Laura Nahmias, "Hydrofracking Protesters at Governor Andrew Cuomo Policy Convention," YouTube video, August 24, 2012, www.youtube.com/watch?v=215KChT52dY; *Waging Nonviolence Blog*, "Thousands Flock to Albany, Delivering Pledge to Resist Fracking," blog entry by Peter Rugh, August 28, 2012.

256 *Following long-held Albany protocol* It was typical for the agency to consult industry, as well as environmental groups, on matters of policy. In one instance, DEC officials gave industry lobbies access to drafts of the SGEIS six weeks before they were made public. "Cuomo Team Gives Drillers Jump Start to Influence Fracking Rules," in *Inside Track*, a blog by Environmental Working Group, June 21, 2012, www.ewg.org/research/inside-track.

257 *In January 2012 two groups representing county health departments* The reports were prepared for the panel of outside experts formed in July 2011 by Martens to advise the administration on practical matters of overseeing shale gas development. The panel meeting was canceled a day before the recommendations were scheduled to be presented and did not meet after that. Jon Campbell, "Leaked Documents Show Fracking Concerns," *Binghamton Press & Sun-Bulletin,* July 1, 2012.

258 *The authors of the SGEIS were against this approach* Danny Hakim, "Gas Drilling Is Called Safe in New York," *New York Times*, January 3, 2013, www.nytimes.com/2013/01/03/nyregion/hydrofracking-safe-says-ny-health-dept-analysis.html.

260 *On February 21, 2012* Anschutz v. Town of Dryden, No. 2011-0902, N.Y. State Supreme Court, Tompkins County, http://dryden.ny.us/Downloads/DrydenCourtDecision.pdf.

260 *Just three days later* Cooperstown Holstein v. Town of Middlefield, No. 2011–0930. NY State Supreme Court, Otsego County. www.scribd.com/doc/82732762/Middlefield-decison

261 *Lawyers might be interested in this dragging on* "Home Rule Adversaries Welcome Date with High Court Ruling," in *Shale Gas Review*, a blog by Tom Wilber, September 6, 2013, http://tomwilber.blogspot.com/2013/09/home-rule-adversaries-welcome-date-with.html.

264 *More than three-quarters* "A 2008 Survey of Land Use Planning & Regulations in NYS," New York State Legislative Commission on Rural Resources, www.dos.ny.gov/lg/publications/Rural_Resource_Survey.pdf.

264 *By the end of 2011* In the fourth-quarter 2011 conference call, Dan Dinges cited the company's operations in northeast Pennsylvania as the "main driver" of a 43.5 percent growth in the company's total oil and gas production in 2011. See Laura Legere, "Marcellus Gas Production in State Continues Steady Climb in 2011," *Scranton Times Tribune*, February 18, 2012, http://thetimes-tribune.com/news/marcellus-gas-production-in-state-continues-steady-climb-in-2011-1.1273478; "Cabot Oil & Gas'

CEO Discusses Q4 2011 Results—Earnings Call Transcript," *Seeking Alpha*, February 21, 2012, http://seekingalpha.com/article/380611-cabot-oil-and-gas-ceo-discusses-q4-2011-results-earnings-call-transcript.

267 *A fire and explosion danger* Tests filed with the Pennsylvania DEP on September 16, 2011, showed methane concentrations in the Sautner well reached explosive levels even though the well was vented. Laura Legere, "High Methane Level Recorded in Dimock Water Well," *Scranton Times Tribune*, October 8, 2011, http://thetimes-tribune.com/news/gas-drilling/high-methane-level-recorded-in-dimock-water-well-1.1215258.

267 *President Obama, whose endorsement* In his 2012 State of the Union Address, Obama touted shale gas and pledged "every possible action to safely develop this energy." Jim Snyder, "Obama Pushes Natural-Gas Fracking to Create 600,000 U.S. Jobs," *Bloomberg*, January 25, 2012, www.bloomberg.com/news/2012-01-25/obama-backs-fracking-to-create-600-000-jobs-vows-safe-drilling.html.

267 *The EPA investigation of the Wind River Reservation* Dominic C. DiGiulio, Richard T. Wilkin, Carlyle Miller, and Gregory Oberley, "Investigation of Ground Water Contamination near Pavilion, Wyoming," U.S. Environmental Protection Agency, December 2011, www2.epa.gov/sites/production/files/documents/EPA_ReportOnPavillion_Dec-8-2011.pdf.

268 *Perhaps the most sustained and influential criticism* The hearings were prefaced with a denunciation, led by Issa, of the EPA's purported eagerness to regulate fracking: "Rising Energy Costs: An Intentional Result of Government Action," Committee on Oversight and Government Reform, U.S. House of Representatives 112th Congress, May 23, 2011, http://oversight.house.gov/wp-content/uploads/2012/02/REPORT_-_Rising_Energy_Costs_An_Intentional_Result_of_Government_Action.pdf.

269 *Michael Krancer was a star figure* A transcript of Krancer's remarks can be found at www.gpo.gov/fdsys/pkg/CHRG-112hhrg75111/html/CHRG-112hhrg75111.htm.

269 *The EPA withdrew its emergency order* Even after withdrawing the order, the EPA stood by its actions as necessary, warranted, and legal. Kathlene Butler, "Response to Congressional Inquiry Regarding the EPA's Emergency Order to the Range Resources Gas Drilling Company," Environmental Protection Agency, December 20, 2013, www.epa.gov/oig/reports/2014/20131220-14-P-0044.pdf.

269 *Cabot released the results of that investigation* "Well Site Evaluations, Areas of Alleged Releases, Susquehanna County, Pennsylvania," prepared by URS Corp. for Cabot Oil & Gas, December 2011, www.cabotog.com/pdfs/FINAL%20Wellsite%20Evaluation%20Report%20-%202011-12-09.pdf.

269 *A follow-up by the DEP* Letter to Phil Stalnaker, Cabot Oil & Gas, from Jennifer Means, Environmental Program Manager for the DEP, March 9, 2012, https://dl.dropboxusercontent.com/u/48182083/drilling/Cabot-%20Dimock%20well%20site%20evals%203-9-12.pdf.

270 *Some values were highlighted in yellow* The EPA results are identified by number and not name. This information is from interviews with Ely, crossed referenced with EPA dataset HW06, www.epaosc.org/sites/7555/files/Dimock%20Chemical%20and%20Bacteriological%20Data%20Weeks%201-5%20and%201st%20Round%20Supplemental%202.pdf.

270 *That same day the EPA issued a one-page press release* "EPA Completes Drinking Water Sampling in Dimock, Pa.," U.S. Environmental Protection Agency press release, July

25, 2012, http://yosemite.epa.gov/opa/admpress.nsf/0/1A6E49D193E1007585257A46 005B61AD.

271 *Hours after the issuance* "Cabot Oil & Gas Management Discusses Q2 2012 Results— Earnings Call Transcript," *Seeking Alpha*, July 25, 2012, http://seekingalpha.com/ article/749791-cabot-oil-and-gas-management-discusses-q2-2012-results-earnings-call-transcript.

271 *By the time of the settlement* The plaintiffs were represented at the settlement by Tate J. Kunkle of the New York–based law firm Napoli Bern Ripka Shkolnik.

272 *On the first anniversary of the settlement* On August 5, 2013, *DesmogBlog* posted a copy of the EPA report, "Isotech Stable Isotope Analysis: Determining the Source of Methane and its Effect on the Aquifer," http://desmogblog.com/sites/beta.desmogblog. com/files/Dimock%20report.pdf.

272 *Staff writer Neela Banerjee reported* Neela Banerjee, "Message Is Mixed on Fracking," *Los Angeles Times*, July 28, 2013, www.latimes.com/nation/nationnow/la-na-epa-dimock-20130728-m-story.html#page=1.

EPILOGUE

279 *The three experts hired* The work of the three consultants—later obtained through the a Freedom of Information Act request—is detailed in correspondence and reports to Nirav Shah in late 2012 and early 2013. They can be found on line at www.toxicstargeting. com/MarcellusShale/documents/2015-1-07/letters-to-DOH-regarding-fracking-public-health-review.

281 *"They're everywhere"* Jon Campbell, "Cuomo: Fracking Protesters Are 'Everywhere,'" *Binghamton Press & Sun-Bulletin*, September 10, 2014.

281 *The primary result was* Sarah Jaffe, "How Zephyr Teachout Taught Democrats a Lesson in Democracy," *The Nation*, September 10, 2014, www.thenation.com/article/181565/ how-zephyr-teachout-taught-democrats-lesson-democracy#.

284 *He held up, one by one* The studies that Zucker mentioned in his oral presentation are cited and summarized in the department's formal report: "A Public Review for High Volume Hydraulic Fracturing for Shale Gas Development," New York State Department of Health, December 2014, www.health.ny.gov/press/reports/docs/high_volume_ hydraulic_fracturing.pdf.

286 *It started when landmen* There were other factors that contributed to the political movement that gave rise to the SGEIS, of course. Several environmental organizations were among the first to recognize the significance of legislation that passed in July 2008 to streamline permitting of unconventional wells in New York. Although Governor Paterson ordered a hold on permitting pending the SGEIS when he signed the legislation, interest in the mechanics of the bill among policy-minded environmental groups became a magnet for organizational involvement in state affairs that, by some accounts, was a forerunner of the anti-fracking movement. Groups that rallied around the bill included the Sierra Club Atlantic Chapter, Catskill Mountainkeeper, Riverkeeper, Delaware Riverkeeper Network, Natural Resources Defense Council, Catskill Center for Conservation and Development, The Wilderness Society and Catskill Citizens for Safe Energy, and Damascus Citizens for Sustainability.

INDEX